KU-466-773

METAMORPHIC TEXTURES

by

ALAN SPRY

Department of Physics, Monash University, Melbourne, Australia

PERGAMON PRESS

OXFORD · NEW YORK · TORONTO · SYDNEY · PARIS · FRANKFURT

U. K.	Pergamon Press Ltd., Headington Hill Hall, Oxford OX3 0BW, England
U. S. A.	Pergamon Press Inc., Maxwell House, Fairview Park, Elmsford, New York 10523, U.S.A.
CANADA	Pergamon of Canada Ltd., P.O. Box 9600, Don Mills M3C 2T9, Ontario, Canada
AUSTRALIA	Pergamon Press (Aust.) Pty. Ltd., 19a Boundary Street, Rushcutters Bay, N.S.W. 2011, Australia
FRANCE	Pergamon Press SARL, 24 rue des Ecoles, 75240 Paris, Cedex 05, France
WEST GERMANY	Pergamon Press GmbH, 6242 Kronberg-Taunus, Pferdstrasse 1, Frankfurt-am-Main, West Germany

Copyright © 1969 Pergamon Press Ltd.

All Rights Reserved. No part of this publication may be reproduced, stored in a retrieval system or transmitted in any form or by any means: electronic, electrostatic, magnetic tape, mechanical, photocopying, recording or otherwise, without permission in writing from the publishers

First edition 1969

Reprinted 1974

Reprinted 1976

Library of Congress Catalog Card No. 68-59126

Printed in Great Britain by A. Wheaton & Co., Exeter

Contents

Preface

METAMORPHIC petrology is at present undergoing a change in emphasis. For the past two decades, metamorphic processes have been considered largely in terms of mineralogical and chemical changes and many petrologists have been so preoccupied with experimental synthesis and the determination of the physical conditions under which minerals or mineral assemblages are stable that the significance of rock textures has been overlooked. There has been an upsurge of interest in textures in the past few years but no comprehensive evaluation of texture-forming processes has been made since Harker's great work of more than thirty years ago.

It is abundantly clear that a study of both texture and mineral assemblage is essential to an understanding of metamorphic rocks and that a great deal of information pertinent to this study is available in the literature on metallurgy, ceramic studies and solid state chemistry or physics.

The purpose of this book is first, to provide definitions, descriptions and illustrations of metamorphic textures for the senior undergraduate student and second, to discuss the fundamental processes involved in textural development at a level appropriate to the graduate student and practising petrologist. The fairly extensive (but by no means exhaustive) bibliography attempts to include references to original definitions, to important examples in geological literature and to the most relevant papers in adjacent fields.

If this book has any underlying theme, it is that an understanding of textures requires that metamorphism be regarded as a series of structural transformations rather than as chemical reactions. These transformations affect *real*, not *ideal* crystals, and to regard minerals as ideal crystals rather than imperfect solids containing structural defects is no more satisfactory in metamorphic petrology than in metallurgy where many processes are considered quantitatively in terms of dislocations. An elementary knowledge of the part that dislocations play in gliding, twinning, nucleation and growth is essential to the petrologist.

Preparation of this book was made possible by the generosity of the Nuffield Foundation in awarding a Dominions Travelling Fellowship at Imperial College, London, in 1965, and in providing further funds in 1966 for preparation costs.

I am very grateful to the following friends who criticized various parts of the manuscript: F. C. Beavis, R. A. Binns, K. A. Crook, D. Flinn, A. W. Kleeman, R. Kretz, C. B. Raleigh, R. L. Stanton, J. Sutton, and R. H. Vernon. Excellent photomicrographs provided by S. Amelinckx, R. A. Binns, J. Christie, A. C. McLaren, P. A. Sabine, W. S. Treffner and R. H. Vernon are acknowledged in appropriate places. My thanks are due to the following bodies for permission to use diagrams for which they hold copyright: American Ceramic Society, *American Mineralogist*, Butterworths, Geological Society of America, Geological Survey of Great Britain, *Journal of Applied Physics*, McGraw-Hill Book Co., M.I.T. Press, Oxford University Press and John Wiley & Sons.

CHAPTER 1

Metamorphism and Metamorphic Processes

METAMORPHISM is the mineralogical and structural (textural) adjustment of (dominantly) solid rocks to physical and chemical conditions which differ from those under which the rocks originated. Weathering and similar processes are conventionally excluded. The type of metamorphism depends on the relative values of temperature (T), confining pressure (P_{con}), pressure or chemical activity or fugacity of water (expressed generally as P_{H_2O}), deformation or directed pressures (P_{dir}) and their variation with time (t). There is only a limited degree of interdependence of these controls and the metamorphic history of a rock is an expression of their mutual interaction with time (Fyfe, Turner and Verhoogen, 1958; Pitcher and Flinn, 1965).

In the simplest possible terms, the whole field of metamorphism may be divided into Thermal (Contact) Metamorphism due to heat, Dynamic Metamorphism due to directed pressures at rather low temperatures, and Regional (also Plutonic) Metamorphism due to heat plus directed pressures. Shock Metamorphism may be regarded as a special type of Dynamic Metamorphism in which the confining pressure and temperature may be high but which is characterized by a very high rate of deformation and very high directed pressures. Burial Metamorphism may be regarded as being transitional between diagenesis and metamorphism (either Thermal or Regional).

Metamorphism is essentially a thermal phenomenon and heat is the most important source of energy allowing mineralogical and textural reconstruction during metamorphism. Energy is required to overcome initial activation energies and to allow sufficient thermal vibration for ion movement and structural change; most metamorphic reactions are endothermic and require energy. Thermal or contact metamorphism takes place with moderate to high T, low P_{con}, variable P_{H_2O} and generally zero P_{dir}. The textures (and mineralogies) of rocks in different aureoles are controlled largely by the relative rates of heat transport and mineralogical transformation.

A rock undergoing metamorphism is subjected to a complex variety of pressures consisting of the confining pressure, intergranular fluid or gas pressure and possibly directed tectonic pressures.

The confining (or load) pressure is due to the weight of the overlying material and is a function of depth and the density of the load. A porous sandstone (with an open framework) whose pores are filled with ground water and which is located near the surface will have an intergranular P_{H_2O} between $\frac{1}{4}$ and $\frac{1}{3}$ of the P_{con}. However, with increasing depth the grains are driven closer together and the intergranular water bears a proportion of the load, the proportion depending on how quickly the water can be driven out. As the permeability decreases due to compaction, a state is reached when the water cannot escape and here $P_{H_2O} = P_{con}$; this is generally regarded as the most common metamorphic circumstance but is by no means universal and in a crystalline rock which is essentially dry, metamorphism takes place with $P_{H_2O} < P_{con}$. Where gas is released in considerable quantities P_{H_2O} may exceed P_{con}; in fact the gas pressure may be sufficiently in excess to lift the roof, disrupt the rock or lower the strength by introducing a fluid between the grains.

The pressure in the pores is not only due to water; carbon dioxide will be present in metamorphosed limestones where P_{CO_2} is more important than P_{H_2O}. The pore-fluid pressure is the sum of the partial pressures of the different gases, i.e. $P_{pore} = P_{H_2O} + P_{CO}$, etc.

Non-directed pressures not only cause mechanical effects such as closer packing, reduction of pore space and increase in density but are important also in controlling mineralogical changes and melting. In general, higher *confining* pressures favour more closely packed atomic structures and thus more dense phases. Regional metamorphism of a basalt at low temperatures gives a greenschist in which sodium is in the feldspar (albite) and iron and magnesium are contained in ferromagnesian minerals such as chlorite and actinolite. Under higher pressures sodium may go into a ferromagnesian to give a glaucophane schist. The behaviour of sodium in eclogites is similarly pressure-controlled.

However, the effects of *directed* pressures are almost entirely textural or structural rather than mineralogical. Harker's suggestion that metamorphic minerals could be divided into *stress* and *anti-stress minerals* and that the formation of *stress minerals* either required or was favoured by directed pressures, and that *anti-stress* minerals could only be formed in non-stress conditions, does not appear to have stood the test of time although there is a little evidence suggesting that directed pressures may control the stability fields of some minerals. There is considerable evidence that directed stresses act almost catalytically in assisting metamorphic changes, probably by lowering activation energies and assisting diffusion; thermodynamically, the directed pressure may be added to the non-directed pressure in calculating total pressure. Directed stresses and movements enable large amounts of energy to be added quickly to a system and to be localized, e.g. along thrust zones.

A rock at a certain depth in the crust is subjected to a confining pressure due to the load of the overlying rocks and if the region is subjected to unequal stresses, the rock will flow in response. While it is flowing it is subjected to a tectonic *overpressure* (Rutland, 1965) so that it is, in effect, confined by a greater pressure than that attributable to the load. Overpressures have been invoked in an attempt to explain discrepancies between the physical conditions for various kinds of metamorphism implied by experiment, and those indicated by the geological evidence, particularly with respect to the occurrence of kyanite.

The *textures* alone of kyanite-bearing rocks indicate that the tectonic overpressure explanation is invalid. Kyanite is most common as a post-tectonic mineral and the kyanite of the most closely studied metamorphic areas occurs as an interkinematic mineral which was formed in the static period between metamorphic phases. Recent (Newton, 1966) experimental work on the aluminium silicates now makes the laboratory conditions much more compatible with those predicted by geology and removes the need for overpressures.

A correct view of metamorphic history requires an understanding of the changes in T, P_{load}, P_{H_2O} and P_{dir} with time and metamorphic rocks contain evidence of variations in all of these controls.

Change of temperature with time is involved in the concepts of progressive, retrograde and repeated metamorphism. Differences in confining pressure between regions allow the recognition of different "trends", e.g. Miyashiro (1961) suggested five trends: high pressure (with jadeite and glaucophane), high pressure intermediate (glaucophane but no jadeite), medium pressure (the kyanite–sillimanite assemblage, i.e. Barrow type), low pressure intermediate (andalusite plus staurolite), and low pressure (Buchan, andalusite plus sillimanite). However, the mineralogical assemblages in the rocks of a single region may indicate different confining pressures at different times. Harker (1939) divided regional metamorphism into two main trends, the normal (Barrow) type and a type with "deficient shearing stress". These two types, commonly referred to as the Barrovian (after Barrow) and Buchan (after the locality), are now attributed by some authors to differences in confining (non-directed) pressures but there is a certain amount of *textural* (not mineralogical) evidence supporting Harker's suggestion that some regional metamorphic rocks are much less deformed than others and it would appear that the various trends may be due only partly to differences in confining pressures and that some differences in amount and rate of deformation are significant.

Variations in P_{H_2O} to give "dry" or "wet" metamorphism may be profound. Water pressures may be high in the early to middle stages of the first metamorphism of sediments, may be lowered by compaction and folding, or temporarily raised by dehydration processes; the subsidiary nature of retrograde metamorphism suggests that P_{H_2O} is low in the late stages. Increased P_{con} and T will both tend to increase P_{H_2O}. The metamorphism of pre-existing crystalline rocks appears to take place at low P_{H_2O} and the granulite facies appears to represent "dry" conditions.

The classical view of regional metamorphism involves simultaneous deformation (constant high P_{dir}) and rising temperature with constant and equal P_{con} and P_{H_2O}. However, the history of most regional metamorphic areas appears to consist of a general rise in temperature with intermittent short periods of deformation; P_{H_2O} and P_{con} vary independently.

The concept of "progressive" metamorphism involving continuous recrystallization and reaction to form successive mineral assemblages stable at progressively increasing temperatures must be examined closely. For instance there is no evidence in many schists that the garnet has been formed by some such succession of reactions as clay → chlorite → biotite → garnet, or that a basic igneous rock has been first broken down to an albite–epidote–actinolite–chlorite assemblage and then built up to andesine–garnet–hornblende aggregate. Progressive metamorphism should occur where the rate of reaction exceeds the rate of heating. Thus it is to be expected in the regional metamorphism of wet sediments but not of dry crystallines. It is not to be expected in all thermal aureoles, especially those containing rocks belonging to the sanidinite facies. The relations of metamorphic to original textures in many rocks suggests that transformations have taken place directly from an indefinite mixed sedimentary material to garnet, sillimanite, etc., or from individual crystals of igneous olivine, pyroxene and plagioclase to metamorphic olivine, pyroxenes and plagioclase.

The Lower Limit of Metamorphism

Rocks (sediments and volcanics) which have been deeply buried but not appreciably deformed may undergo mineralogical changes at comparatively low temperatures. These changes belong to a group involving cementation, lithification, diagenesis and incipient metamorphism and it is a matter of concern to find a satisfactory definition of the boundary between lithification and metamorphism. Packham and Crook (1960) suggested that the boundary be made on the basis that processes are diagenetic until the original fabric is extensively modified. Coombs (1961, p. 213) disagreed with this and emphasized the transitional or progressive nature of the process from diagenesis to metamorphism. Coombs (1961, p. 322) defined *burial meta-*

morphism as "reconstitution without obvious relation to igneous intrusions and commonly of regional extent; incipient, extensive or incomplete. The fabric of silicate rocks is not modified by development of a schistosity.

"The metamorphism appears to follow burial and is not accompanied by significant penetrative movements. Diagenetic processes in the most restricted use of that term, that is, occurring essentially at the temperature of deposition, are excluded. Burial metamorphism has been observed to produce mineral assemblages conventionally ascribed to the zeolite facies, the greenschist facies and perhaps the glaucophane schist facies. Many slates and slightly sheared greywackes such as those of the Ch. 1. subzone of Otago mark transitions from the products of burial metamorphism to those of regional metamorphism."

Coombs (1954, 1960) stressed the transitions from undeformed burial metamorphic rocks to deformed regional metamorphics but it would appear that the Glaucophane Schist and Greenschist Facies rocks should be regarded as genetically distinct from the Zeolitic Facies of burial metamorphism.

The study of the Zeolite Facies has been almost entirely mineralogical and the textures have been largely ignored. Future work may well show that Coombs' transitions from the Zeolitic Facies (without schistosity) through the Prehnite–Pumpellyite Metagreywacke Facies (with or without schistosity) to the Greenschist or Glaucophane Schist Facies (with schistosity) can be subdivided on a textural basis.

The retention of abundant original grains and textures and also of complex mineral assemblages suggests that equilibrium, both chemical and textural, has not been achieved generally, although it may have been achieved in limited regions of matrix or cement. The textures are dominated by palimpsest (igneous or sedimentary) features but have been little studied. Reference may be made to the papers of Coombs *et al.* (1959), Crook (1960, 1961, 1963), Coombs (1954, 1960, 1961) and Packham and Crook (1960).

METAMORPHIC TEXTURES

The texture of a rock involves the size of the component crystals (both absolute and relative to each other), their shape, distribution and orientation. Textures can be divided into two main categories:

Intergranular (between grains); concerned with grain boundaries, the size and shape of crystals, preferred orientations, compositional layering etc.
Intragranular (within grains); concerned with zones, twins, kinks, sub-grain structure, exsolution intergrowths, inclusions, etc.

The origins of textures are most easily understood when regarded as functions of three variables: crystallization, deformation and time. The process of crystallization includes the *recrystallization* of existing minerals and the *crystallization* of new minerals. Inasmuch as crystallization can generally be regarded as a *positive* process in which minerals are "built up" and converted to a lower energy condition, deformation is a *negative* process in which existing crystals are strained, broken up and converted to a higher energy condition. The time factor is extremely important because both crystallization and deformation are slow processes and textures must be interpreted as much in terms of kinetics and interrupted processes as in terms of thermodynamics and an equilibrium arrangement. Minerals and mineral textures may persist metastably through a number of metamorphic episodes.

The texture of a metamorphic rock may be subdivided into three possible elements:

(1) *Relict:* original pre-metamorphic features which have not been obliterated by the metamorphism.
(2) *Typomorphic:* the characteristic texture produced by the metamorphism.
(3) *Superimposed:* alteration or modification textures due to later events which are not part of the metamorphism proper.

Metamorphic rocks which have undergone deformation and crystallization may contain evidence of various chronological relationships, e.g. crystallization of a given mineral may be said to be:

(1) Pre-tectonic, if it took place before deformation.
(2) Syntectonic, if it took place during deformation.
(3) Post-tectonic, if it took place after deformation.

The three main kinds of typomorphic textures, in *descriptive* terms, are:

(1) Granular (granoblastic).
(2) Foliated.
(3) Porphyroclastic (cataclastic).

The term "granoblastic" is used in its normal sense and "foliated" means possessing a fissility or dimensional preferred orientation, not a compositional layering. The *genetic* term "cataclastic" is generally used rather than the *descriptive* "porphyroclastic" or "mortar texture" and it is used here in the normal sense of mechanically crushed or dynamically metamorphosed. The term "cataclasis" is generally taken to indicate mechanical fragmentation without any recrystallization; it should be realized that such an ideal process is rare geologically and that most rocks accepted as cataclastic have evidence of considerable recrystallization in their fine-grained matrices.

Elements of all these textures, viz. granoblastic, foliated and cataclastic, may occur in a rock and it is not always clear to what extent each is pre-, syn-, or post-tectonic, is typomorphic or is superimposed.

Some rocks (e.g. peridotites, eclogites and granulites) are found as tectonic inclusions, i.e. as fragments in an environment which is foreign to that in which they were formed and into which they have been emplaced by tectonic processes. It is therefore necessary to distinguish between:

(1) *Exotic* textures: those which were formed elsewhere and which have persisted as relicts with respect to the existing environment.
(2) *Transit* textures: those which were produced by the process of emplacement.
(3) *In situ* textures: those which belong to the existing environment and which have been imposed on the host rocks and superimposed on previous textures in the inclusion.

The exotic character of mineral assemblages is suggested by their having compositions incompatible with the metamorphic conditions *in situ* or by having isotopic ratios indicating a greater age than that of the metamorphism. The exotic nature of enclosed structures or textures such as compositional layering, mineral foliation, lineation and lattice preferred-orientation is indicated by a greater complexity than that of the host or by discordance with that of the host.

Transit mineralogies and textures are typically concentrated at the boundaries of the tectonic inclusion where the foliation and lineation are parallel to the boundary. *In situ* textures and minerals may be concentrated at the boundaries because of the necessity of accession of chemical agents such as water but may occur sporadically throughout the body. Such textures are superimposed on older exotic or transit features (if present) but characteristically have orientations related to those of structures in the host rocks, i.e. are concordant with the host rock.

These distinctions are not always easy to make because of the tendency for tectonic inclusions to form lenses parallel to the foliation of the host rocks and to contain internal structures parallel to the long axis of the lens. Thus exotic, transit and *in situ* features may have similar orientations. Structural relations may be difficult to interpret because of the great contrast in physical behaviour between competent inclusions and incompetent country rocks.

Mineral assemblages constituting granoblastic to foliated textures are formed syn- and post-tectonically so that the textures and the minerals are contemporaneous and in equilibrium. The mineral assemblages constituting cataclastic rocks which have not been recrystallized are older than the texture and must be considered separately.

Relict Textures

Original textures found in metamorphic rocks include:

Igneous: ophitic, intergranular, intersertal, porphyritic, amygdaloidal and spherulitic.
Sedimentary: bedding (both as a compositional layering and as a structural feature accompanied by ripple marks, cross-bedding, flute casts, graded bedding, etc.), pebbles and fossils.
Metamorphic: banding, foliation, lineation, folds and porphyroblasts.

Metamorphic textures may be subdivided as follows:

(1) *Due to deformation:* intergranular textures such as foliated, lineated, mortar, flaser- and augen-structure and intragranular textures such as strained, polygonized, bent, twinned, kinked, exsolved, and polymorphically transformed crystals.

(2) *Due to crystallization:* intergranular textures such as granoblastic, porphyroblastic (maculose, knotted, spotted), heteroblastic, idiotopic, spherulitic and decussate textures; intragranular textures such as poikiloblastic and zonal textures; various intergrowths, overgrowths and reaction rims.

(3) *Due to a combination of crystallization and deformation:* foliated, lineated, layered, granoblastic, porphyroblastic and poikiloblastic textures.

Original textural or structural features which persist after metamorphism are extremely important in indicating the original nature of a rock. Premetamorphic minerals or textural features are said to be *relics* (noun) or *relict* (adjective), or to be *palimpsests* or *palimpsest textures.* The words "relic" and "relict" have come to be used differently from everyday English where both tend to be used as nouns and where "relict" is obsolete except in a legal sense. The degree of persistence of an original feature depends mainly on its size and the degree by which it differs from its surroundings, e.g. bedding marked by shaly parting is easily obliterated but bedding marked by alternations of contrasting composition (sandstone and shale) is not; fine-grained sediment (lutite) is more easily recrystallized than coarser (arenite); small fossils or pebbles disappear more quickly than large ones; calcitic fragments or fossils in a calcareous matrix are easily destroyed but a calcareous fossil in an argillaceous matrix may be replaced by wollastonite (pseudomorphed) and be recognizable even after high-grade metamorphism. Vitroclastic and flow textures in acid volcanics are easily destroyed but large phenocrysts of quartz or feldspar may retain their characteristic shape even in strongly sheared rocks (porphyroids).

The Greek root "blastos" is used in two ways:

(1) The prefix "blasto" refers to a relict texture which has been modified by metamorphism but which is still recognizable, e.g. blastoporphyritic (i.e. modified porphyritic, original phenocrysts still remain in a metamorphosed igneous rock), blasto-ophitic, blastopsammitic, etc.

(2) The suffix "blast" or "blastic" refers to a true metamorphic texture (e.g. porphyroblastic texture contains large crystals formed during metamorphism and is comparable with the porphyritic texture of igneous rocks). Porphyroblastic, granoblastic and poikiloblastic textures are examples. The term "granoblastic" is widely used as the metamorphic equivalent of "granular" in igneous rocks but departures from this convention are very common.

Original inhomogeneities may be sufficiently large to control the mechanical and chemical behaviour of a rock during metamorphism. Bedding, particularly in sediments with contrasting compositional layering, is important in controlling the direction of mechanical movements during dynamic and regional metamorphism. The strength of a rock in compression is greater across than along the bedding but its permeability and diffusivity may be an order of magnitude greater along than across the bedding. Thus elongate and flaky minerals (e.g. amphiboles, epidotes, micas, etc.) show a distinct tendency to grow preferentially along bedding or pre-existing foliation. Such textures which imitate or "mimic" an original texture are said to be *mimetic*. Nucleation of certain new minerals may take place preferentially on premetamorphic mineral grains which are dispersed along the bedding thus the bedding layering may be accentuated, not destroyed, by metamorphism.

It must be emphasized, however, that not all banding, layering, "striping" or foliation in schists, gneisses, amphibolites, quartzites and marbles is necessarily relict bedding. It may be an entirely metamorphic structure due to metamorphic differentiation or it may be relict bedding which has been folded, transposed and rotated around to coincide with a metamorphic foliation.

THE "INTERGRANULAR FLUID"

The role of the intergranular fluid in metamorphism and metasomatism has been discussed extensively (Harker, 1939; Grout, 1941; Ramberg, 1952; Read, 1957; Fyfe *et al.*, 1958).

The intergranular fluid only concerns us at present in the way in which it might affect the textures of metamorphic rocks and we are relieved of making any decision as to its probable presence or absence by the conclusion that there are very few signs of its presence in the textures and it does not appear to have a *significant* control over their formation. The presence of an intergranular fluid is considered probable because water is abundantly present in sediments before metamorphism, is liberated by dehydration processes, may be added by metasomatism and once present in a rock is very difficult to remove.

Due to the relatively low thermal conductivity of rocks, heated water may be an efficacious medium for transferring heat. Experimental evidence shows that the rate of chemical reaction between solids may be increased by orders of magnitude on the addition of small quantities of water to the system. The majority of metamorphic changes involve H_2O as an active phase and its abundance, chemical activity or pressure control the thermodynamics and kinetics of many reactions.

It seems clear, however, that the metamorphism of some crystalline rocks at least is essentially "dry". Ramberg (1952) and others have questioned the conventional idea of a complete intergranular film of water adhering to all grain boundaries and allowing reaction to take place by solution and redeposition in the fluid phase. There is certainly very little room for water along the grain boundaries of many crystalline metamorphic rocks. If water is the "intergranular fluid", it will be a compressed superheated fluid under most metamorphic conditions, i.e. it may be comparatively dense like a liquid but have insignificant surface tension like a gas. Little is known of the internal structure, degree of disorder, adsorption and wetting powers, etc., of the hypothetical "intergranular fluid" and no serious attempt has been made to estimate its detailed behaviour (Fyfe *et al.*, 1958, p. 36).

For most metamorphic conditions the viscosity of intergranular water would be minute at about 2×10^{-3} poises. The surface tension at 20°C is 72·8 dynes/cm and is higher than that of all the common liquids except mercury. Solutions of inorganic salts have higher surface tension than pure water but dissolved gases reduce the surface tension. The surface tension of water decreases with increase of temperature until at and above the critical temperature it is reduced to zero. The *critical temperature* of a fluid is defined as the temperature above which a gas cannot be converted into a liquid by increase of pressure. The *critical pressure* is the least pressure which will liquefy the gas at the critical temperature. The critical point is defined by the critical temperature plus the critical pressure and for H_2O is 374·0°C, 224·9 kg/cm^2 when it has a density of 0·4 g/cm^3 (Turner and Verhoogen, 1960).

Compaction of a water-bearing sediment drives out the water by reduction of pore space until the grains come in contact. Under sufficiently high stresses the grains will either deform plastically or will recrystallize to further reduce the pore space until all the grains are separated by a mere film of fluid. If the fluid wets the solid surfaces it will remain at the boundary and not be driven out, but if its surface tension is high the liquid will reduce its surface area by contracting into globules, leaving the intergranular surfaces dry.

The surface tension per unit area of an intergranular boundary depends on a variety of factors and some boundaries have an even lower energy than the solid–liquid interface and thus will not be wetted. The interfacial tensions are temperature-dependent but the solid–liquid tension will decrease much more per unit increase of temperature.

Thus high temperatures will favour wetting of grain boundaries, but the wetting behaviour of an intergranular fluid cannot be predicted at moderate metamorphic temperatures because the surface tensions of minerals are not known. However, some guesses can be made. The surface tension of water is very low so that a silicate–water interface will also be low, e.g. the surface energy of the intergranular boundary between mica flakes may be about 1600 ergs/cm^2; a figure of 1170 ergs/cm^2 has been given for the mica–water interface, so that intergranular water penetrates an aggregate of mica and mica sheets may be split with a pin and a drop of water. The water penetrates along the cleavage very quickly and splits the flakes apart. The strong wetting power is no doubt due to chemical absorption of ions in the water on loose bonds on the mica surface. This kind of process may be important in allowing water to wet a variety of silicates (Fyfe *et al.*, 1958, p. 42; De Vore, 1963). The introduction of strongly polarizable cations to a surface and the asymmetrical field orientation brings about a considerable increase in the contact angle (Eitel, 1954, p. 343). Hydroxyl ions on the crystal surface improve wettability with water because hydrogen bonds "protrude" from the surface.

A considerable interchange of oxygen between silica and high-pressure stream takes place at metamorphic temperatures and pressures (Spitzer and Ligenza, 1961) and it is suggested that water reacts with silica to form internal silicon hydroxide (silanol) groups. The demonstration (Griggs and Blasic, 1965) that H_2O at very high pressures markedly reduces the strength of quartz suggests that water enters the silica structure and breaks bonds between the tetrahedra.

Thus it can be seen that H_2O under metamorphic conditions is a mobile tenuous fluid. It has a strong wetting ability towards rock-forming minerals and once present in a rock it is difficult to see how even extreme metamorphic conditions could completely remove it from grain boundaries. However, this should not be taken to indicate that such adsorbed water would be present in sufficient bulk or in an appropriate condition to act as a solvent as in the classical view of the role of the intergranular fluid. Nevertheless, the presence of H_2O molecules could well affect the magnitude of the grain boundary energies and their relative mobilities.

It is difficult to find petrographic evidence as to whether an intergranular fluid phase was present at the time of recrystallization of most metamorphic rocks. Aggregates whose grain boundaries are completely wetted by the intergranular fluid might behave similarly to those which are completely dry, but although the end result might be the same, the rates of various processes would be different. On the other hand, an aggregate containing some coherent, low-energy, dry (solid–solid) boundaries and some high-angle, high-energy, wet (solid–liquid) boundaries might not be equigranular with equal interfacial angles at triple-points and the grain size and grain shape would probably be very irregular. An intergranular liquid might enhance nucleation and crystallization rates and in this way might increase the mobility of the grain boundary. On the other hand, processes which belong to strongly coherent or highly coincident, solid–solid boundaries (slip, twinning, strain-induced boundary migration, polygonization) would be depressed.

Textural readjustments involving solution and precipitation of minerals in a pore water phase take place under small strains during diagenesis by the process known as "pressure solution". This process converts sandstones to quartzites, recrystallizes limestones and evaporites, and operates during diagenesis, burial and metamorphism to produce stylolites, indented pebbles and minor recrystallization.

Consider a porous sandstone whose intergranular cavities are filled with water. The load of overlying material is transmitted through the grain contacts so that each grain contains regions which are strained and are under high pressure and other regions adjacent to pores which are less strained and at lower pressure. This inhomogeneity results in preferential solution of the quartz in the pore water at the regions of high pressure (contacts) and deposition in regions of low pressure (pores). The sandstone may become less porous, more dense and develop a siliceous cement (Carozzi, 1960, pp. 7–36). The quartz grains become angular with a low sphericity almost entirely due to pressure solution (Heald, 1955) and grain boundaries may be sutured. The grains are commonly not fractured at contacts but may penetrate into each other. A similar process occurs in conglomerates (Keunen, 1943) where pebbles penetrate into each other at contacts to produce deep indentations ("pitted pebbles").

In certain common, but not well-understood circumstances, irregular, sutured or dentate surfaces of discontinuity (*stylolites*) are produced through limestones and sandstones by pressure solution. The stylolitic surfaces may cut across grains, fossils or bedding (Plate XII). Stylolites in cross-section have an irregular, wave-like form in which the amplitude (which may be up to about 5 cm or down to microscopic size) is greater than the wavelength, and the crests and troughs are sharp. Some stylolitic surfaces are composed of columnar projections with striated sides.

Contraction in the direction perpendicular to the stylolite (thus parallel to the load) is demonstrated by the reduction in dimensions of objects of known shape such as fossils or ooliths (Plate XII). Soluble material is removed from the stylolite surface but an insoluble residue may remain to mark it as a dark line in thin section. Heald (1955, p. 105) has shown an order of increasing tendency to pressure solution from (least) zircon and pyrite; sphene, tourmaline and collophanite; micas, sericite and clays; feldspar; quartz; calcite; to hematite (greatest).

Pressure solution may be partly controlled by the surface energy of the interface (Correns, 1926, 1949; Voll, 1960, p. 533). Ernst and Blatt (1964) have suggested that when the load-pressure exceeds the pore solution pressure there will be solution at grain contacts, etc., but where the two pressures are equal the quartz grains will be enlarged by overgrowths.

Overgrowths due to secondary enlargement of quartz commonly give a rim which is in optical continuity with the central grain. In some instances the sandstone contains grains which had been strained and were undulose prior to deposition in the sandstone and these have unstrained secondary rims. Where the rock has been strained after cementation, both centre and rim may show identical undulose behaviour. However, Ernst and Blatt (1964) have given experimental evidence suggesting that an imperfect rim with a curved lattice and undulose extinction may be formed on a strained nucleus grain because the defects of the core are extended out into the epitaxial rim.

The development of crystal faces is much more strongly favoured at a solid–liquid interface (e.g. crystal in a melt or solution) than a solid–solid interface. Smith (1964, p. 46) considered that "growth of well-formed idiomorphic crystals in a sand or shale matrix can only occur if capillarity carries liquid continuously to all parts of the growing surface. The interfering particles will be perpetually displaced, not by the growing crystal, but by the layer of liquid carried into the interface which perpetually prevents contacts."

Idiotopic texture (in which most crystals are idioblastic) is uncommon; indeed it is commonly claimed that a crystal cannot develop crystal faces in the solid state when surrounded by its own species. It is perhaps significant that the texture is most common in minerals with comparatively high solubilities; an intergranular solvent may be necessary for its formation. Carozzi (1960, p. 421) considered that this texture is primary in sedimentary anhydrite and is not a recrystallization texture.

Solid diffusion. Solid diffusion (inasmuch as it influences textures) involves a number of concepts (Christian, 1965).

1. *Self-diffusion* in the simplest case is the movement of a unit of a certain composition through a crystal lattice of the same composition.

2. *Volume diffusion* is the movement of atoms or ions through a lattice. This may be the simple self-diffusion of metals discussed above or the more complex case of ions of a certain species migrating through a lattice containing a variety of ions of different sizes or charges and in various configurations.

Diffusion is difficult at low temperatures because the ions of the lattice may form such a tight mass that penetration is impossible. The crystal lattice may be regarded as a fairly closely packed aggregate of large ions (oxygen in most rock-forming minerals) with the smaller cations fitting between, although even the perfect lattice would have spaces along which small atoms can move.

Volume diffusion may result from mechanical causes, e.g. Nabarro–Herring stress-induced diffusion, or from chemical causes. The simplest of these is controlled by an order–disorder relation.

Self diffusion in a solid leads to change of shape and texture, but differential migration leads to changes in composition and is important in metasomatism. The lattice is regarded as a rigid framework (of oxygen ions or SiO_4 tetrahedra in silicates) through which independent diffusion of different interstitial cations takes place.

The effect of pressure on diffusion is not clear. Read (1957, p. 237) stated that "high pressures *promote* solid diffusion in that they decrease the pore space and lessen pore magma", but Girifalco (1964) considered that "increased pressures *decrease* ionic diffusion because the repulsive forces between atoms increase with decreasing distance".

Diffusion through polycrystalline aggregates is aided by the presence of point defects and various kinds of two- and three-dimensional defects including edge and screw dislocations, cleavages, cracks, grain boundaries and crystal surfaces. The mean jump frequency of an atom at such defects is greater than that for an atom in the interior so that diffusion may be expected to be easier.

Volume diffusion in the complex silicates which are the main metamorphic minerals appears to be very difficult and slow because of the complexity of the ideal structures and the nature of the bonding (summary in Fyfe *et al.*, 1958, p. 60). Volume diffusion will thus be important only at high grades of metamorphism. Diffusion along dislocations or dislocation arrays, and along grain boundaries, is thus the more likely mechanism in dry metamorphic rocks but the extreme sluggishness of all dry processes is one of the main arguments for the existence of a fluid phase in metamorphic rocks.

The kinetics of grain boundary diffusion is profoundly influenced by the nature of the boundaries themselves. The activation energy is lower and the rate of diffusion higher in the following boundaries (in order): highly coherent boundaries (such as low-angle tilt and twin boundaries), semi-coherent boundaries, boundaries with regions of alternate good and bad fit, high-angle incoherent boundaries, boundaries with large intergranular spaces, open boundaries occupied by an intergranular fluid phase.

Diffusion may differ in rate in different directions within a single grain boundary if there is a preferred orientation of linear dislocations along it.

Lattice and grain boundary diffusion are both strongly dependent on the concentration of defects, and as deformation produces defects it might be expected that the rate of diffusion might increase during deformation. A literature survey, however, shows many conflicting results as to whether contemporaneous deformation does, or does not, influence the rate of diffusion (Fyfe *et al.*, 1958, p. 64; Guy and Philibert, 1961). Diffusion has been enhanced by two or three orders of magnitude in certain metals under certain conditions but no change has been found in others. Plastic deformation of ionic solids produces point defects (Comagni, Chianotti and Manara, 1960) and thus diffusion should be promoted by deformation, but Riggs (1964) concluded that an increase in diffusion in metals was only appreciable at high strain rates and that it might increase to a maximum and then decrease.

CHAPTER 2

Metamorphic Minerals and the Solid State

Most metamorphic rocks consist of less than half a dozen major minerals out of a total selection of a couple of dozen mineral groups. The mineral assemblage depends on the bulk chemical composition of the rock and on the physical conditions of metamorphism. The view that a metamorphic rock is a complex multicomponent chemical system in which the phases (minerals) constitute the most stable assemblage having been reduced to a minimum number depending on the bulk composition is the basis of the Phase Rule treatment and the Facies classification of metamorphic rocks. However, metamorphism cannot be adequately described in terms of stable chemical end-products alone. It is a process which takes place over a period of time and involves numerous processes which take place at differing rates to approach or achieve a lowest energy state.

A change in texture is a change in the relative size, shape and distribution of crystalline solids and this takes place by the transformation of one group of crystals to another. A metamorphic rock is a polycrystalline assemblage in which each crystal is surrounded by crystals of similar or different composition, structure or orientation, and from which it is separated by grain boundaries. Metamorphism involves the breaking and reforming of bonds, translation and rotation of ionic units and the diffusion of ions to form new assemblages or orientations of crystals.

The behaviour of a mineral during metamorphism, i.e. the way it becomes mechanically deformed or the way it grows, depends largely on its atomic structure. However, the behaviour of a pure metal with its identical atoms closely packed and weakly held by metallic bonds to give a structure of high symmetry is not markedly different from that of a simple ionic solid (halite) with moderate ductility or a brittle homopolar compound (sulphide ore mineral) or a rock-forming silicate consisting of different kinds of atomic units or ionic groups strongly held by a complex network of bonds of mixed ionic and homopolar character in an open structure of low symmetry.

An excellent modern approach to the structure of solids in general, and of minerals in particular, has been given by Fyfe (1964). Other more detailed treatments are given by Azaroff (1960), van Bueren (1961), Eitel (1958, 1965) and Smith (1963) and only a summary of the more important principles necessary for an understanding of mineralogical changes in metamorphism will be discussed here.

Crystallography conventionally regards crystals as *ideal*. Crystal lattices of minerals are described in terms of unit cells containing ions and ionic groups, such as silica-tetrahedra or alumina-octahedra, built up by various symmetry operations. However, as long ago as 1914 Darwin appreciated that *real* crystals were not *ideal* and that crystalline solids contain imperfections or regions where the regular pattern breaks down so that certain atoms are not properly surrounded by their neighbours (Gray *et al.*, 1957).

REAL CRYSTALS—THE DEFECT SOLID STATE

Real solids are not perfect and contain many imperfections. Most quantitative metallurgical explanations of phenomena such as gliding, twinning, grain boundary movement, nucleation, growth, diffusion and creep involve dislocation mechanisms and it is clear that such an approach is essential in understanding metamorphic textures. Extensive details are given in texts such as Cottrell (1953, 1964a, b), Read (1953), van Bueren (1961), and Amelinckx (1964).

Dislocations cause considerable distortion in the lattice. An edge dislocation can be regarded as an extra plane inserted in the lattice which is compressed in the region of the extra plane but is in tension opposite to it. Bending occurs around an *edge* dislocation and twisting or shearing around a *screw*. The energy of the dislocation is the sum total of the energies of all the displaced atoms and thus a certain amount of strain energy is involved in each dislocation. A reduction in strain will occur when dislocations interact and cancel each other or disappear in other ways and as the energy is proportional to the length there is an energy gradient tending to straighten them. Dislocations are visible under the electron microscope because of the diffraction effects of the strained region along and around them.

Dislocations are present in all real crystals. They may be present *ab initio* from original growth or may be produced during the deformation of the crystal. Dislocations can be produced at grain boundaries or inclusions, etc., and at special sources.

The *dislocation density* is defined as the total length of all dislocation lines contained in unit volume. As it is expressed as a length per unit volume it has the dimensions of l^2 and is expressed as a number per cm^2. If the dislocations are parallel straight lines, the density is the number of dislocations cutting unit area normal to the lines. Metal crystals can be grown with densities as low as 10^2 per cm^2, and "whiskers" may have very low dislocation densities. Low values (10^0 to 10^2) have been suggested for high-quality quartz crystals (McLaren, personal communication) and for artificial sapphire crystals by Conrad, Stone and Janowski (1964) with $10^6/cm^2$ before deformation and 5×10^7 after deformation. Griggs and Blasic (1965) estimated a density of $10^9/cm^2$ for some synthetic quartz. Strongly deformed metal crystals have densities up to $10^{14}/cm^2$ and annealed metals may be as low as $10^8/cm^2$. The latter density indicates approximately one dislocation per square micron and this is very low.

Dislocations are not uniformly distributed in deformed crystals (Christian, 1965, p. 293) and a pronounced substructure may develop. This consists of cells of relatively low dislocation density with slightly different orientations, separated from each other by regions of high dislocation density. The dislocation tangles form walls and the cells act as "preformed" nucleii during annealing.

Emergent screw dislocations which allow a crystal growth on a spiral growth ramp are not uncommon in crystals precipitated from liquids, e.g. synthetic quartz (Joshi and Vagh, 1966), but appear to be rare in metamorphic crystals.

CHAPTER 3

Grain Boundaries

THE size and shape of the constituent crystals in a metamorphic rock reflect its history. Each individual grain is surrounded by, and separated from, adjacent crystals, by a grain boundary which exists because the grains on either side are different in composition, structure or orientation (Smith, 1948, 1953a, b, 1954, 1964; McLean, 1957). The texture of a rock can be pictured in terms of the configuration of the grain boundary system, and changes in texture (e.g. grain growth) in a single-phase polycrystalline aggregate can be considered in terms of movements of grain boundaries. Some effects of grain boundaries are due to the difference in composition or structure across them, e.g. part of the difference in mechanical behaviour between a single crystal and polycrystalline aggregate is due to the different orientations of the crystals. However, other effects are due to the boundary itself being a region which is physically distinct from the crystals.

Interphase boundaries (between crystals of different compositions) differ considerably in behaviour from *self-boundaries* (between crystals of the same com-

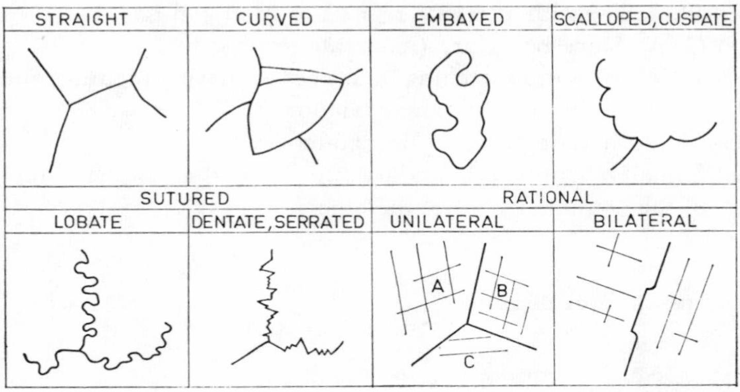

FIG. 1. Grain boundary shapes

position) which may have a considerable degree of *coherence* or correspondence of structure. The orientation of a self-boundary has five degrees of freedom; the relative orientation of the two adjacent lattices can be related to rotation about three axes, leaving two degrees of freedom for the orientation of the boundary.

Grain boundaries can be classified on their shape (straight, curved, sutured, dentate, serrated, cuspate, scalloped, irregular, etc.) (Fig. 1) or on their origin (impingement, mobile, arrested, etc.) and these are discussed further when considering crystal shapes in Chapter VI.

Boundaries are best classified on their structure as this controls their energy, mobility and ultimate shape. The boundary may have a *symmetrical* or *asymmetrical* orientation with respect to the adjacent lattices. The term "coherent" is used in its metallurgical sense and does not necessarily imply anything about the *adhesion* of the boundary.

STRUCTURAL CLASSIFICATION OF GRAIN BOUNDARIES (Fig. 2)

I. Self-boundaries

(1) Coherent (interface in a rational direction for both lattices plus a close match of lattices).
- (a) Twin boundary.
- (b) Coincidence boundary (good fit).
- (c) Rational boundary, bilateral.

(2) Semi-coherent (interface rational for both lattices but only a proportion of sites match).
- (a) Low-angle tilt boundary.
- (b) Boundary with alternate regions of good and bad fit.
- (c) Coincidence boundary (moderate to poor fit).

(3) Non-coherent (no matching of lattices or sharing of lattice sites).
- (a) Ordinary misfit, high-angle boundary.
- (b) Non-coherent, twin, or kink boundary.
- (c) Unilateral rational boundary (i.e. rational for a crystal on one side of the boundary but not the other).

II. Interphase Boundaries

(1) Coherent.
- (a) Martensite transformation boundary.
- (b) Stacking fault boundary.
- (c) Exsolution lamella boundary.

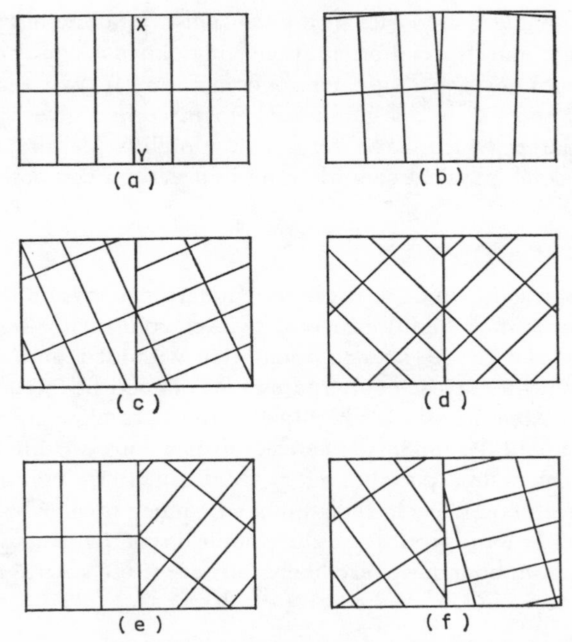

FIG. 2. Models of grain boundary *XY* at the atomic level. (a) Coherent contact of two perfect parallel crystals. (b) Symmetrical, low-angle tilt boundary. (c) Coincidence boundary. (d) Coherent twin boundary. (e) Asymmetrical, high-angle, unilaterally rational boundary. (f) Asymmetrical, high-angle boundary

(2) Semi-coherent.
  (a) Unilateral rational boundary.
  (b) Epitaxial boundary.
  (c) Topotactic junction.
  (d) Fortuitous, concidence boundary.
(3) Non-coherent.
  (a) Normal moderate- to high-angle grain boundaries.

The boundary may thus range from a coherent twin boundary where there is a minimum structural disturbance and a geometrical relationship between the adjacent lattices, to a disordered region a few atom diameters wide in which the structure is more like that of a liquid. As the misorientation of the two lattices increases, so the structure changes from one of good "fit" with regularly distributed dislocations to one with poor "fit" where the lattices are unrelated and the boundary is rather open. Between this are special configurations which happen to show some atomic correspondence. Many boundaries must consist of "islands" of good atomic matching separated by regions in which the matching is poor as suggested by Mott (1948).

Not all high-angle boundaries in a given polycrystalline aggregate have the same energy, and "special orientations" may have a significant number of coincident sites and thus a greater coherence. A 40° twist around $\langle 111 \rangle$ in aluminium, and 22° or 38° around $\langle 111 \rangle$ in copper gives special coincidence boundaries which have very high mobility (despite their lower energy) under some physical conditions and for certain compositions.

GRAIN BOUNDARY WIDTH

The width of a solid–solid (dry) grain boundary depends on the degree of boundary coherence. A highly coherent twin or coincidence boundary has zero width and the lattices change orientation without a gap. Moderately coherent boundaries may have alternate areas of good fit (zero width) and poor fit (small space). In metals a highly discordant, random grain boundary may be from about 1 atomic diameter if dislocations are interspersed, to perhaps 3 diameters for a poor boundary (McLean, 1957, 1965). The width of a random grain boundary in silicates may be much greater because of the size of the atoms or ionic groups and the complexity of the structure. Regions of bad fit on quartz boundaries are likely to range up to several multiples of the diameter of an (SiO_4) tetrahedron (4 to 10 Å).

GRAIN BOUNDARY AREA

It is not easy to calculate accurately the grain boundary area in a rock because of the irregular shape of the grains but a reasonable, simple approximation can be made:

Consider a cube of rock of side a centimetres composed of cubes of side D centimetres situated within a very large volume of rock. The surface area of each cube face is D^2 and the total surface area of each crystal is $6D^2$. The total number of cubes is $(a/D)^3$. The total grain boundary area is thus

$$\tfrac{1}{2}6D^2 \cdot \frac{a^3}{D^3} = \frac{3a^3}{D}$$

the area being halved because the unit cubes are in contact with each other. If $a = 1$ cm, then the grain boundary area in 1 cm$^3 = \frac{3}{D}$ cm^2.

Alternatively, if a traverse of length L across a polished surface or thin section crosses n grain boundaries $\left(\text{the grain size is } \frac{L}{n}\right)$, then the grain boundary area in a cube of side L is

$$A = 3nL^2$$

The grains in a real rock depart appreciably from a cubic shape and resemble tetrakaidodecahedra in a granoblastic polygonal texture. This form has a hexagonal cross-section in most directions. Measurements of grain boundary areas using interlocking hexagonal networks do not differ appreciably from $\frac{3}{D}$ cm^2/cm^3. Irregular (sutured, dentate) boundaries will increase the figure slightly and a mean grain diameter should be taken for elongate crystals.

The grain boundary area for a rock with grain diameter 0·001 mm (i.e. 10^{-4} cm) is thus about 3×10^4 cm^2/cm^3, with diameter 0·01 mm (10^{-3} cm) is 3×10^3 cm^2/cm^3. Recrystallization of a chert with grain size 0·001 mm to a quartzite (1·0 mm) involves the loss of almost the whole grain boundary area (3×10^4 cm^2/cm^3). Recrystallization of sandstone with 1 mm crystals to a quartzite with 3 mm (3×10^{-1} cm) crystals involves only a loss of 2 cm^2/cm^3. An aggregate of clay particles 1μ (10^{-4} cm) across has a grain boundary area of 10^5–10^6 cm^2/cm^3.

Grain boundary motion. The change of size or shape of a crystal may be considered in terms of the movement of its boundary surface. Very little is known of boundary motion in rock-forming minerals but those in metals have been considered in terms of the following (Rutter and Aust, 1960):

(1) Motion of dislocations, where the boundary is regarded merely as an array of dislocations.
(2) Group process theory, i.e. transfer of groups of atoms across the boundary,
(3) Volume diffusion of the boundary.

The *mobility* of a grain boundary refers to the rate of movement of a boundary between two crystals of the same species, under a given driving force. The boundary may move during annealing recrystallization from an unstrained into a strained grain by *strain-induced boundary migration* where the driving force is the difference in strain energy. The boundary may be moved during deformation by *stress-induced boundary migration* where the driving force is stress difference in different parts of a crystal which is being deformed. A boundary between two grains of the same species but with differing orientations may move under the action of stress by transforming the structure at the boundary from that of the phase ahead of the boundary to that of the advancing phase. The change is analogous to a martensitic transformation.

Mobility is enhanced by increase of temperature and lack of impurity phases but mostly depends on the orientation of the boundary with respect to the two lattices and the mutual orientation of the two lattices. A good fit or high coherence means a low interfacial energy (Fig. 2) and low mobility; high-angle, high-energy boundaries are most mobile. Twin boundaries or crystal faces have low energies and are particularly stable and immobile. The true low-angle tilt boundary is an exceptional case and is probably rare; it has a low angle and low energy but a high mobility.

Low-energy boundaries (such as those parallel to rational lattice planes, or at least straight if irrational) are stable and suggest thermodynamic equilibrium (this is also suggested by equal-angle triple-points). Many straight boundaries in regionally metamorphosed rocks have unequal triple points and are not in equilibrium. This is true also for curved boundaries which are typical of textures frozen while in the process of mutual adjustment.

Serrated, dentate and scalloped boundaries are also non-equilibrium "frozen" mobile boundaries, formed in some instances by minor late stage deformation. The indentations probably indicate movement approximately normal to the boundary (Plate VIId).

A number of origins are possible for the various kinds of irregular boundaries:

(1) Most dentate boundaries in moderately deformed rocks occur between slightly undulose and strained quartz (Macgregor, 1950; Voll, 1960) and have been attributed to strain-induced boundary migration (Flinn, 1965, p. 55). Deformation of a quartzite with regular grain boundaries produces strain which is moderately evenly distributed within the grains but is inhomogeneously distributed along the boundaries. Incipient recrystallization results in slight boundary movements from less-strained into more-strained regions so as to reduce the total amount of strained crystal). The "teeth" on the boundary represent greater grain boundary movement into small regions of higher strain.

(2) Some dentate boundaries in well-crystallized quartzites might possibly consist of rational (rhombohedral) faces which could be stable (Voll, 1960).

(3) The irregular, somewhat blurred and vaguely "grainy" margins of some quartz crystals in mylonites, gneisses, granulites and quartzites with mortar texture, differ from normal sutured boundaries and are probably due to crushing and recrystallization (Turner, 1948, p. 252).

(4) A certain amount of minor irregularity will be produced during normal grain boundary movement. Small irregularities may be caused by dislocations, slip planes, cracks and cleavages enhancing or hindering movement and large irregularities may be due to pinning of the boundaries by inclusions.

(5) A certain amount of grain boundary irregularity in slightly deformed sandstones appears to be stylolitic and caused by pressure solution.

(6) It is possible that quartz boundaries can move by the process of stress-induced boundary migration. A boundary between two crystals of the same phase may be moved from one into the other by the application of stress.

PLATE I. (a) Growth-mosaic substructure in vein quartz (×20). (b) Quartz phenocryst in porphyry with growth zoning and growth-mosaic substructure in the upper edge (×12). (c) Detail of substructure in (b) (×26). (d) Deformation substructure in quartz phenocryst of sheared porphyritic rock (×15)

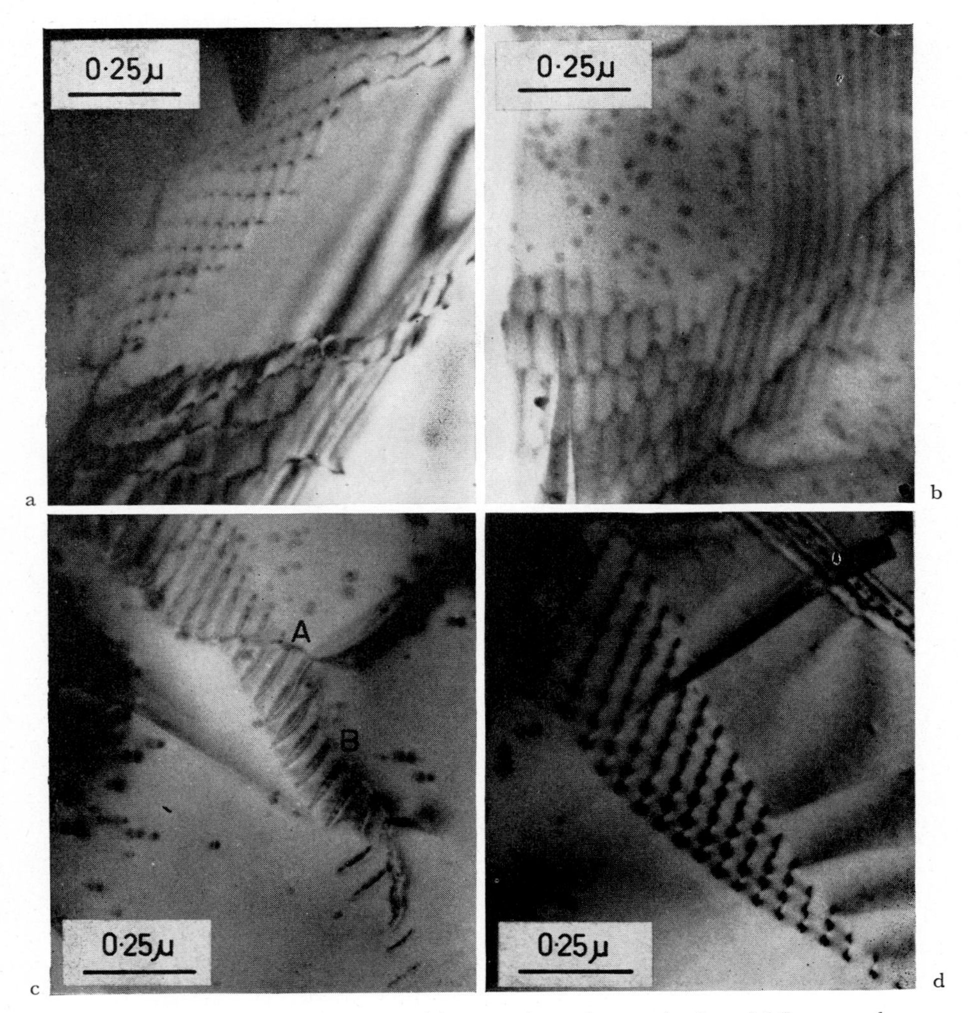

PLATE II. Dislocations in quartz (electron photomicrographs from McLaren and Phakey, 1965). (a) Hexagonal network of screw dislocations. (b) Hexagonal dislocation network. (c) Sub-grain boundary composed of basal edge dislocations. (d) Array of simple edge dislocations forming a tilt boundary

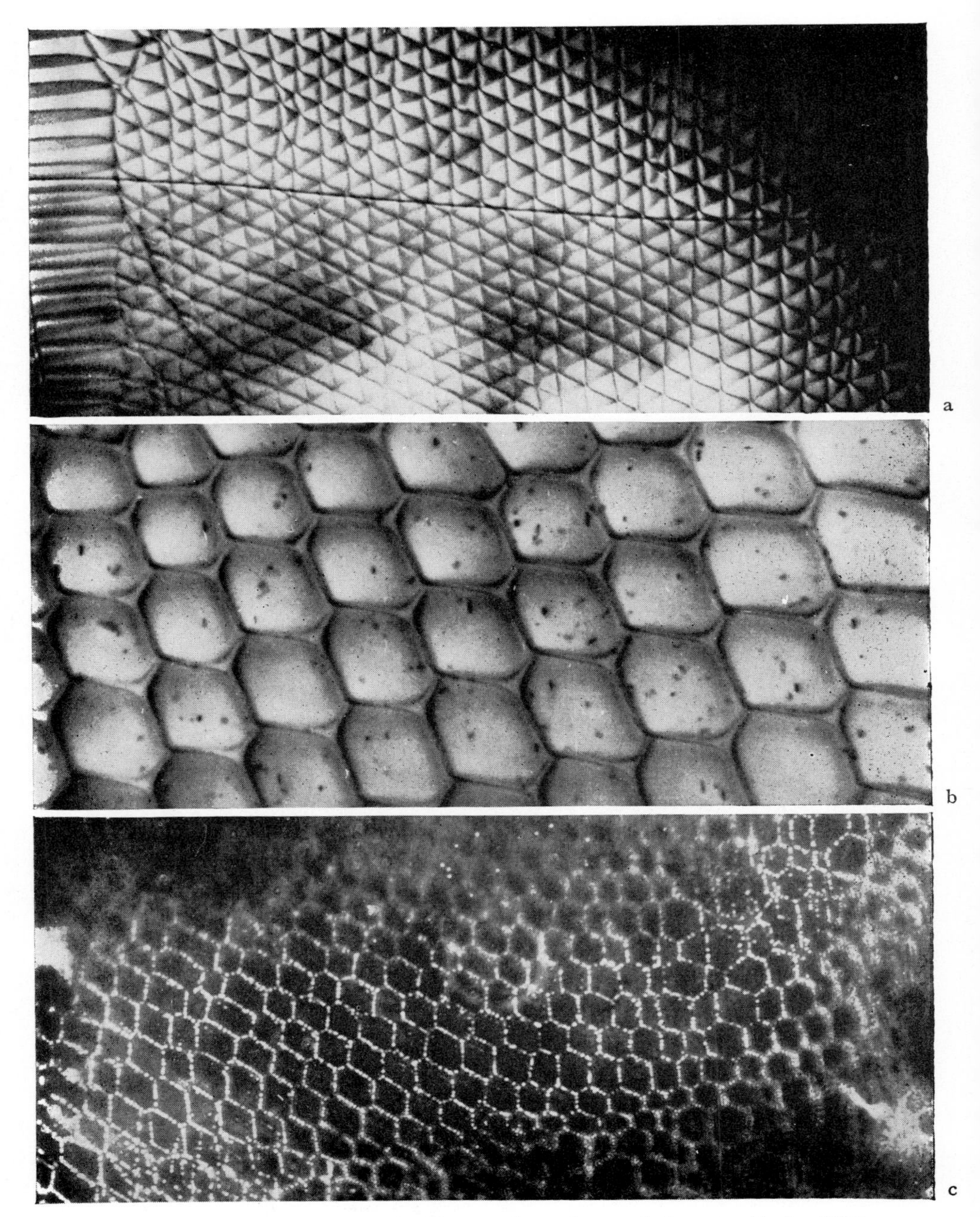

PLATE III. Dislocation networks (electron photomicrographs from Amelinckx, 1964). (a) Network of nodes in molybdenum sulphide ($\times 20{,}000$). (b) Network in tin disulphide ($\times 30{,}000$). (c) Twist boundary in potassium chloride formed by a hexagonal network

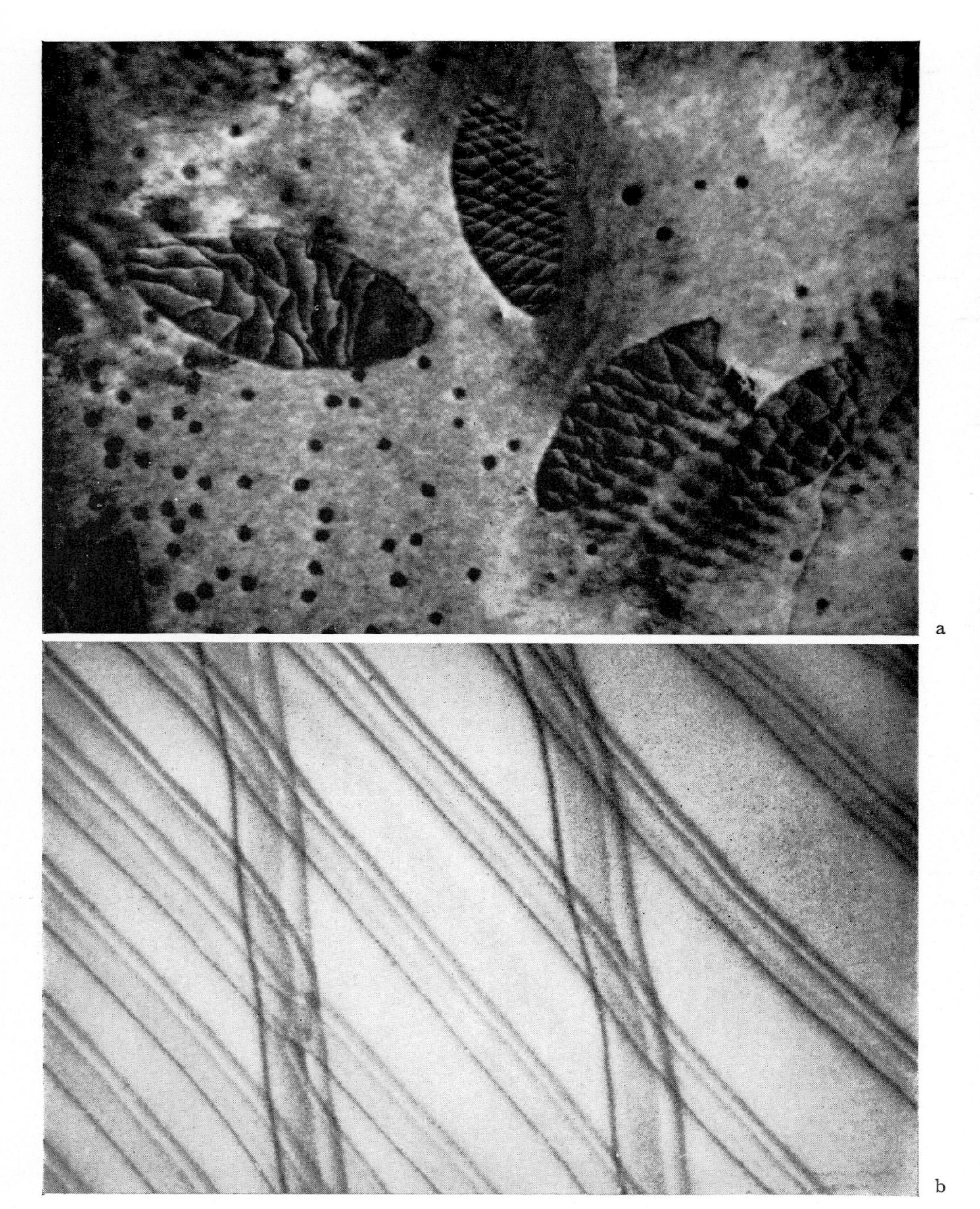

Plate IV. (a) Epitaxial dislocation networks between oriented overgrowths and chromium bromide substrate (×50,000). (b) Stacking fault ribbons in talc (electron photomicrographs from Amelinckx, 1964)

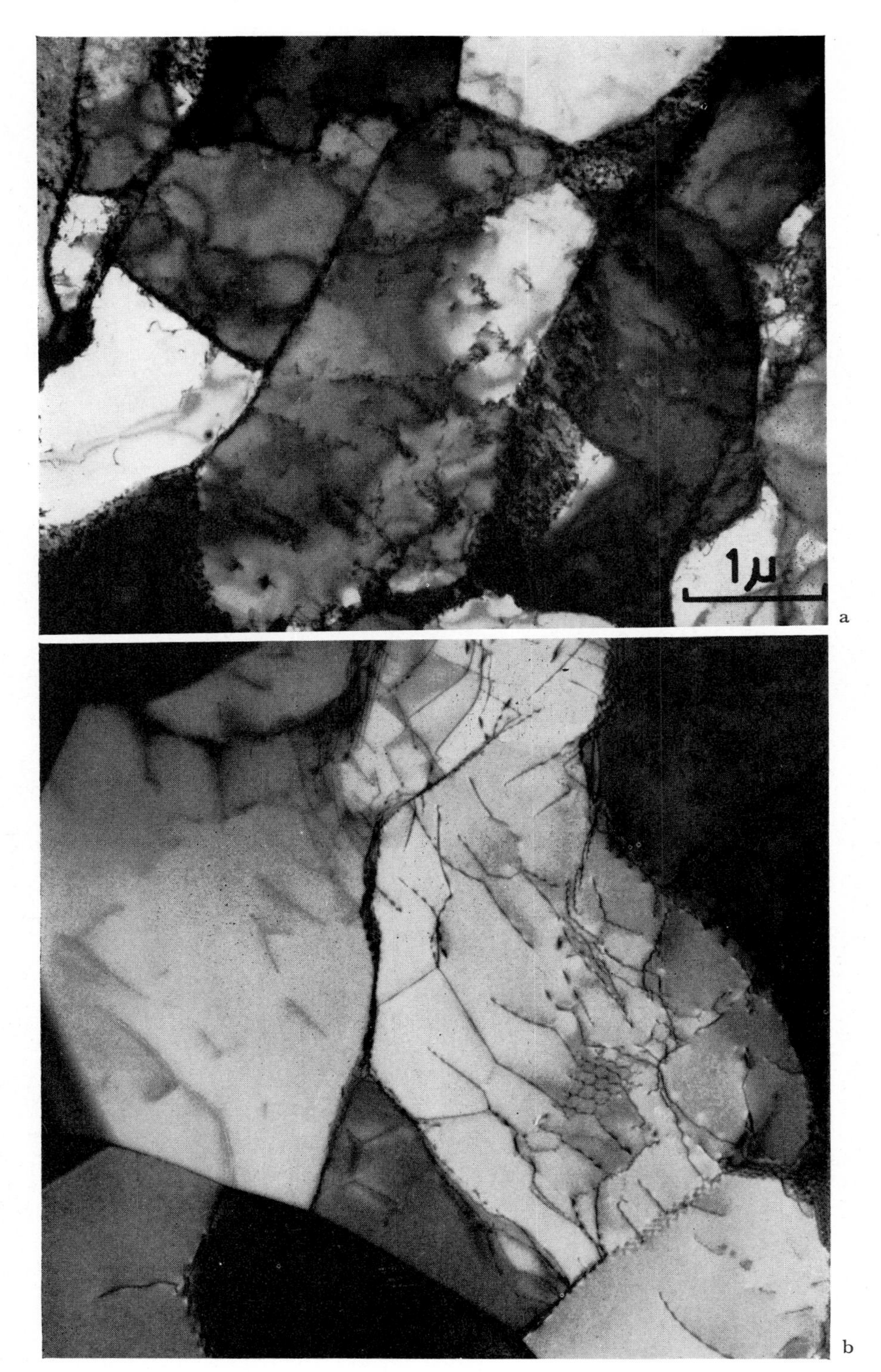

PLATE V. (a) Cell structure in rolled chromium sheet. Cells are 2 to 4μ across and are bounded by dislocation tangles. (b) Annealing of (a) produces sub-grain boundaries mainly of hexagonal dislocation networks (electron photomicrographs from McLaren, 1964)

PLATE VI. (a) Recrystallized quartz in hematite chert. The quartz has nucleated on the hematite (opaque and thus black on this photograph) and has extended away with columnar impingement texture (×40). (b) Comb structure in a quartz vein. Columnar impingement texture in quartz nucleated at the fissure walls (×10). (c) Quartz–siderite grain boundaries. Siderite boundaries are rational; irrational quartz boundaries meet siderite boundaries almost at right angles (×30). (d) Spherulitic topaz (Mt. Bischoff) under crossed polarizers showing an extinction cross and a similarity with bow-tie structure (×35)

PLATE VII. Grain boundaries in polycrystalline quartz. (a) Fine-grained quartz in unmetamorphosed chert (×30). (b) Partial recrystallization of chert to granular quartzite (×30). (c) Granoblastic-polygonal texture in granulite facies quartzite; note straight boundaries and triple-points (×30). (d) Sutured quartz boundaries due to slight post-crystalline deformation in quartzite (×30)

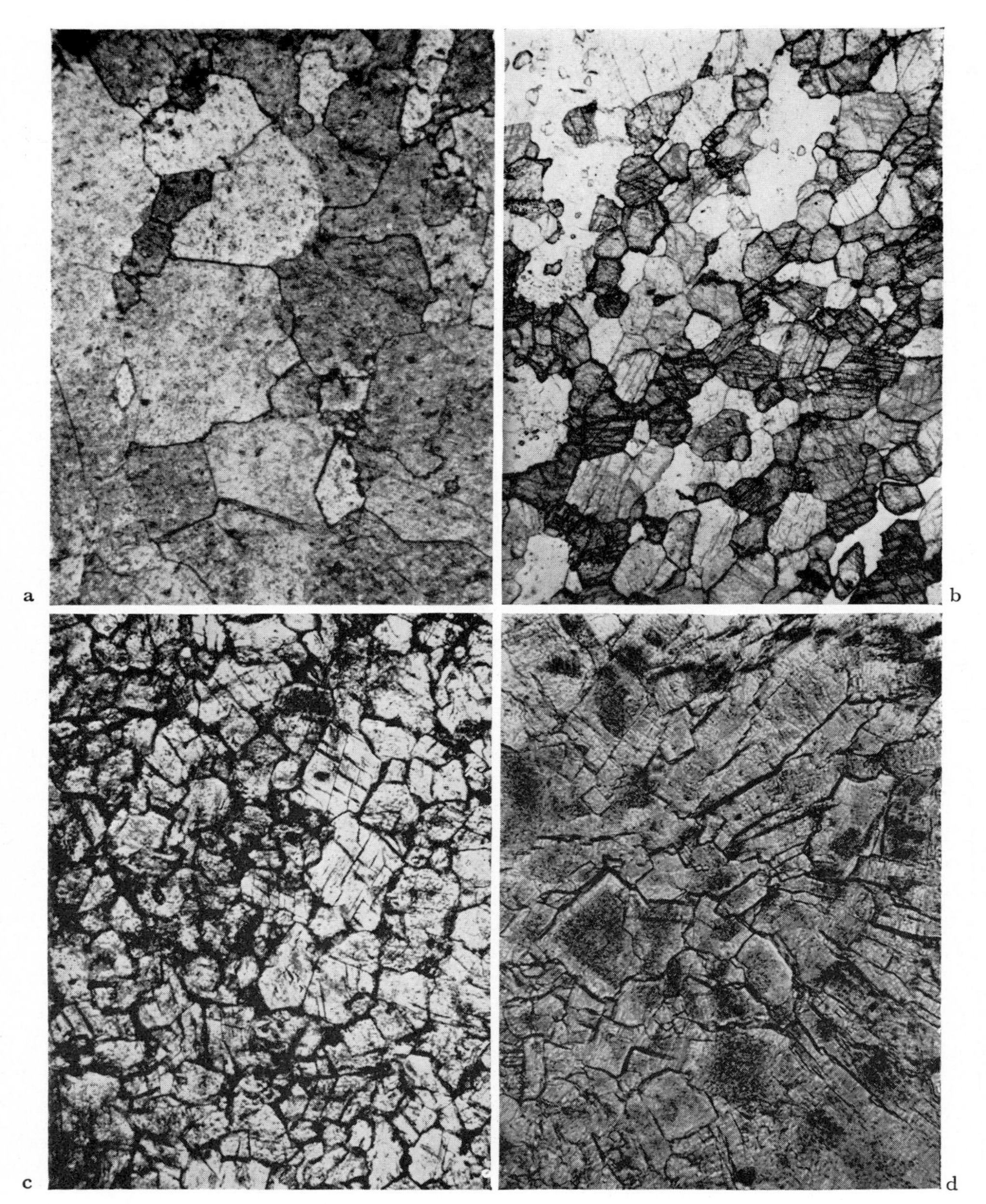

PLATE VIII. Grain boundaries in single-phase polycrystalline aggregates. (a) Granoblastic-polygonal texture in dolomite with straight to slightly curved boundaries and triple-points (×55). (b) Fine-grained marble containing calcite and quartz (low relief, no cleavage). Granoblastic-polygonal texture with straight boundaries and triple-points; some calcite boundaries are rational (×40). (c) Dolomite; granoblastic texture with unequal angle triple-points; many boundaries are rational (×100). (d) Idiotopic texture in anhydrite. Most grain boundaries are rational (×120)

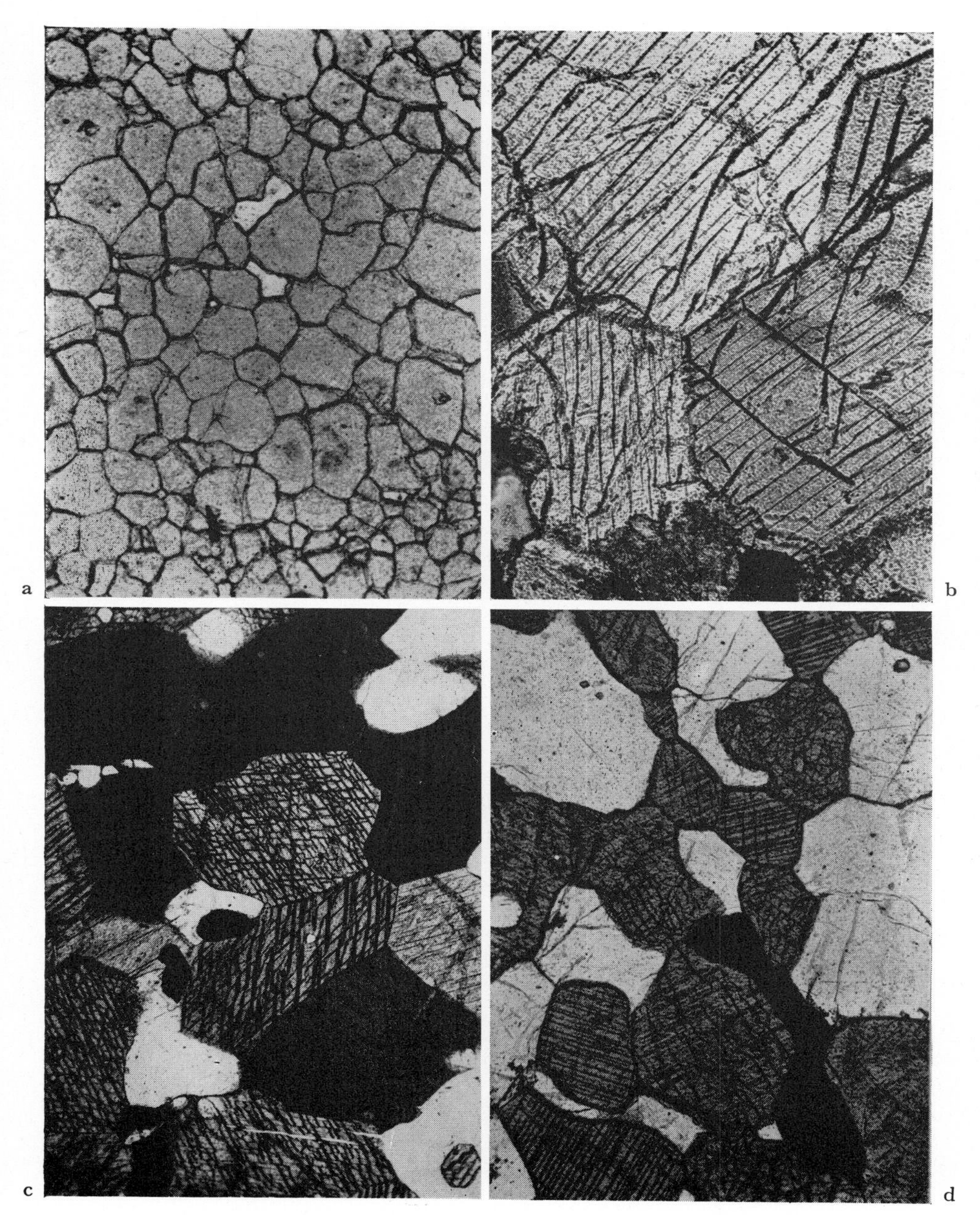

PLATE IX. Grain boundaries in polycrystalline aggregates. (a) Granoblastic-polygonal garnet, Broken Hill (×30). (b) Triple-point between pyroxene crystals in eclogite, Steinbach (×30). Boundaries tend to be rational. (c) Hornblende–plagioclase relations in amphibolite, Broken Hill. Grain boundaries are straight to slightly curved, some are parallel to {110} of hornblende (×30). (d) Pyroxene–plagioclase relations in granulite, Broken Hill (×30). (Photographs (a) and (c) by courtesy of R. H. Vernon and (d) by R. A. Binns)

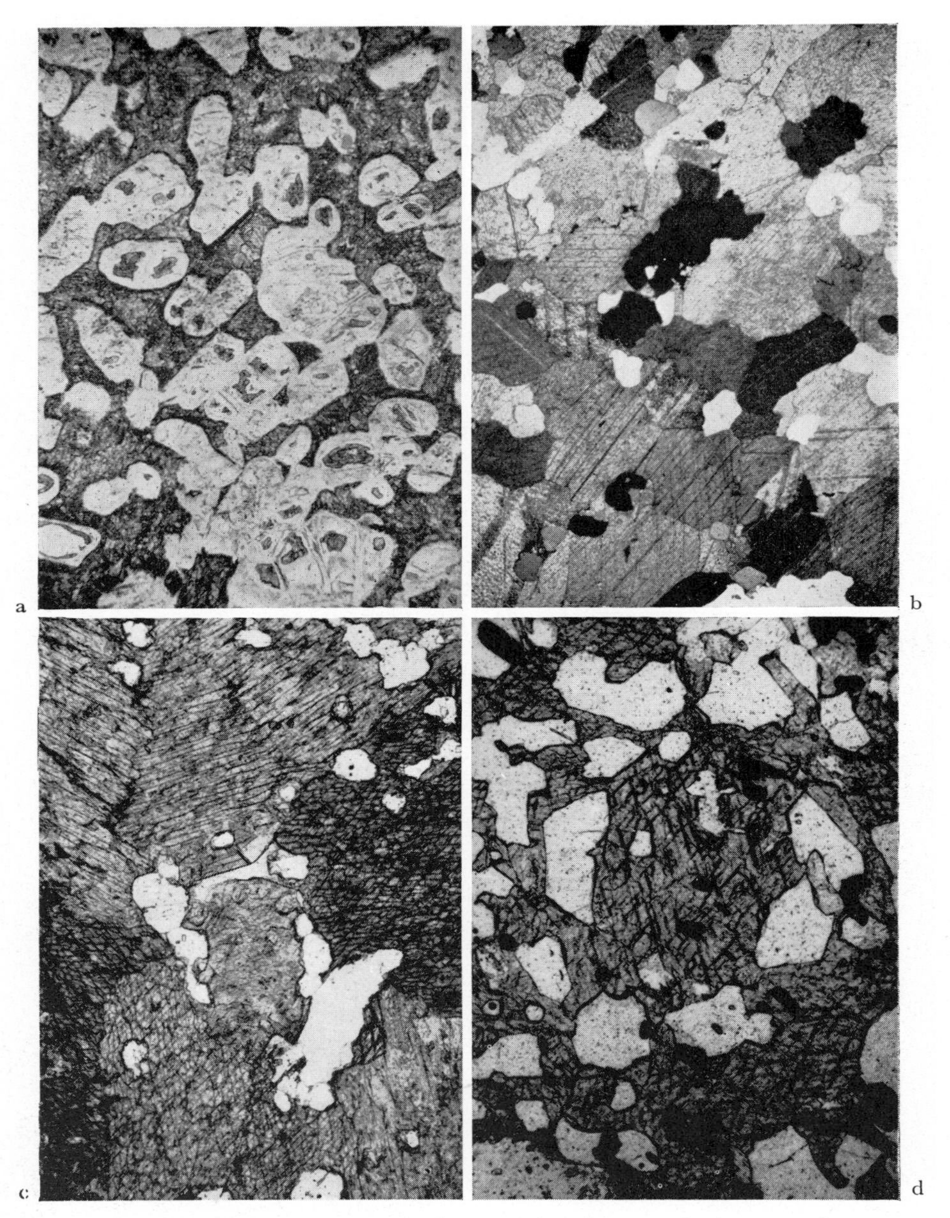

PLATE X. (a) Forsterite (serpentinized) crystals in calcite matrix show a tendency towards idioblastic form with many curved boundaries ($\times 35$). (b) Round crystals of quartz (low relief) at calcite triple-points in marble ($\times 40$). (c) Quartz in amphibolite forms round crystals at hornblende grain boundaries and triple-points ($\times 20$). (d) Quartz in amphibolite with extensive post-tectonic crystallization. Many grain boundaries are unilaterally rational (i.e. rational for hornblende but irrational for quartz) ($\times 35$)

GRAIN BOUNDARY ENERGY AND TENSION

The importance of surfaces in controlling the thermodynamics and kinetics of metamorphic changes has been emphasized by Verhoogen (1948), McLean (1957, 1965); De Vore (1959), Voll (1960), Stanton (1964), Rast (1965) and Lacey (1965b). The surface properties of solids, however, are complex and the nature of surface energy and tension will be developed here by analogy with liquids.

The forces of attraction between the atoms within a body of liquid act equally in all directions but the atoms at the surface are asymmetrically attracted towards the body of the liquid because of the lack of neighbouring atoms above the surface and are less closely packed than in the interior. The asymmetrical pull on the surface atoms causes a state of strain which is called the *surface tension* (Willows and Hatschck, 1923; Seitz, 1943; Smith, 1948; McKenzie, 1950; van der Merwe, 1950; Shuttleworth, 1950; De Vore, 1963; Cottrell, 1964b).

The tendency for the surface to shrink is shown by the way in which isolated drops of liquid are drawn into a spherical form. The surface tension may be regarded as that force necessary to deform the surface so that the interatomic spacing becomes equal to that within the interior. When the term is used without qualification it should mean the tension in the boundary layer between a liquid and its vapour in equilibrium. A liquid can exist as a body because the cohesion between the molecules is greater than their tendency to separate under thermal motion. The net attraction between atoms is most closely fulfilled in the interior and the surface atoms are less tightly held than those within. Thus the surface atoms are at a higher energy level. Since bond energies are negative, the energy of a surface atom is higher than that of one in the interior by approximately its share of missing energy. The behaviour of the surface can be understood in terms of *surface energy* because the energy is proportional to the area of surface, thus stability is increased by any decrease in surface area. The difference between the energy of surface and of interior atoms is the *surface free energy* which depends on the work required to form unit area of surface by a reversible isothermal process. Since the creation of surface requires work, the process must be accompanied by a force which is the surface tension. Surface tension (expressed in dynes/cm) and surface energy (expressed in ergs/cm^2) are numerically equal in liquids. They equal zero at the critical temperature of the liquid and increase with decrease of temperature.

The relationship that the surface tension equals the surface free energy requires easy interchange of material between the surface and the interior and breaks down if the viscosity of the system becomes so high that the rearrangement of molecules (when the surface is extended) is slower than relaxation of the shear stresses within.

A solid crystal has a surface energy and surface tension which is *similar* to those of a liquid but not *exactly* the same. Differences between the surface properties of a solid and of a liquid are decreased by any condition promoting mobility of the atoms (i.e. high temperature, presence of a solvent at the surfaces, etc.). Compared with a solid crystal, a liquid at rest has negligible viscosity, rigidity and anisotropy, and the atoms are able to move freely. The atoms in a crystalline solid are comparatively immobile and each one is firmly bonded to its neighbour so that the crystal lattice is rigid and anisotropic.

The following treatment of the meaning of surface energy is by Shuttleworth (1950). A crystal is imagined to be cut by a plane normal to the surface but extending just below the surface. In order that the surface on

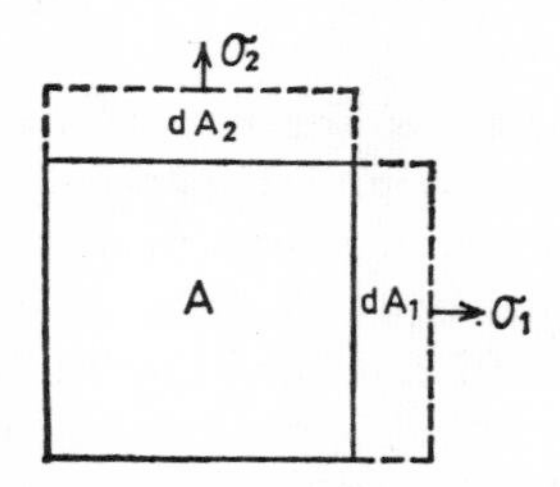

FIG. 3. Surface free energy defined by the work done to increase unit area A of crystal surface by amounts dA_1 and dA_2 against surface stresses σ_1 and σ_2 isothermally and reversibly

each side of the cut shall remain in equilibrium and that no additional stresses shall occur in the volume of the crystal, it is necessary to apply equal and opposite forces in the plane of the surface to the atoms on each side of the cut. The total force per unit length of cut is the surface stress across the cut. A crystal face of area A is then supposed to be deformed so that the area is increased by amounts dA_1 and dA_2 against surface stresses σ_1 and σ_2 (Fig. 3).

If this deformation is reversible and takes place at constant temperature (isothermal) the work done against the surface stresses will be equal to the increase in total surface (Helmholtz) free energy.

$$\sigma_1 \, dA_1 + \sigma_2 \, dA_2 = D(AF_s)_1$$

where F_s is the surface free energy per unit area.

For a liquid, an isotropic solid, or a crystal face with a three-fold or higher axis of symmetry, Shuttleworth considered that $\sigma_1 = \sigma_2 = \gamma$ (a surface tension) and that the equation above becomes

$$\gamma = \frac{dAF_s}{dA}$$

$$= F_s + A\left(\frac{dF_s}{dA}\right)$$

where $$dA = dA_1 + dA_2$$

In a liquid, the atoms are mobile and F_s is not a function of area, so that $\frac{dF_s}{dA}$ is zero and $\gamma = F_s$. Atoms in solids are not mobile, and F_s is a function of surface area, i.e. the term $\frac{dF_s}{dA}$ is finite and γ does not equal F_s at low temperatures. However, it is customary and simpler to use γ to refer to surface tension, surface energy and intergranular energy in crystalline aggregates. The terms F_s and $\frac{dF_s}{dA}$ may have comparable magnitudes, and although F_s must always be positive, the surface tensions of inert gas crystals and some alkali halides may be negative at absolute zero, and some homopolar solids may be zero.

The thermodynamic definition of *total* surface energy per unit area (E_s) as opposed to the surface *free* energy per unit area (F_s) is given by $E_s = F_s + TS_s$ where S_s is the surface entropy per unit area. The total surface energy is the total potential energy of the molecules in 1 cm^2 of surface, in excess of the energy they would have if they were within the interior of the crystal.

Anistropy of surface energy is marked in rock-forming silicates which have low symmetries and complex structures. This anisotropy is extremely important in controlling all textures dependent on surface energies and is much more marked in silicates than in most metals and ionic salts. Although measurements have not been made to determine the magnitude of surface energies in different directions in silicates, experiments on reactivity of surfaces may give some indication. Attack of quartz by hydrofluoric acid is 100 times faster on (0001) than on (1011) (Ernsberger, 1952).

The surface energy increases with decrease in radius of curvature of the surface so that the energy is reduced if curvature of the surface is removed. Surfaces tend to straighten or shorten themselves by migrating towards their centres of curvature. The rigidity of a crystal is such, however, that although the surface energy is greater on the edges and corners this driving force cannot pull the crystal into a spherical shape. The total free energy of a crystal consists of two parts, first the surface energy and second the very much greater internal energy. The internal energy is defined by the work necessary to completely disrupt the crystal and separate the atoms to such distances that they no longer affect each other and is the factor concerned with the stability of the crystal as a mineral or chemical substance. The surface energy is concerned with its shape.

The total free energy of a given weight of small crystals is greater than that of the same weight of large crystals. If small and large crystals are placed in the same enclosure under appropriate conditions, the larger will grow at the expense of the smaller. If a large crystal is crushed, the finely divided powder will have a higher surface free energy and be less stable; the energy difference is not sufficient, however, to allow a finely divided solid to spontaneously reconstitute itself into a single crystal because crushing is not a thermodynamically reversible process. Nevertheless, if a fine-grained aggregate is heated sufficiently to produce ionic mobility, then an increase in grain size results in a reduction of total surface area and hence surface free energy. Recrystallization tends to produce a grain boundary configuration in which the surface energy is at a minimum. The total intergranular surface area will be as low as possible and the orientation of the grain boundaries will tend to reduce their surface energy.

The shape of the interface between two liquids has been related to their relative liquid–air tensions, but care must be observed in considering that the grain boundary energy between two anisotropic crystals is approximately equal to the difference between their absolute or specific surface energies against air or other mutual fluid (Antonow's Rule). This is only really applicable to liquids but may be a useful first approximation for the maximum possible value for a solid–solid interface. The energy of self-boundaries varies from almost zero up to a maximum depending on the orientation. Interphase boundaries between different minerals are less orientation-dependent because they are generally incoherent, but the relative orientations are still significant.

Interpretation of metamorphic textures may be in terms of surface *tension* or of surface *energy*, but although surface tension models are useful in introductory studies, the surface tension of metamorphic crystals should not be thought of as similar to that of a liquid surface; confusion may be created by envisaging tiny forces pulling along grain boundaries. A crystal face is planar and immobile because it is a stable *low-energy* boundary (compared with other possible external surfaces) not because it is a strong *high-tension* boundary. It is more fruitful to consider surface tension merely as the work force involved in passing from one energy level to another and to consider grain boundaries only in terms of interfacial energies.

The basic reason for the existence of a surface energy lies in the asymmetry of the bonding. The layer of atoms at a free (solid–vapour) surface lacks an adjacent layer outside of the crystal, so the atoms are incompletely coordinated and in a higher energy condition than those in the interior. Those bonds belonging to surface atoms which should extend outwards must turn around and try to attach themselves to adjacent atoms. The greater the degree of mutual satisfaction obtained, the lower the energy of the surface. High surface energies will be favoured by a low density of atoms per unit area of surface, a short distance to the nearest neighbour atoms below, a high proportion of bonds which would make a large angle with the surface if normally coordinated and not cut by the surface, high bond strength, a high proportion of covalent rather than ionic bonding, a type of structure which does not allow the surface atoms to rearrange themselves from the normal to the surface condition, and the presence of defects such as solid solution atoms and dislocations of various sorts. The energy of the free surface is thus controlled by the atomic structure, packing, bonding, etc., just as hardness and strength are controlled.

The energy of a solid–solid grain boundary will show a similar control but the energy will be lowered if a certain degree of bond fulfilment is possible from one crystal to the next across the boundary. The greatest bond sharing and thus the greatest coherence and least grain boundary energy will occur between crystals of the same species. The energy of self-boundaries ranges from near zero for low angle tilts, through very low values for special coincidence and coherent twin boundaries, to a maximum figure. Smith (1948, Fig. 6) has shown, however, that the energy of self-boundaries can be greater than that of interphase boundaries.

Dislocation Models of Grain Boundaries

A grain boundary is a discontinuity in the periodicity of the lattices of adjacent crystals and is a kind of lattice imperfection (Read and Shockley, 1950) and it is useful to consider boundaries in terms of *arrays* or collections of imperfections. The boundary may lie between crystals of the same species (self-boundary) or between different species (interphase boundary) and although much of the following refers to self-boundaries (particularly subgrain boundaries) the same principles hold for semi-coherent interphase boundaries, particularly those which show atomic matching, e.g. oriented intergrowths and overgrowths.

The simplest possible boundary (between twins in a perfect face centred cubic metal) is a planar dislocation itself but may be hardly a boundary at all. The (111) twin boundary (Fig. 4) is a special orientation with high coincidence, low mobility and very low energy (in copper it has 1/30th the energy

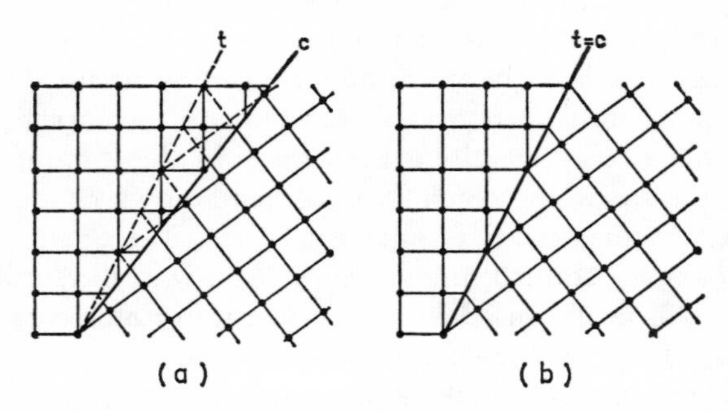

FIG. 4. Detail of twin structure in hypothetical cubic lattice. (a) Irrational composition plane *c* does not coincide with twin plane *t*. (b) Rational composition plane

of a (111) face). Twinning of this kind is easily achieved in pure metals because it only involves a slight change in the packing sequence of successive (111) layers; i.e. the sequence goes from *ABCABC* to *ABCACBAC*.

A simple model of a *low-angle tilt boundary*, i.e. a narrow transitional boundary between slightly differently oriented segments of the same lattice, is shown in Fig. 5 (Read, 1953, p.157). The segments can be regarded as having been rotated by equal and opposite amounts about the axis Z perpendicular

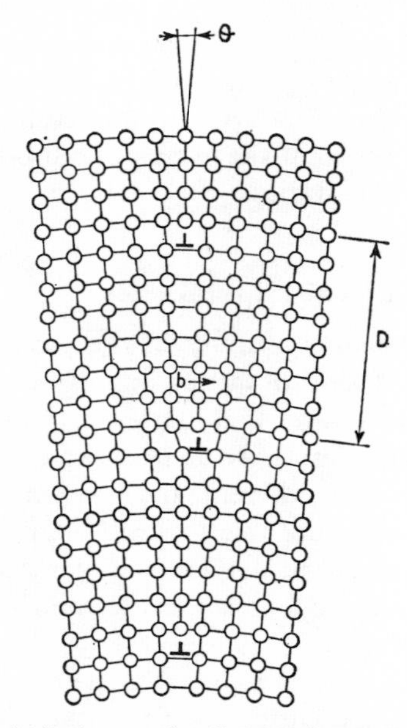

FIG. 5. Dislocation model of a low-angle, tilt boundary which is marked by a line of edge dislocations every eighth atom layer (after Read, 1953, p. 157). θ is the angle of tilt, b is the Burgers vector, D is the spacing of the dislocations

to the section and differ in orientation by the angle $\theta = \frac{b}{D}$ where b is the Burgers vector and D is the spacing of the dislocations parallel to the boundary. The grain boundary consists of a sheet of parallel edge dislocations of the same sign parallel to Z; D is large as long as θ is less than about 5°.

The grain boundary has five degrees of freedom with regard to the crystal lattices of the adjacent grains. Three degrees of freedom due to possible rotations of the two lattices on the X, Y and Z axes and two degrees of

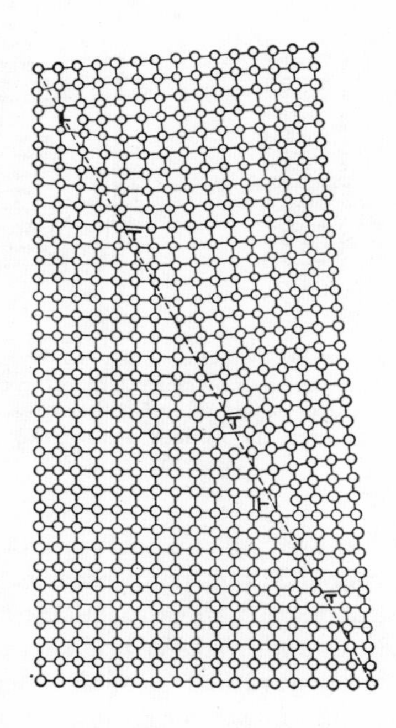

FIG. 6. Dislocation model of a low-angle boundary marked by edge dislocations of two kinds (after Read, 1953)

freedom of orientation of the boundary surface with respect to the crystals. Rotation about one axis, e.g. Z can be accomplished with one set of edge dislocations; rotation about an axis between Z and Y requires an additional set of edge dislocations parallel to Y (Fig. 6). Rotation about Z requires a crossed grid of screw dislocations and such a boundary is called a *twist* boundary (Fig. 7). An asymmetrical boundary between lattices with different atomic spacing requires extra dislocations with Burgers vectors inclined to the first to take up the misfit.

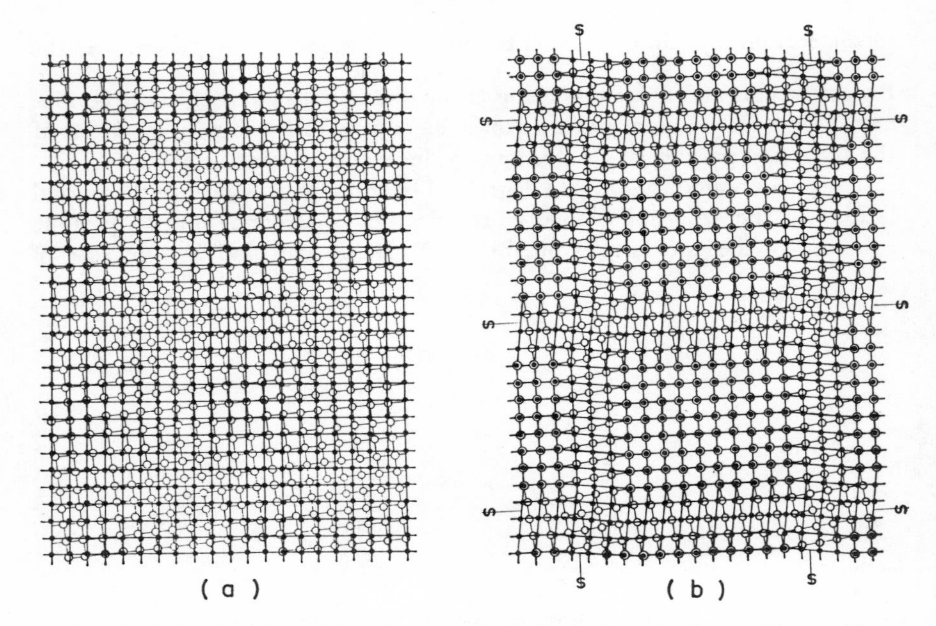

Fig. 7. Twist boundary (viewed perpendicularly to boundary). (a) Two cubic lattices (one shown by open, the other by closed, circles) in contact have a difference in orientation. (b) The lattices adjust to each other to give "islands" of good fit separated by a network of screw dislocations marked *S–S* (after Amelinckx, 1964, Fig. 36)

The simple tilt model in Fig. 5 allows coherence between lattices differing in orientation by up to about 5° but the fit becomes progressively poorer as the angle increases. At about 10° an extra dislocation can be arranged in every sixth plane and the fit improved. As θ is increased there will be certain angles at which moderately good fit is possible (special coincidence boundaries) and others where the fit is poor. Although only simple boundaries have been adequately explained by dislocation models, the same explanation has been generally extended to random high-angle boundaries and these may possibly be represented by arrays of complex and partial dislocations (Hornstra, 1960).

If a grain boundary is regarded as an array of dislocations then its energy is the sum of the dislocation energies.

The interfacial energy of the low-angle tilt boundary is given by the formula (Cottrell, 1953, p. 95):

$$\gamma = \frac{\mu b}{4\pi(1-v)} \cdot \theta(A - \log \theta)$$

where γ = surface energy of unit area of the boundary,
θ = the angle between the lattices,
μ = shear modulus,
b = Burgers vector of the dislocations,
v = Poisson's ratio,

and A is a constant depending on the energy at the centre of the dislocation.

The surface energy-orientation relation of a hypothetical crystal is illustrated in Figs. 2 and 8. The surface energy of two crystals of the same species in contact varies from zero when the lattices are parallel and coherent up to

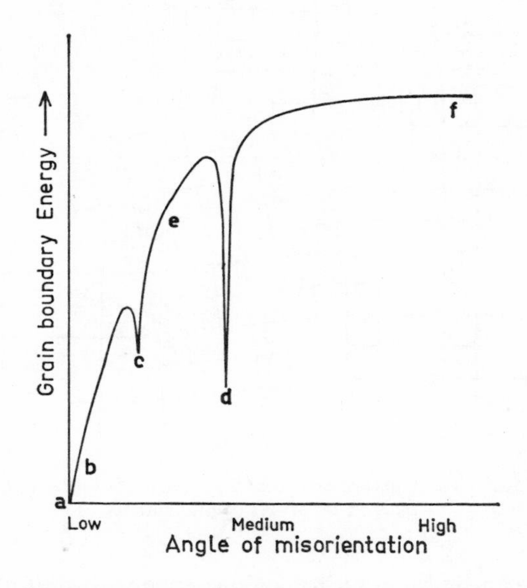

FIG. 8. Relation between angle of misorientation of the lattices of adjacent crystals and the energy of the grain boundary. Low-energy cusps occur at twin and special coincidence orientations. Points a, b, c, d, e and f correspond approximately to the orientations lettered similarly in Fig. 2

a maximum with a high-angle misfit boundary. "Cusps" indicating low-energy orientations exist for the twin orientation and for various special orientations which happen to have a moderate lattice coincidence across the boundary (Moment and Gordon, 1964).

Most solids are anisotropic with respect to surface energy to an even greater extent than to optical or mechanical properties; a cubic solid may have much lower surface energies on rational crystal faces than on random surfaces.

Mosaic structure in a crystal (Fig. 9) is a kind of sub-grain structure (or grain substructure) in which small adjacent blocks differ slightly in orientation. The mosaic appearance is due to sectioning across *lineage structure*, and the crystal is internally partitioned by discontinuities in such a way that the entire structure is continuous but branched (Buerger, 1934). The lattice of

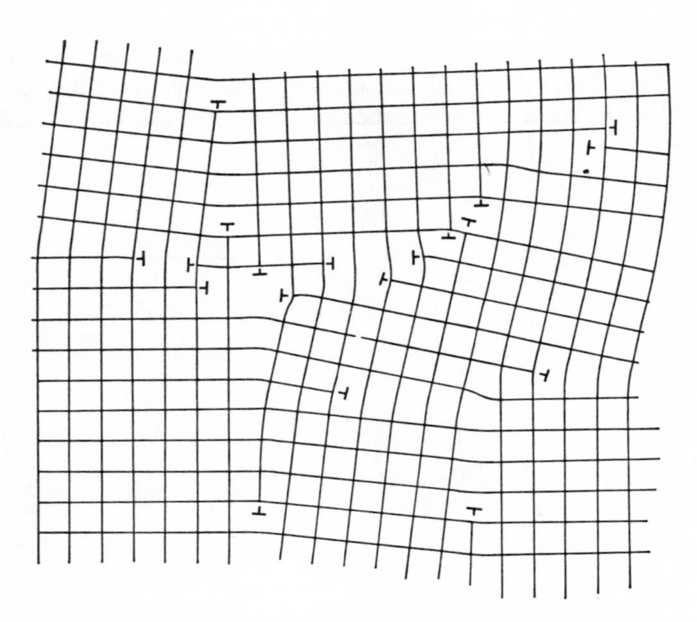

FIG. 9. Mosaic substructure produced by lineage. Sub-grain boundaries marked by arrays of dislocations

each sub-grain can be brought into coincidence with that of its neighbours by a rotation of a few degrees about a certain axis and the sum of these rotations is zero for the whole crystal. The sub-grain boundaries are low-angle twist or tilt boundaries and can be regarded as consisting of arrays of dislocations. Mosaic-type substructure may be found in undeformed crystals deposited from a liquid (Seitz and Read, 1941, p. 106) or in deformed crystals (Beck, 1954, Fig. 11). The boundaries of sub-grains grown from a melt or solution are roughly parallel to the growth direction (Doherty and Chalmers, 1962) because the dislocation arrays of the sub-grain boundaries are continuously produced at the liquid–solid interface during growth. A mosaic-type substructure can be produced by growth, deformation or polygonization. Lineage in quartz was described by Frederickson (1955) and is shown in Plate Ia; a similar structure is common in galena.

All crystal structures are built up by the repetition of units like the bricks in a wall. These units may be sufficiently regular in some simple structures to fit together without distortion or gaps but the "bricks" in a complex rock-forming silicate are so irregular that some misfit occurs between each. Addition of successive units to a nucleus causes increased distortion and bending of the structure until it is necessary for a defect to be produced before adding further units. The lattice strains involved in building large crystals of the serpentines are compensated by considerable curvature and defect structure (Pauling, 1930; Roy and Roy, 1954; Eitel, 1958, p. 135).

Chrysotile, gibbsite, endellite, halloysite, etc., are made up of sheets of silica alternating with magnesia and aluminia but the slight misfit between the ideal lattice vectors in the sheets causes strain and curvature. The total strain between the sheets depends on their lateral extent and produces a tubular structure in chrysotile. It is the basic reason why large crystals of clay or serpentine do not form. Despite the large surface area of an aggregate of clay particles and hence a large surface free energy, this condition has lower energy than a large crystal in which there would have to be very large lattice strain energy.

Surface Tension/Energy Values

The review of surface energy measurements of solids by Inman and Tipler (1963) is very useful and is mainly concerned with work since 1952; earlier work is covered by Shockley *et al.* (1952) and Gomer and Smith (1952).

Surface and grain boundary energies are so important in controlling the size and shape of crystals in metamorphic rocks that some quantitative knowledge is essential in understanding metamorphic textures. However, at present no *accurate* methods are known to determine surface or grain boundary energies of rock forming minerals.

The surface tensions of some liquids are as shown in Table 1 (Bikerman, 1958).

TABLE 1

Liquid	Against	Temperature °C	γ dynes/cm
Water	Air	20	72
Mercury	Vapour	16	484
Sodium (molten)	CO_2	90	290
Sodium chloride	Nitrogen	908	106
Gold (molten)	Air	1200	1120
Silicate glass (1)		1400	273
Silicate glass (2)		694	282
$CaSiO_2$		1570	389

The surface tension or surface free energy of a solid is measured in an atmosphere thought to be inactive. The tension is different for surfaces between a solid and a vacuum, a vapour, a liquid or another solid. An approximation made by Taylor is very useful and has some experimental support; he stated that the surface free energy of a solid metal at the melting point is $\frac{1}{8}$ greater than that of the liquid just above the melting point, and that the grain boundary free energy is $\frac{1}{3}$ that of the surface. The surface tension is zero at the critical temperature. It is essential to distinguish between the *surface energy* of a solid against a liquid or vapour and the *grain boundary energy* between crystalline solids. The grain boundary energy of a crystal will differ according to the nature of its surroundings.

The surface tension of liquid iron at 1600°C is about 1700 dynes/cm, of δ iron at the melting point about 1900, the grain boundary energy is about 500 ergs/cm^2 for δ iron (1/4 of surface energy) and about 600 ergs for γ iron (1/3 of surface energy). The surface energy of copper varies as follows: theoretical absolute (2350–2500 ergs/cm^2), solid surface against a vacuum (350–1400), against argon (1650), against liquid copper (177), and against liquid lead (425), a copper–copper grain boundary (646), a coherent twin boundary (19), and non-coherent twin boundary (440 ergs/cm^2) (McLean, 1957).

The surface tension of a liquid can be measured by virtue of its definition as the force per unit length to break the surface layer and Kuznetzov (1957, pp. 5–10) attempted to determine the surface tension of solids by scratching a surface.

An alternative definition of the surface energy of a liquid depends on the view that the surface atoms are in a higher energy condition and that work must be done to move an atom from within the liquid to the surface. Thus the surface energy depends on the work required to form unit surface area. This method has been applied, possibly validly, to mica cleaved reversibly in a vacuum; it is of very doubtful value when applied to cleaving minerals in air (Obreimoff, 1930; Gilman, 1959; Brace and Walsh, 1962) and its use by Kuznetzov (1957, p. 34) in splitting, abrading, drilling (p. 111) and crushing (pp. 116–17) is invalid. All such methods fail because the process is not isothermal and reversible. Such processes are valid for liquids where they are reversible and the work done is almost entirely involved in the surface layer, the body of the liquid requiring very little energy for its distortion. The process is not valid for a solid where only a very small and unknown proportion of work is expended on breaking or enlarging the surface. The greatest proportion of work is done on disrupting the subsurface layers of the solid and thus energy is lost irreversibly as heat and sound. The energy values obtained by Kuznetzov are more related to internal free energy (the total internal free energy being that required to completely disrupt the lattice and

TABLE 2. VALUES OF SURFACE ENERGY (ERGS/CM²) OR TENSION (DYNES/CM)

Mineral or compound	Face	γ (calculated)	γ(experimental)	Reference
Aragonite	(010)	270		Gilman, 1960
Barite	(001)	480		Gilman, 1960
	(against water)		123	Walton, 1965
Calcite	$(10\bar{1}1)$	78		Bikerman, 1958
		1000		Paterson, 1959
		190	230	Gilman, 1960
CaO		1310		Bikerman, 1958
		980		Walton, 1965
Diamond	(100)	7050		Gilman, 1959
	(110)	5500		Gilman, 1959
Fluorite	(111)	1100		Gilman, 1959
		540	450	Gilman, 1960
Forsterite	(010)	370		Gilman, 1959
Galena	(100)	625		Gilman, 1959
	(100)	233		Gilman, 1959
	(110)	165		Gilman, 1959
Glass			1214	Berdennikov, 1933 (in Kuznetzov, 1957)
			540	Berdennikov, 1933
Graphite	(0001)	119		Good *et al.*, 1958
	(100)	27		Gilman, 1959
	(110)	2340		Gilman, 1959
Gypsum			370	Kutznetzov, 1957
Hematite	(0001)	980		Gilman, 1959
$MnCO_3$		144		Bikerman, 1958
MgO	(298°K)	1000		Jura and Garland (in De Vore, 1959)
	(100)	1310	120	Gilman, 1960
	(110)	2330		Gilman, 1959
	(100)	1362		Dent, 1929
	(100)	1389		Walton, 1965
Muscovite	(in vacuum)		5000	Obreimoff, 1930
	(in air)		375	Kuznetzov, 1957
	(in air)		2442	Lazarev (in Kutnetzov, 1957)
	(by grinding)		31	Kuznetzov, 1957
	(in water)		1170	Kuznetzov, 1957
Orthoclase	(001)	200	7770	Brace and Walsh, 1962
$PbCO_3$			112	Walton, 1965
	(grinding)		27	Kuznetzov, 1957
$PbSO_4$	(against water)		79	Walton, 1965
Pyrite	(100)	1550		Gilman, 1960
Quartz	$(10\bar{1}1)$	760	410	Brace and Walsh, 1962
	$(\bar{1}011)$	450	500	Brace and Walsh, 1962
	$(10\bar{1}0)$	480	1030	Brace and Walsh, 1962
Rutile	(110)	1420		Brace and Walsh, 1962
Sphalerite	(110)	360		Gilman, 1959
Spinel	(111)	1780		Brace and Walsh, 1962
$SrCO_3$			92	Kuznetzov, 1957
$SrSO_4$	(against water)		86	Walton, 1965
$ZnCO_3$		245		Bikerman, 1958

move all atoms apart). The surface energy is very much less than the internal energy.

The determination of surface energy on theoretical grounds involves the assumption that the surface free energy per unit area can be expressed as the sum of the energies of the individual broken bonds. This involves a knowledge of the atomic structure of the material and the type, energy and distribution of the bonds. The method may be comparatively simple for face-centred and body-centred metals (Mackenzie, Moore and Nicholas, 1962) but can become complex (Benson and Dempsey, 1962; Benson, Freeman and Dempsey, 1963; Shewmon, 1963). Calculations are for ideal, not real, crystals.

Determinations of the surface energies of the various alkali halides are of interest because they form structurally simple ionic crystals. Van Zeggeren and Benson (1957) gave values of 188 ergs/cm^2 for (100) and 445 ergs/cm^2 for (110) in NaCl, but calculated and experimentally determined values range from 87 to 210 ergs/cm^2 for (100) and from 252 to 445 ergs/cm^2 for (110). The range of values is such that none can be accepted with confidence.

Surface Energy and Grain Boundary Shape

Consider a system of two immiscible liquids in contact, one resting on the other. The interface between them will be curved because of its tension. If the surface tension of a liquid is taken as that at the liquid–vapour interface, then the interfacial tension between the two liquids A and B is $\gamma_{AB} = \gamma_{A\text{-vapour}} - \gamma_{B\text{-vapour}}$. A pressure difference occurs across the curved surface such that the pressure is greatest on the side nearer the centre of curvature.

If two immiscible liquids are shaken together, the one with the greater surface tension will be broken up and will form spherical globules within the other. There is a certain amount of attraction between the molecules within each liquid (self-cohesion) and a certain amount between molecules of the different liquids across the interface (adhesion). The liquids will be miscible if these attractions are equal (the liquids will "wet" each other), and will be immiscible if the self-cohesion is greater than the adhesion. The surface tension of the interface between the immiscible liquids will shorten the interface so that it tends to be spherical and thus have the smallest surface and lowest surface energy for a given volume. It has been considered that as the surface tension of a liquid is about half the value of its cohesion, the tendency for shortening will be greater in the liquid with the higher absolute surface tension and that the surface will become convex towards the liquid with the lower tension.

A slightly curved interface results when one immiscible liquid is carefully poured on to the surface of another. When a bubble of a third immiscible liquid is placed at the interface, the boundaries between the three will rotate so that the three surface tensions balance (Fig. 10). The three interfacial angles will depend on the magnitude of the three interfacial tensions, γ_{12}

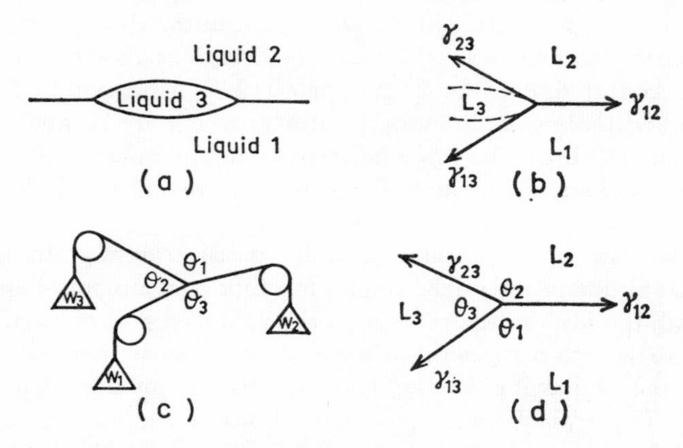

FIG. 10. (a) Three immiscible liquids in contact (after Smith, 1948). (b) Surface configuration at the triple-point where three liquids (L_1, L_2, L_3) meet is due to the surface tensions (γ_{12}, γ_{13}, γ_{23}) between the liquids. (c) Mechanical analogue. Three weights W_1, W_2, and W_3 are supported by three strings tied together at one point. At equilibrium, the angles θ_1 and θ_2 and θ_3 depend on the relative weights. (d) The interfacial angles (θ_1, θ_2, θ_3) at the triple-point depend on the surface tensions γ_{23}, γ_{13}, and γ_{12} of the three liquids

(between liquid no. 1 and liquid no. 2), γ_{13} (liquid 1/liquid 3), and γ_{23} (liquid 2/liquid 3).

Thus

$$\gamma_{12} = \gamma_{23} \cos \theta_2 + \gamma_{13} \cos \theta_1$$

or alternatively

$$\frac{\gamma_{12}}{\sin \theta_3} = \frac{\gamma_{23}}{\sin \theta_1} = \frac{\gamma_{13}}{\sin \theta_2}$$

The same reasoning applies if an oil bubble (of liquid 3 above) is floated on the surface of water (of liquid 1) with air being analogous to liquid 2. The mutual boundaries of three gas bubbles in contact (but separated by a thin liquid film) will meet at a triple-point with equal angles because the surface tensions along the interfaces are equal.

Recrystallization of thin films of polycrystalline metals produce granular aggregates with polygonal outlines and intergranular boundaries meeting at triple-points. The interfacial angles tend to be equal (120°) when the boundaries are perpendicular to the viewing plane, but there is a small spread of values.

Recrystallization of single-phase polycrystalline aggregates of various ceramic and rock-forming minerals (oxides, calcite, quartz, etc.) gives a similar result but complexity is introduced because triple-points cannot be viewed exactly at right angles. The shape for equal solid grains fitting together is a distorted tetrakaidodecahedron (Smith, 1948; Kingery, 1960, p. 410). Experiments show that random sections through such an aggregate show a marked preference for equal angle (120°) triple-points. It is rarely possible to find three grain boundary surfaces which are cozonal, i.e. intersect in a line which can be oriented parallel to the microscope axis. Solid angles may, however, be measured in transparent minerals with the universal stage.

The explanation of the meaning of triple-points, straight grain boundaries and the magnitude of interfacial angles in solids is commonly based on the analogy with liquids and the concept of the balanced pull of interfacial tensions. Although such a mechanism is simple and instructive, the concept of grain boundaries moving in response to tensions pulling along them is misleading. The grain boundary configurations at equilibrium are formed by diffusion in response to the requirement for lowest total surface energy.

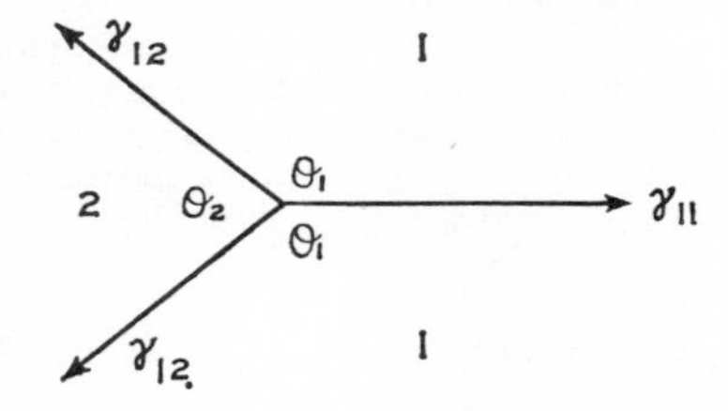

FIG. 11. Three crystals of two species (1 and 2) in contact at a triple-point

The surface tension model for grain boundary configurations is identical with that for liquids and is considered first. Interfacial energies and angles for three different crystals in equilibrium at a triple-point (Fig. 10d) are given by the equation:

$$\frac{\gamma_{12}}{\sin \theta_3} = \frac{\gamma_{23}}{\sin \theta_1} = \frac{\gamma_{13}}{\sin \theta_2}$$

If three crystals of two species (1 and 2) are in contact at an equilibrium point (Fig. 11):

$$\frac{\gamma_{11}}{\sin \theta_2} = \frac{\gamma_{12}}{\sin \theta_1}$$

i.e.

$$\frac{\gamma_{11}}{\gamma_{12}} = \frac{\sin \theta_2}{\sin \theta_1}$$

or

$$\gamma_{11} = 2\gamma_{12} \cos \frac{\theta_2}{2}$$

i.e.

$$\frac{\gamma_{11}}{\gamma_{12}} = 2 \cos \frac{\theta_2}{2}$$

Thus relative values of grain boundary energies can be obtained from measurements of interfacial angles at triple-points.

For example, Vernon (1968) has found that θ_2 is 135° for a crystal of garnet against two crystals of quartz and θ_2 is 90° for a crystal of quartz against two crystals of garnet.

Thus

$$\frac{\gamma_{\text{quartz-garnet}}}{\gamma_{\text{garnet-garnet}}} = 0{\cdot}71$$

and

$$\frac{\gamma_{\text{quartz-garnet}}}{\gamma_{\text{quartz-quartz}}} = 1{\cdot}31$$

and

$$\frac{\gamma_{\text{garnet-garnet}}}{\gamma_{\text{quartz-quartz}}} = 1{\cdot}9$$

Polygonal textures of many anisotropic rock-forming silicates in which a large proportion of the grain boundaries are rational depend on the achievement of a low total surface energy. The interfacial angles in aggregates of hornblende, plagioclase, pyroxene, wollastonite are not equal although there is a recognizable tendency for the angles to be equal in some monomineralic aggregates. The careful, accurate measurement of interfacial angles in polymineralic aggregates is difficult but Stanton (1964) and Vernon (1968) have had some success.

A tendency to form equal interfacial angles in monomineralic aggregates will follow from the general reduction of total surface energy by reducing the total grain boundary area. The shape of grain boundaries of grains of equal size and properties is a space-filling problem and interlocking bodies such as the tetrakaidodecahedron will give many 120° interfacial angles.

Grain boundaries meeting at a triple-point may rotate about the point of intersection or move the triple-point and so change from high-energy orientations, relative to the adjacent lattices, to positions which achieve lower energy. Movement of the boundary takes place by changing the orientation of successive surface units from that of one crystal to that of the other. The boundary will move to a position which achieves maximum coherence and fit between the two crystal lattices.

Consider three grain boundaries meeting at a point. For simplicity it will be assumed that the three crystals are of the same species and that the grain

TABLE 3. VALUES OF INTERFACIAL ANGLES AND RELATIVE GRAIN BOUNDARY ENERGIES (after Stanton, 1964; Vernon, 1968)

Phase A	Phase B	Interface C	θ	γ_{A-B}/γ_C	γ_{A-A}/γ_{BB}
Quartz	Orthoclase	Or–Or	105°	0·82	
Quartz	Plagioclase	Pl–Pl	110°	0·87	1·0
Quartz	Plagioclase	Qu–Qu	110°	0·87	
Quartz	Garnet	Ga–Ga	90°	0·71	0·5
Quartz	Garnet	Qu–Qu	135°	1·31	
Orthoclase	Garnet	Ga–Ga	95°	0·74	0·6
Orthoclase	Garnet	Or–Or	130°	1·18	
Plagioclase	Hornblende	Ho–Ho	100°–105°	0·8	0·8
Plagioclase	Hornblende	Pl–Pl	120°	1·00	
Plagioclase	Clinopyroxene	Cl–Cl	110°–115°	0·9	1·0
Plagioclase	Clinopyroxene	Pl–Pl	115°	0·93	
Hornblende	Clinopyroxene	Cl–Cl	115°–120°	0·97	1·0
Hornblende	Clinopyroxene	Ho–Ho	115°	0·93	
Orthoclase	Ilmenite	Or–Or	130°	1·18	
Quartz	Calcite	Ca–Ca	115°	0·93	0·8
Quartz	Calcite	Qu–Qu	130°	1·18	
Quartz	Apatite	Ap–Ap	70°	0·61	0·4
Quartz	Apatite	Qu–Qu	145°	1·66	
Galena	Sphalerite	Sp–Sp	102°	0·79	0·6
Galena	Sphalerite	Gal–Gal	134°	1·28	
Chalcopyrite	Sphalerite	Sp–Sp	108°	0·85	0·75
Chalcopyrite	Galena	Gal–Gal	125°	1·10	
Sphalerite	Pyrrhotite	Pyr–Pyr	107°	0·84	1·0
Sphalerite	Pyrrhotite	Sp–Sp	106°	0·83	
Galena	Chalcopyrite				0·8
Galena	Pyrrhotite				0·6
Chalcopyrite	Pyrrhotite				0·8

boundary energy is independent of orientation. The total surface energy of the system is the surface energy per unit area multiplied by the total area. Three alternative configurations are considered (Fig. 12a): (1) boundaries AO, BO and CO meeting at O at angles of 100°, 100°, 160°, (2) boundaries AP, BP and CP meeting at P at angles of 120°, (3) boundaries AQ, BQ and CQ meeting at Q at angles of 90°, 90° and 180°.

The total length of $CP + AP + BP$ is less than that of the other configurations, thus the lowest total surface energy will be achieved by altering interfacial angles until equal angles are achieved.

Consider three different species, 1, 2 and 3 in Fig. 12b in contact at a triple-point O. The interfacial or grain boundary energies per unit area of the three boundaries are γ_{13} greater than γ_{23} greater than γ_{12}. The least energy condition will be achieved by movement of the triple-point to such a position that the sum of the boundary energies (i.e. area $\times$ unit surface energy) is a minimum, i.e. $OC.\gamma_{13} + OB.\gamma_{23} + OA.\gamma_{12}$ is least.

If the triple-point moves (e.g. from O to P in Fig. 12a), then a length OP of new 1–3 surface (and an area dA_1) is produced at the expense of some 1–2 (dA_2) and 2–3 (dA_3) surface; the total surface area may be increased or decreased, i.e. $dA = dA_1 + dA_2 + dA_3$ is positive or negative. The change in energy will be $dA_1 . \gamma_{13} - dA_2 . \gamma_{12} - dA_3 . \gamma_{23}$.

If $dA_1 . \gamma_{13}$ is greater than $dA_2 . \gamma_{12} + dA_3 . \gamma_{23}$, then work must be done to move the triple-point to P. If $dA_1 . \gamma_{13}$ is less than $dA_2 . \gamma_{12} + dA_3 . \gamma_{23}$, then

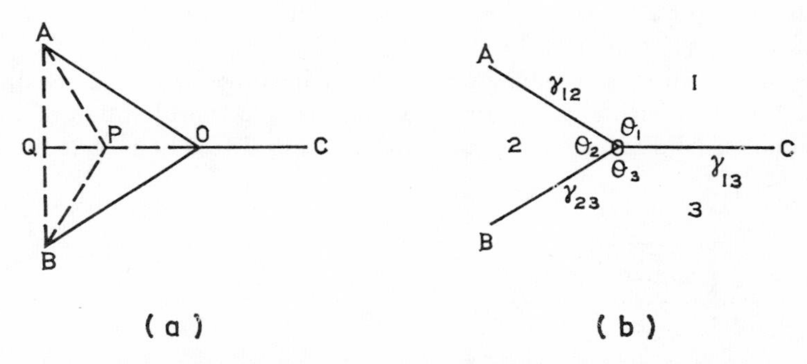

FIG. 12. Surface-energy model for grain boundaries of three species (1, 2 and 3) at a triple-point (see text)

energy is released by movement of the triple-point from O to P. The lowest total free surface energy (i.e. the equilibrium texture) will be achieved when $A_1\gamma_1 + A_2\gamma_2$, etc., is least.

The interfacial angle method is based on a bubble analogy and may be valid for liquids; it may be reasonably valid for metals of high symmetry, it is reasonable for some sulphides but it is not valid for silicate minerals of low symmetry and considerable anisotropy of structure.

The equation

$$\frac{\gamma_{23}}{\sin \theta_1} = \frac{\gamma_{13}}{\sin \theta_2} = \frac{\gamma_{12}}{\sin \theta_3}$$

at a triple-point is only valid if γ is a function of the angle between the interfaces but is independent of the orientation of each interface relative to the lattices of the adjacent grains, ϕ (Moment and Gordon, 1964). When γ is a function of both θ and ϕ, the condition of equilibrium is much more complicated (Herring, 1951), and involves six variables, the energy for each boundary γ and the $\frac{d\gamma}{d\phi}$ for each. The terms $\frac{d\gamma}{d\phi}$ are called *torque terms* and when they are present relative values of γ, measured from triple-points, will have a large range of values. R. Vernon (*personal communication*) has pointed out the range of interfacial angles may reflect the anisotropy of γ. It is suggested that the discrepancies between values of γ calculated from two- and from three-phase aggregates might be due to ϕ.

Moment and Gordon (1964) annealed tri-crystals of NaCl at 750° for 12 days to equilibrate and found a considerable orientation dependence of surface energy consistent with independently determined values of γ (i.e. the surface energy of {100} is lowest). The angles at triple-points in silicon tri-crystals depend on the orientations of the lattices and vary from 90° to 135° (Dunn, Daniels and Bolton, 1950).

The role of surface tension in relation to wetting may be conveniently regarded here. Wetting can be demonstrated by the consideration of a drop of liquid on a solid surface in contact with air (Fig. 13). The mobile liquid to

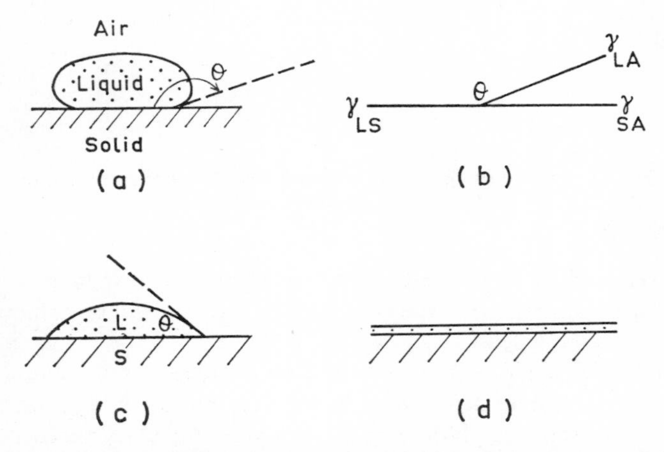

FIG. 13. (a) A drop of liquid on a solid surface. (b) The angle of contact θ depends on the relative magnitudes of the surface tensions. A high contact angle in (a) does not favour wetting; a lower angle (c) allows greater wetting; an angle of zero in (d) allows complete wetting

air interface rotates to a stable position so that it meets the solid surface at an angle θ. The net horizontal force per centimetre towards the left is equal to that towards the right, so that

$$\gamma_{SA} = \gamma_{SL} + \gamma_{LA} \cos \theta \qquad \text{(Davies and Rideal, 1961, p. 35)}$$

The size of the "wetting angle" depends on the relative values of γ_{LA} and γ_{SA}. The equation cannot be satisfied if γ_{SA} is greater than $\gamma_{LS} + \gamma_{LA}$ because θ approaches zero and the liquid phase spreads over (or wets) the solid. The size of θ depends on the relative magnitudes of the adhesion of the liquid to the solid and the self-adhesion of the liquid. The *wetting coefficient* W_{LS} (liquid to solid) is

$$W_{LS} = \gamma_{SA} + \gamma_{LA} - \gamma_{SL}$$

The air–liquid–solid model discussed above involves a solid surface which is inert and of such high strength that it is unaffected by the tiny surface tensions. The component of the liquid–air tension, normal to the solid face, cannot pull it sideways.

This model is not satisfactory when applied to three solids in contact. For instance, Fig. 14 shows a grain boundary between crystals of two species

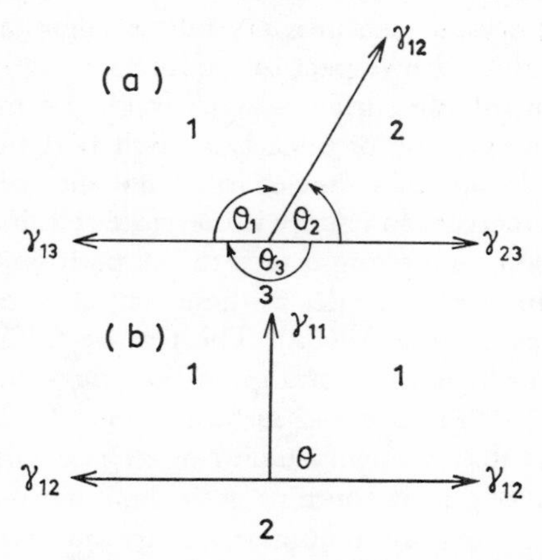

Fig. 14. Interfacial angles at a straight grain boundary of phase 3 which is met by a grain boundary between two phases 1 and 2 in (a) or two crystals of one phase in (b) (see text)

and/or of two crystals of the same species meeting a crystal face of another species. An analogy with the air–liquid–solid model would suggest that the crystal face has a very high boundary tension compared with that of the boundary which meets it. In fact, the tension of the crystal face would have to be very much greater to allow the interfacial angle of 180° to be retained. If the species 1, 2 and 3 were quartz, feldspar and mica respectively then the maximum surface energy for the surface between mica and either quartz (γ_{mq}) or feldspar (γ_{mf}) might be 1600 ergs/cm^2 and that between quartz and feldspar perhaps about 200 ergs/cm^2. Applying the equation

$$\gamma_{qf} = 2\gamma_{mq} \cos \frac{\theta}{2}$$

$$200 = 2 \times 1600 \cos \frac{\theta}{2}$$

and $$\theta = 173° \, 36'$$

A deviation of $6\frac{1}{2}°$ by a plane surface of mica would surely be noticeable but is not seen. It is necessary to increase the value of γ for mica against quartz or feldspar to 3000 ergs/cm^2 and reduce that for a quartz-feldspar surface to 50 ergs/cm^2 to increase the angle to within $\frac{1}{2}$ degree of 180 degrees so that the interface would appear straight, but these figures do not seem reasonable.

It is thus doubtful whether the air–liquid–solid model is a useful analogy for a straight crystal face in a crystalline aggregate. The strength of the solid face in the air–liquid–solid model is perhaps eight orders of magnitude greater than the tension in the air–liquid interface which meets it. The surface tension of the straight crystal face in a crystalline aggregate does not have a comparably large value with respect to the intersecting interfaces.

An explanation involving surface *energies* might be more satisfying. It is suggested that the anisotropy of mica is so high that the surface energy of (001) is very much lower than that of irrational surfaces nearly parallel to (001). A very great increase in energy is therefore required to move a triple-point from its position on a straight face to the position which would result in the appropriate interfacial angles as these vicinal faces would have much higher surface energy (Kretz, 1966a). The texture fulfils the final thermodynamic requirement for lowest total grain boundary energy.

The total energy (both lattice and surface energy) for a given mass is least for a single crystal so that a polycrystalline aggregate will never be in a least energy condition as long as it contains grain boundaries. The driving force for recrystallization is the grain boundary energy and this force progressively decreases as the grain size increases until it is too small to overcome the activation energy for boundary movement. Thus even the granular polygonal texture has a kinetic control. However, as De Vore (1959) has pointed out, the average surface energy of each real grain in a polycrystalline aggregate of anisotropic crystals depends on the proportion of low- to high-energy contacts and a position may be reached when an increase in grain size must replace lower- by higher-energy boundaries as new grains come in contact and thus growth will cease.

CHAPTER 4

Metamorphic Reactions

A ROCK undergoes structural and mineralogical (and chemical) changes during metamorphism in an attempt to reach equilibrium with the new conditions. This generally results in *recrystallization* which is the reconstitution of existing stable phases, and *crystallization* (neomineralization) which is the formation of new phases resulting from the destruction of unstable ones.

The evolution of textures in metamorphic rocks requires an understanding of changes in minerals regarded as units of crystal structure and not merely as chemical compounds. Metamorphic processes will be regarded here as a series of *structural transformations*. Some mineralogical transformations may be comparatively simple, e.g. twinning or minor polymorphic change, or the recrystallization of, e.g., quartz. Others are more complex in that they involve considerable structural changes (e.g. andalusite to sillimanite) and many are much more profound in that they involve a change in chemical composition and structure (e.g. chlorite to garnet).

These changes are regarded as transformations from an *unstable* or *metastable* phase to a *stable* phase in *equilibrium* with the new conditions; the changes must overcome a certain *activation energy* and take place at a certain *rate* under a certain *driving force*.

Metamorphic reactions may be studied by either *thermodynamic* or *kinetic* methods. These may be very complicated but a great deal can be understood even with the few elementary principles outlined here.

"Kinetics deals with the rate of reaction and with the explanation of the rate in terms of the reaction mechanism. Chemical kinetics with its dynamic viewpoint may be contrasted with thermodynamics with its static viewpoint. Thermodynamics is concerned only in the initial and final states of a system; the mechanism whereby one system is converted to another and the time required are of no importance. Time is not one of the thermodynamic variables. The most important subject in thermodynamics is the state of equilibrium, and consequently thermodynamics is the more powerful tool for investigating the conditions at equilibrium. Kinetics is concerned fundamentally with the details of the process whereby a system gets from one state to another and with the time required for the transition. Equilibrium can also be treated in principle on the basis of kinetics as that situation in which the rates of forward and reverse reactions are equal" (Frost and Pearson, 1961).

The kinetics and thermodynamics of metamorphic reactions have been discussed at length by Ramberg (1952), Fyfe, Turner and Verhoogen (1958) and Turner and Verhoogen (1960), but the student can proceed a long way by remembering three very simple concepts.

(1) The lower the total energy of a system, the greater the stability, i.e. metamorphic transformations proceed so as to reduce the total energy.
(2) Mineralogical transformations are impeded by an energy barrier which is called the activation energy. Energy must be available to surmount this barrier before the process takes place. The higher the activation energy, the slower the process.
(3) Entropy is not a major factor in most metamorphic processes (see Fyfe *et al.*, 1958, p. 25). Entropy may be equated with atomic disorder and an increase in temperature increases the disorder. Reactions proceed in the direction of increased entropy.

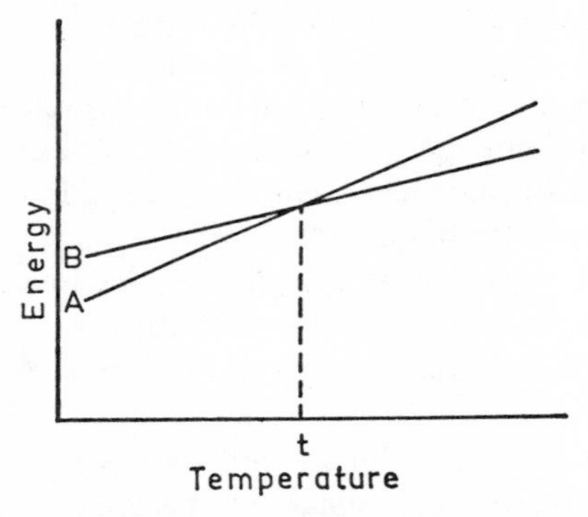

FIG. 15. Energy levels of phases or assemblages (denoted by *A* and *B*) at different temperatures; *t* is the equilibrium temperature. *A* is more stable below and *B* more stable above *t*

Figure 15 shows the relation between two materials *A* and *B* which may interchange reversibly, i.e. *A* and *B* can be polymorphs of one mineral, or a mixture *A* which can change isochemically into a mixture *B* (e.g. *A* = calcite + quartz, *B* = wollastonite plus carbon dioxide). The energy of *A* is less than that of *B* below a temperature *t*, thus *A* is more stable. Above *t*, *B* is more stable. The energy difference between *A* and *B* depends on the temperature; it is zero at *t* where *A* and *B* are in equilibrium and becomes larger with departure (above or below) from *t*. The energy difference itself does not overcome the activation energy as this is achieved at local points by thermal fluctuations. The energy difference only indicates the direction of reaction—not the rate.

MacKenzie (1965) emphasized the difference between equilibrium and metastability. Equilibrium is synonymous with "reversible reaction", i.e. under certain conditions $dG = 0$ (at constant P and T) or $dF = 0$ (at constant V and T). A substance is stable if after any disturbance it tends to return to its original state. However, a substance can be in a condition of "false equilibrium" or "metastable equilibrium", if the velocity of reaction is zero.

The mineral assemblage in a rock is considered to be "in equilibrium", i.e. the mineral phases are "in equilibrium" with each other, if they appear to be stable and not tending to break down individually or to react with each other. The assemblage need not, however, be in a lowest energy (most stable) condition but may be in a metastable equilibrium. No experimental, mineralogical or petrological criteria can tell unequivocally that an assemblage is stable, and metastable assemblages are known to persist throughout geological time because the reaction rates are infinitely slow and the activation energy barrier cannot be overcome.

Metamorphic reactions are strongly time-dependent and the *kinetics* of reactions are extremely important.

CHAPTER 5

Mineral Transformations

TRANSFORMATIONS are changes of atomic arrangement in a solid (including bonding changes, atomic rotation or translation, ordering and disordering) but in the strict sense excluding chemical changes and gliding.

Textures can be best understood by considering them as transformations in the solid state whereby atomic lattices are changed, disrupted or built up. This concept is contrasted with the chemical or phase rule treatment which is suitable only for an understanding of mineral assemblages. A great deal of information is available on the processes of solid state transformations involving inversion, nucleation and growth, lattice strain, twinning, diffusion, reaction, etc., and many of the basic principles are directly applicable to metamorphic processes.

The strict principles of Willard-Gibbs (Christian, 1965) and even the more general application of Eitel (1958) are thus considerably extended for this petrological treatment and care must be taken that conclusions made about transformations (*sensu stricto*) by metallurgists are not applied rigidly to metamorphic transformations which are only loosely analogous.

Following Christian (1965) a very wide range of processes is classified as transformations in Table 4. Most transformations are *heterogeneous*, in that, at an intermediate stage, the assembly can be divided into microscopically distinct transformed and unchanged regions (Christian, 1965, p. 4). The process involves some kind of nucleation plus growth.

The simplest transformation is the inversion between high- and low-temperature polymorphs, e.g. low quartz, high quartz, tridymite and cristobalite. Such transformations are divided into two main categories: *displacive* (Buerger, 1948), involving only slight structural change, and *reconstructive*, involving breaking of bonds, significant changes of bond angle, and rebuilding of the structure. Changes of intermediate complexity are termed *semi-reconstructive*. Sosman (1927) used the terms rapid and sluggish respectively for displacive and reconstructive changes in quartz. The change from high- to low-temperature quartz involves slight structural readjustment, is essen-

TABLE 4

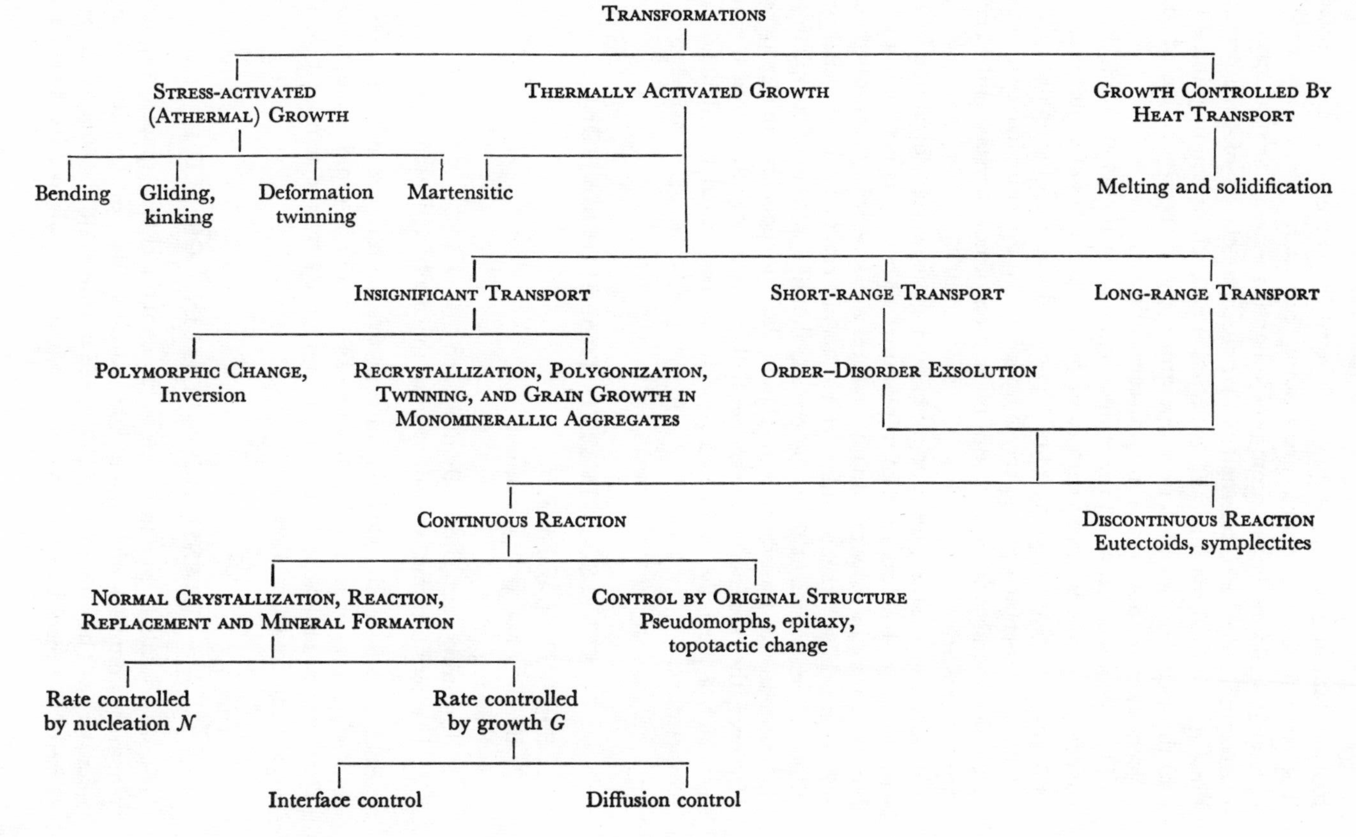

TRANSFORMATIONS
STRESS-ACTIVATED (ATHERMAL) GROWTH
THERMALLY ACTIVATED GROWTH
GROWTH CONTROLLED BY HEAT TRANSPORT
Bending
Gliding, kinking
Deformation twinning
Martensitic
Melting and solidification
INSIGNIFICANT TRANSPORT
SHORT-RANGE TRANSPORT
LONG-RANGE TRANSPORT
POLYMORPHIC CHANGE, Inversion
RECRYSTALLIZATION, POLYGONIZATION, TWINNING, AND GRAIN GROWTH IN MONOMINERALLIC AGGREGATES
ORDER–DISORDER EXSOLUTION
CONTINUOUS REACTION
DISCONTINUOUS REACTION
Eutectoids, symplectites
NORMAL CRYSTALLIZATION, REACTION, REPLACEMENT AND MINERAL FORMATION
CONTROL BY ORIGINAL STRUCTURE
Pseudomorphs, epitaxy, topotactic change
Rate controlled by nucleation N
Rate controlled by growth G
Interface control
Diffusion control

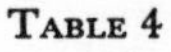

tially instantaneous, and cannot be delayed by quenching, i.e. is displacive. The transformation quartz–cristobalite, however, involves considerable change in bond angle and atom position and is slow, i.e. is reconstructive.

Metamorphism is concerned with two main classes of transformations, those due to deformation (i.e. stress-activated such as bending and gliding) and those due to elevation of temperature (e.g. thermally activated such as inversion, exsolution, reaction and recrystallization).

This classification, and in fact the usefulness of the whole transformation concept, depends to a certain extent on the existence of some relationship between the original minerals and those resulting from the change although much metamorphic crystal growth does involve the complete destruction of original minerals with the *de novo* nucleation of new crystals and their subsequent growth.

At any temperature, the amount of homogeneous transformation increases with time until a condition of minimum free energy is reached. The rate, however, may be slow and many textures are kinetically controlled in that they represent a "frozen" stage which has not reached a lowest energy condition. Given sufficient time, the process should continue to completion and the amount of transformation should be independent of temperature. However, the rate increases exponentially with temperature thus the degree by which a partially completed process has proceeded will depend on the temperature.

STRESS-ACTIVATED TRANSFORMATIONS

1. Slip (glide, translation).
2. Bending and polygonization.
3. Kinking.
4. Twinning (also thermally activated).
5. Martensitic transformations (also thermally activated).

Such transformations are due directly to superimposed stresses. They tend to be displacive or semi-reconstructive and thus to be rapid and diffusionless. The velocity and kind of change is influenced by the temperature but is largely controlled by stress.

TRANSLATION (GLIDING OR SLIP)

Some crystals in metamorphic rocks contain closely spaced, sub-parallel planar structures which appear as parallel lines or lamellae under the microscope. They have been interpreted as due to slip because of the importance of this mechanism in the deformation of metals and ionic crystals and because similar structures have been produced experimentally in some minerals.

Slip (glide, translation) occurs when one portion of a crystal slides over another without loss of cohesion (Barrett, 1952; Gilman, 1961). It is a process of indefinite magnitude which takes place such that corresponding crystal elements in each portion are similarly orientated before and after the movement. It is a rapid (almost displacive) diffusionless process which produces displacements in the lattice (and accompanying change in external shape) but which requires no change in structure or packing and thus is not a transformation in the strict sense.

Slip takes place along a *slip plane* (glide plane) in a direction parallel to the *slip direction*. The intersection of glide planes with the outer surface of the crystal produces *slip lines* and closely spaced assemblages of such lines are called *slip bands*.

CRYSTALLOGRAPHIC VECTOR INDICES

The conventional crystallographic nomenclature used in discussing glide and dislocations is as follows:

parentheses for the slip plane, e.g. (*hkl*), in Miller indices.
braces for the planes of a form, e.g. {*hkl*}, in Miller indices.
square brackets for a direction, e.g. [*uvw*], in vector indices.
angle bracket for a set of equivalent directions, e.g. ⟨*uvw*⟩.

The indices of a direction may be explained as follows (Fig. 16): a direction is outlined by the movement of a point from the origin to a given position which has coordinates of *u*, *v*, *w* on the three rectangular cartesian coordinate axes *X*, *Y*, *Z*. The direction is written [*uvw*] where *u*, *v* and *w* are the smallest integral values possible. The *X* axis *(AB)* has indices [100], the *Y* axis *(BC)*

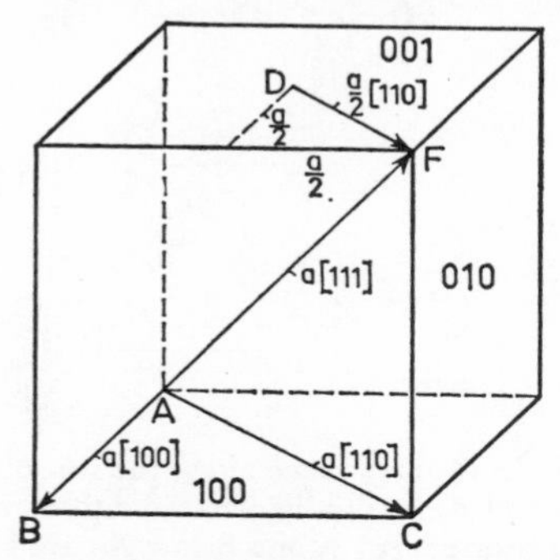

Fig. 16. Direction of a line (i.e. vector index) within a crystal. The index consists of a distance plus a direction, e.g. *AF* is one lattice vector *a* in a direction [111], thus the index is *a* [111]. The *X* axis is parallel to *AB*, the *Y* axis to *BC* and the *Z* axis to *CF*.

indices [010], the diagonal *AC* of the *XY* face has indices [110] and the body diagonal [111].

Directions are only perpendicular to the planes with the same Miller indices in the cubic system and for a few planes and directions in other systems. Directional indices are not Miller indices and reciprocals are not used.

A glide system is described in terms of the slip plane, e.g. (010) plus the slip direction ⟨001⟩, e.g. (010) ⟨001⟩.

The *direction* of a slip is practically always a close-packed lattice direction, e.g. ⟨110⟩ in face-centred cubic metals, [111] in body-centred cubic metals and [$2\bar{1}\bar{1}0$] in hexagonal metals. The slip directions in ionic crystals are along lines of similar ions or at least of ions of similar charge, e.g. [110] in NaCl, but this rule is not obeyed in all compounds with mixed bonds. The *plane* of slip is generally, but not always, a close-packed plane, e.g. (111), in face-centred cubic metals and (0001) in hexagonal. Body-centred cubic metals do not slip on a clearly defined crystallographic plane and (110), (112) and (123) planes have been reported.

Slip begins on a given system (i.e. slip plane plus slip direction) when the component of the imposed shear stress, resolved parallel to the slip direction and plane, reaches a critical value. The observed yield strengths of the softest crystals are about 10^3 less than the calculated strength, but near perfect single crystals in "whiskers" (long, thin single crystals almost free of defects) approach the theoretical strength. The low strengths are due to dislocations assisting glide (Fig. 17). It is emphasized that a consideration of dislocations

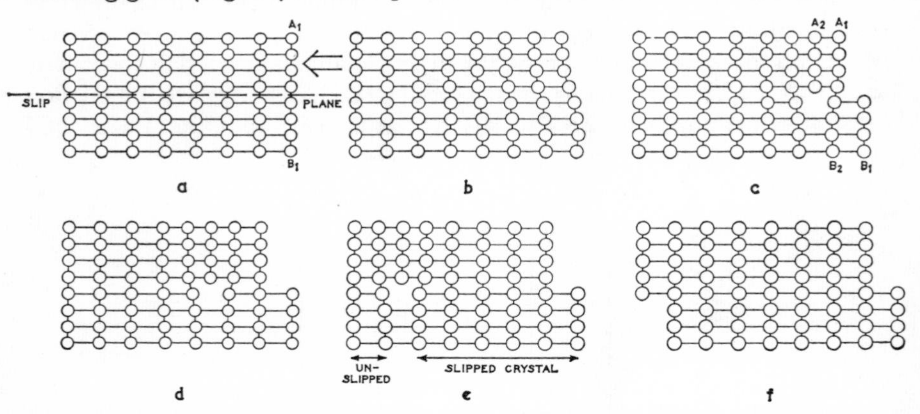

FIG. 17. Translation glide (slip) takes place along the slip plane by movement of an edge dislocation. (a) An atom A_1 is initially connected to B_1 in the unstrained lattice. (b) The lattice becomes distorted. (c) Atoms below A_1 are joined to those above B_2 giving an extra layer A_2 and a slipped region between B_1 and B_2. (d) and (e) The dislocation moves through the crystal. (f) The whole is slipped by one unit after the dislocation has traversed the whole crystal

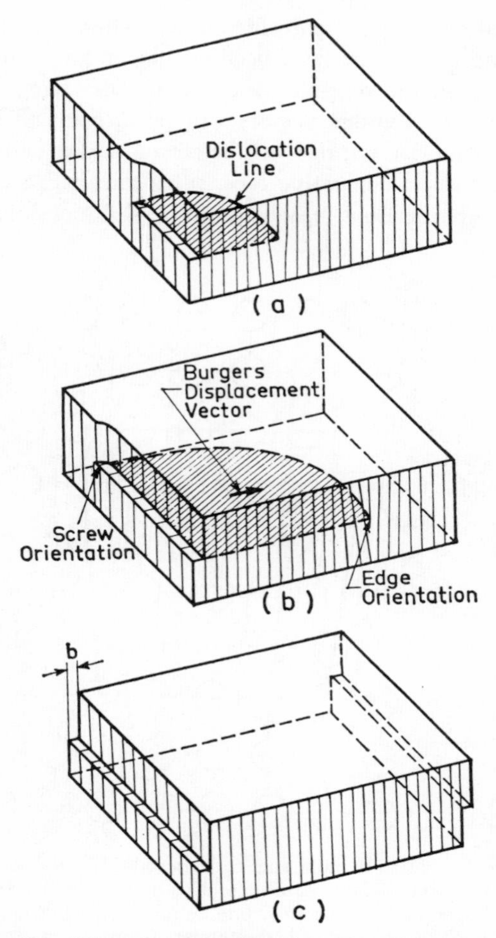

Fig. 18. Complex line dislocation with edge and screw orientation. (a), (b) and (c) show succession positions as the dislocation sweeps through the crystal to allow a unit of slip *b* (after Gilman, 1961, Fig. 5)

is essential in understanding gliding, as well as many other texture-controlling processes. Slip is permitted by *edge*, *screw* and a whole range of more complex dislocation types (Fig. 18). Dislocations are generated at grain boundaries, cracks, interior defects or special sources and sweep through the crystal assisting the slip. They move along the slip planes but edge dislocations may climb, and screw dislocations may undergo cross-slip and move on to intersecting slip planes. The dislocations may interfere with each other, "pile-ups" or "tangles" are formed, and work-hardening of metals occurs when dislocation movement has been hindered.

In passing from metal to ionic to silicate (ionic-homopolar) structures, the symmetry decreases, the unit cells become bigger and more complex, the bonding stronger and the possible dislocations more irregular and difficult to visualize. Thus the Burgers vectors will be larger and the tendency to dissociate into partial dislocations will be strong and even edge dislocations may contain positive and negative ions, silica-tetrahedra or alumina-octahedra, and the edge insertions may be of unit cell dimensions (Fig. 19).

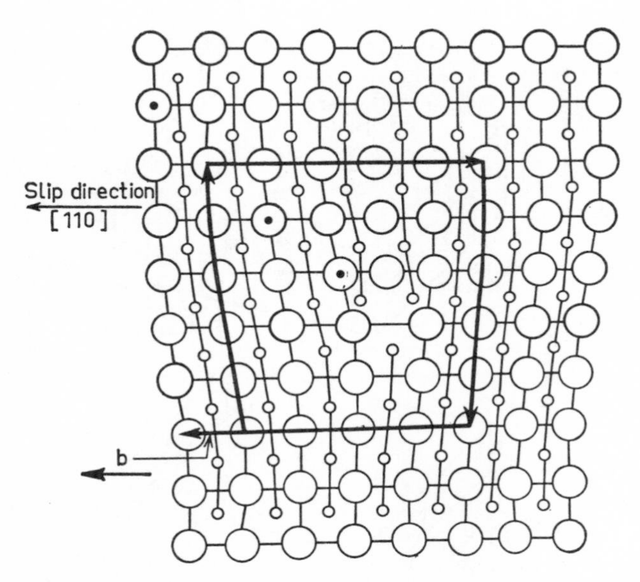

Fig. 19. An edge dislocation in halite structure; the edge "insert" can be regarded as a double layer of Na and Cl ions parallel to (110) or as a single layer of Na plus Cl shown by dotted open circles. The solid line outlines a Burgers circuit with a misclosure of *b*, the Burgers vector

The plastic behaviour of metals during cold working and the parts played by gliding and dislocations have been very extensively documented. Excellent summaries are given by Cottrell (1953, 1964), Read (1953), Friedel (1964) and Amelinckx (1964). A knowledge of the behaviour of metals is important in understanding the textures produced by the deformation of ore minerals such as galena, hematite, etc., which show many similarities with the metals, and to a lesser extent the behaviour of ionic minerals with strong glide properties (calcite, dolomite, gypsum, rock salt). Analogies between metals and common rock-forming silicates are of considerable importance (Gilman, 1961).

The mechanism of slip is by no means always simple even in metals and the correlation of slip planes with close-packed lattice planes, and of slip directions with close packed directions, does not hold for all metals and is of doubtful validity in low symmetry complex silicates. The observed second order pyramidal slip on $\{11\bar{2}2\}\langle 11\bar{2}3\rangle$ in zinc (h.c.p.) is hard to understand because the observed slip direction is not parallel to a direction of close packing and the atom arrangement parallel to this direction is highly puckered (Bollman, 1961). The required Burgers vector is relatively large thus necessitating a relatively thick edge insertion with unusually large terminal distortions. Similar relations have been demonstrated in beta uranium and in rutile and are probably common in minerals.

Slip mechanisms and systems in minerals of geological interest have been given by Buerger (1930). The glide behaviour of calcite, both in single crystals and polycrystalline aggregates has been well documented (Turner and Weiss, 1963, p. 343) at various temperatures and pressures and with various kinds of strain.

Translation takes place on $\{10\bar{1}1\}$ parallel to $(f_2{:}f_3)$ in a positive sense. This mechanism is dominant at 20°C, is the sole mechanism at 300°–400°C and is possible at high temperatures. Translation takes place on $\{02\bar{2}1\}$ parallel to $(f_1{:}f_2)$ in a few experiments at low temperatures and high pressures, but it becomes dominant at between 500° and 600°C and continues to 800°C at 5kb. Twin gliding takes place on $\{01\bar{1}2\} = e$ in a positive sense over the complete range of investigated conditions 20° to 800°C, 1 to 10kb. Etch pits reveal the dislocations associated with gliding in calcite (Keith and Gilman, 1960).

Raleigh (1965b) demonstrated experimentally that slip takes place in kyanite, enstatite and olivine at 5 kb and 700–800°C and is accompanied by kinking. The choice of glide plane is such that the strong silicon–oxygen bonds remain unbroken and the glide direction is that direction for which the Burgers vector for a unit edge dislocation in the plane is least, thereby minimizing the length of unit translation during slip.

Deformation of quartz by gliding has been postulated many times and although demonstrated experimentally by Christie *et al.* (1964) is not accepted as a major factor in the deformation of natural quartz. The operation of slip is proved by slip lines on the outside of crystals or by the internal rotation of some marker. The slip planes may or may not be visible under the microscope and not all parallel fractures or lamellae are due to slip.

Deformation lamellae have been considered to be the visible expressions of slip planes. Their nature and origin have been surveyed by Christie and Raleigh (1959), Turner and Weiss (1963), Friedman (1964) and Carter and Friedman (1965). The origin of lamellae produced experimentally by

Carter, Christie and Griggs (1964) has been established by Christie, Griggs and Carter (1964).

Textural features due to slip in a variety of crystals appear under the microscope as:

1. *Slip planes:* visible as lines, straight to curved, without measurable thickness resembling cleavage, e.g. Raleigh (1965).
2. *Deformation lamellae:* straight to curved parallel thin layers ("lamella" is derived from the diminutive of lamina and means a very thin plate, scale or film). Lamellae may be *continuous*, i.e. pass across a number of crystals of different orientation ("Tuttle" lamellae of Rast, 1967), or *discontinuous*, i.e. restricted to individual grains.
3. *Deformation bands:* these are layers differing in orientation from adjacent parts of the crystal. Some deformation lamellae may be very thin deformation bands. Kink bands and deformation twin bands are special cases. The "*c* lamellae" in calcite are thin $(01\bar{1}2)$ twins.

Discontinuous deformation lamellae of two kinds are important textures in quartz. Less important are the "Boehme" lamellae which consist of planar arrays of minute cavities or inclusions. More important are the "normal" lamellae which are thin, planar to curved features occurring in closely spaced parallel sets. They are lenticular and do not transgress grain boundaries, are approximately perpendicular to strain shadows and deformation bands and are visible in plane polarized light, between crossed polarizers or in phase-contrast. They have slightly different refractive index and birefringence from the host quartz.

Christie *et al.* (1964, p. 739) concluded that lamellae are comprised of a region of higher refractive index and birefringence on one side of a plane discontinuity and lower index on the other side. They suggested that lamellae are caused by basal slip or deformational Brazil twins (McLaren *et al.*, 1967) and that their optical character is due to an array of edge dislocations lying on the basal plane. The dislocations would be of the same sign (those of opposite sign would cancel each other out) and would result in a region of compression on the "extra layer" side of the slip plane and a region of tension on the other. The part of the crystal in compression would have a higher R.I., and that in tension, a lower R.I. than unstrained crystal.

Quartz consists of a compact network of SiO_4 tetrahedra each of which is bonded at the four oxygen corners to adjacent tetrahedra. There are no *simple* close-packed layers or lines that could be expected to be *simple* slip planes or directions. However the atomic pattern parallel to (0001) is regular and repeats itself at intervals of 1 unit cell parallel to *a* and *c* axes. The slip system proven experimentally (slip plane (0001), slip direction $\langle 11\bar{2}0 \rangle$, i.e. **a**) has the shortest and simplest atomic movements. The simplest dislocation to assist glide would be an edge dislocation perpendicular to $\langle 11\bar{2}0\rangle$ and lying in (0001) (Fairbairn, 1939). The dislocation is large, involving a unit cell sized insertion and has a Burgers vector $\mathbf{a}\ [11\bar{2}0]$ where the unit cell dimension $\mathbf{a} = 4{\cdot}91$ Å (Plate II).

The strong linear atomic features parallel to **a** and the absence of a layered structure suggests that slip may not take place along single (0001) planes but that slip planes would have a strong tendency to wander from level to level giving a kind of pencil glide. It is perhaps more likely that screw dislocations parallel to **a** operate with abundant cross-slip. Plastic deformation of quartz takes place under experimental conditions at very high temperatures and pressures by basal glide but the process occurring in rocks is not necessarily the same. Deformation of quartz has taken place in rocks which have been only slightly deformed at low temperatures and pressures and natural deformation seems to take place more easily than experimental deformation. The "softening" effect of high P_{H_2O} (Griggs and Blasic, 1965) may help but natural deformation may take place over long periods by a creep process not involving simple slip. As deformation is produced geologically by apparently small strains, it is suggested that slip is aided by dislocation movement. The structure of quartz is such that simple perfect dislocations would involve large Burgers vectors which would result in large lattice strains.

Direct observation of quartz under the electron microscope confirms conclusions made previously by various workers as to the probable dislocation systems and their relations to slip, grain boundaries, and sub-grain boundaries (Willis, 1952; Arnold, 1960; Dienes, 1960; McLaren and Phakey, 1965a, b; and Bursill and McLaren, 1964). The smallest Burgers vector for unit slip is $\langle 11\bar{2}0\rangle$ and other possible, but energetically less favourable vectors are $\langle 0001\rangle$ and $\langle 11\bar{2}1\rangle$. Christie *et al.* (1964) and Carter *et al.* (1964) found slip on $\{0001\}\ \langle 11\bar{2}0\rangle$ mainly but also on $\{10\bar{1}0\}\ \langle 0001\rangle$. Observations of the structure of quartz suggest $b = \mathbf{a}\langle 11\bar{2}0\rangle$ but the relation $\mathbf{a}\{2\bar{1}\bar{1}0\rangle + \mathbf{a}[\bar{1}2\bar{1}0] \rightarrow \mathbf{a}[11\bar{2}0]$ is possible. The interaction of a set of screw dislocations with Burgers vector $b = \mathbf{a}[\bar{1}2\bar{1}0]$ will give a set of three-fold dislocation nodes as in Plate III.

Edge dislocations form symmetrical tilt boundaries as predicted by Carter *et al.* (1964) parallel to [0001]. The spacing between the dislocations is 500 Å and as $b = 4{\cdot}91$ Å, this gives a tilt angle of $\theta = b/h = 0°\ 36'$. Sub-boundaries are formed by edge dislocations (McLaren and Phakey, 1965b, Fig. 3).

Examination of the structure suggests that for slip on (1011) the Burgers vector is $|\mathbf{a} + \mathbf{c}|\ \langle 11\bar{2}1\rangle$. Two dislocations with Burgers vectors $|\mathbf{a} + \mathbf{c}|\ \langle 11\bar{2}1\rangle$ and $|\mathbf{a} + \mathbf{c}|\ \langle 11\bar{2}1\rangle$ may interact to produce a third dislocation with vector $\mathbf{a}[\bar{1}2\bar{1}0]$ to produce a hexagonal network.

The role of high P_{H_2O} in allowing plastic deformation of quartz and other silicates is indicated by the behaviour of synthetic quartz which has a water

TABLE 5. GLIDE AND TWIN GLIDE IN SOME ROCK-FORMING MINERALS

Mineral	Slip		Twinning		Reference
	Plane	Direction	Plane K_1	Direction n_1	
Anhydrite	(001)	[010)			Mugge (1898)
Barite	{001}	[100]	{110}		Grigorev (1965)
	{011}	[011]			Grigorev (1965)
Brucite	(012)	[100]			Veit (1922)
	(0001)				Mugge (1898)
Carbonates					
Aragonite	(010)	[100]			Veit (1922)
Calcite	$\{01\bar{1}2\}$	$[\bar{1}\bar{2}31]$			Mugge (1898)
	$(10\bar{1}1)$	$[f_2:f_3]$	$\{01\bar{1}2\}$	$[e:r_2]+$	Turner *et al.* (1954)
	$(02\bar{2}1)$	$[f_1:f_2]$			Turner *et al.* (1954)
Dolomite	{0001}	[0100]	$\{02\bar{2}1\}$	$(02\bar{2}1):(2\bar{1}\bar{1}0)$	Johnson (1902)
Magnesite	(0001)	[0100]			Higgs and Handin (1954)
	$(10\bar{1}1)$	$[r_1:f_2]$			Higgs and Handin (1954)
Rhodochrosite	(0001)	$(01\bar{1}0)$			Veit (1922)
Siderite	(0001)	[0100]			Johnson (1902)
Chalcopyrite	{111}		{110}		Grigorev (1965)
Corundum	(0001)	$[11\bar{2}0]$			Gibbs (1959)
	(0001)	$[\bar{1}\bar{2}10]$			Gibbs (1959)
	(0001)	$[\bar{2}110]$			Gibbs (1959)
	(1120)	[1100]			Gibbs (1959)
	$(\bar{2}110)$	$[01\bar{1}0]$			Gibbs (1959)
	$(1\bar{2}10)$	$[\bar{1}010]$			Gibbs (1959)
	$(10\bar{1}0)$				Klassen-Neklyudova (1964)
			$\{10\bar{1}1\}$		Grigorev (1965)
Fluorite	(001)	[110]			Veit (1922)
Galena	{100}	⟨110⟩			Buerger (1928)
					Lyall and Paterson (1964)
	{001}	⟨100⟩			Mugge (1898)
	{110}	$\langle 1\bar{1}0\rangle$			Lyall and Paterson (1964)
	{001}	⟨110⟩			Lyall and Paterson (1964)
Graphite	(0001)	[2110]			Mugge (1898)
	(0001)	$[10\bar{1}0]$			Grigorev (1965)
Gypsum	(010)	[301]			Mugge (1898)
	(010)	[001]			Mugge (1898)
Halite	{110}	$\langle 1\bar{1}0\rangle$			Johnson (1902)
	{001}	$\langle 1\bar{1}0\rangle$			Buerger (1930)
Hematite	{0001}	$[10\bar{1}0]$	$\{10\bar{1}1\}$		Grigorev (1965)
Kyanite	(100)	[001]			Mugge (1898), Raleigh (1965b)
Magnetite			{111}		Grigorev (1965)
Mica	(001)	[100]	(001)		Mugge (1898)
	(001)	[110]			Mugge (1898)
	(135)				Ramberg (1952)
	(205)				Ramberg (1952)
	(104)				Ramberg (1952)

TABLE 5—*continued*

Mineral	Slip		Twinning		Reference
	Plane	Direction	Plane K_1	Direction n_1	
Molybdenite	(0001)	[21$\bar{3}$0]			Mugge (1898)
	(0001)	[10$\bar{1}$0]			Grigorev (1965)
Olivine	(100)	[001]			Raleigh (1965)
	(110)	[001]			Raleigh (1965)
	(010)	[100]			Raleigh (1965)
Periclase	(110)	[1$\bar{1}$0]			Mugge (1898)
Pyroxenes					
Enstatite	(100)	[001]			Mugge (1848), Griggs *et al.* (1960) Raleigh (1965)
Diopside	{100}	[001]	(100)	[001]	Raleigh and Talbot (1967)
			(001)	[100]	
Clinopyroxene	{100}	[001]			Talbot *et al.* (1963)
Pyrrhotite	{0001}	[10$\bar{1}$0]			Grigorev (1965)
Quartz	{0001}	⟨11$\bar{2}$0⟩			Carter *et al.* (1965)
	{10$\bar{1}$0}	⟨0001⟩			
			{101$\bar{0}$}	[1$\bar{2}$10]	
Rutile	{110}	001	{101}		Grigorev (1965)
Sphalerite	(111)	?			Mugge (1898)
	{111}	[112]	{111}		Grigorev (1965)
Stibnite	(010)	[001]			Mugge (1898)
Sylvite	(110)	[1$\bar{1}$0]			Mugge (1898)
Wolframite	(010)	?			Mugge (1898)

content of 0·1% (about 10 times that of natural quartz). The weakness is probably due to the formation of Si–OH (Silanol) groups by the breaking of Si-O-Si bonds and their replacement by Si-OH-HO-Si. These silanol groups could align themselves along existing dislocations, but are in fact defects themselves and, like edge and screw dislocations, could move through the structure with successive rupture and healing of bonds allowing glide.

The relations of dislocations to slip under various physical conditions have been determined for a number of natural or synthetic minerals. Most are not rock-forming or even significant in metamorphic rocks, however they constitute a group with structures and bonding (ionic or ionic-homopolar) which are intermediate in character between metals and rock-forming silicates. Phases include corundum (Gibbs, 1959, Kronberg and Wilson, 1962; Voruz *et al.*, 1963; Stephens and Alford, 1964; Conrad *et al.*, 1964; Conrad, 1965); spinels (Hornstra, 1960); rutile (Hirthe and Brittain, 1962); fluorite (Burn and Murray, 1962); magnesium oxide (Day and Stokes, 1966; Hulse *et al.*, (1963); galena (Lyall and Paterson, 1966).

BENDING AND POLYGONIZATION

Crystal lattices may be easily bent within the elastic limit with suitable small strains but permanent bending is more difficult. Permanent bending results in a change in lattice and optical directions across a crystal so that it does not extinguish as a whole when viewed between crossed polarizers under the microscope. A region of extinction sweeps across the crystal as the stage is rotated and the crystal is said to exhibit *undulose, undulatory* or *strain extinction.* Undulose extinction is common in quartz, feldspar, calcite, olivine, and

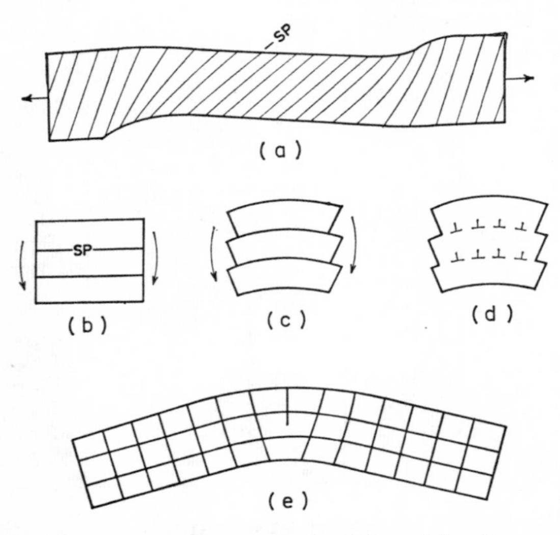

FIG. 20. (a) A single crystal is experimentally deformed by slip on a single set of oblique planes (SP) with accompanying bending. A crystal (b) may be bent by slip on one set of planes (c), or by slip by movement of a series of dislocations of the same sign (d). (e) Bending of a crystal is permitted by "insertion" of an extra layer, i.e. an edge dislocation

mica where the bending may exceed twenty degrees. As the radius of lattice curvature becomes smaller (i.e. the crystal is more sharply bent) the undulose crystal may be composed of narrow regions of slightly different orientations (*deformation bands*). The radius of lattice curvature is comparable with, or larger than, the width of the reoriented zone in undulose extinction but where the radius is smaller than about half the width of the reoriented zone (Carter *et al.*, 1964) the regions are called deformation bands.

A layered structure with good glide properties will bend plastically by differential slippage (bend gliding) of layers or by the formation of edge dislocations of one sign along the glide plane. Figure 20 shows a structure in which the sheets have slid over each other like a pack of cards; (d) shows a sheet structure in which differential slip has taken place on each layer by a series of edge dislocations which allow slip at low stress values and retain structural continuity throughout the whole stack of sheets during bending; (e) shows a layered structure in which bending is caused by the "insertion" of an extra layer of atoms to give an edge dislocation. If the glide packet is bent around a radius r the number of dislocations required is $1/rb$ (Cottrell, 1953, p. 29), where b is the Burgers vector of the dislocations.

Many deformed rocks contain crystals which have been twisted or bent (quartz, mica, calcite, gypsum, amphibole, pyroxene, olivine, kyanite, etc.). The bending causes undulatory (undulose) extinction, anomalous optical properties, and curved cleavage, curved twin planes or curved crystal faces.

Bending of mica (or gypsum) can be explained with the simple model above. Mica has a layered structure with (001) the glide plane, and the a axis the glide direction. Glide can be caused by bending around b.

Silk and Barnes (1961) and Akizuki (1966) have shown that micas contain dislocations, some of which lie in the cleavage and others which could be edge dislocations perpendicular to the cleavage. The optic directions in a crystal depends on the lattice orientation and as the (001) plane is bent, so the mica will develop undulose extinction. Mechanical deformation of muscovite causes slip along the basal plane and this may result in a restacking of the layers to produce twinning or stacking faults with a resulting change in the optics (lowering of $2V$). Bending lowers the symmetry of the quartz lattice from hexagonal to monoclinic or triclinic and thus the $2V$ changes from 0° to perhaps 20° with strain. Deformation of biotite may also cause an increase in $2V$. Slip along (001) causes either twinning or stacking faults so that cross-sections show strips with differing birefringence.

The mechanism for plastic bending in quartz is not known but may be partly due to inhomogeneous (0001) slip. Undulose extinction with accompanying deformation lamellae is common in quartz which has been naturally deformed even at quite low temperatures but has only been produced experimentally (Carter, Christie and Griggs, 1964) at very high temperatures and pressures. Natural undulose extinction forms bands approximately parallel to [0001] with almost perpendicular deformation lamellae.

A *deformation band* across a crystal is a layer which differs in orientation from the remainder and which is due to deformation. It is visible under the microscope as a comparatively sharply bounded band differing in birefringence or extinction from the host. It differs from a twin in lacking a precise symmetrical relation with the host.

A deformation band in calcite was defined by Turner *et al.* (1954) as a layer in which the crystal lattice is progressively rotated away from that of the host crystal during deformation; it is generally inclined at a high angle to the glide planes. Carter, Christie and Griggs (1964, p. 170) defined deformation bands in quartz as bands whose orientation differ slightly across clear junctions with an abrupt change in orientation.

KINKING

Kink bands may be regarded as a special type of deformation band. A *kink band* in a crystal is a sharply defined layer in which the orientation differs from that of the adjacent regions by a low to moderate angle (up to about 60°), the orientations being related by a single rotation about an axis which lies in the boundary (Orowan, 1942, Urusovskaya, 1962). Kinking is due to slip along closely spaced surfaces (such that the slip direction is perpendicular to the kink axis) and is due to shortening approximately parallel to these surfaces. The change of orientation is abrupt and may or may not be accompanied by a fracture (Fig. 21 and Plate XI). Very small kink bands

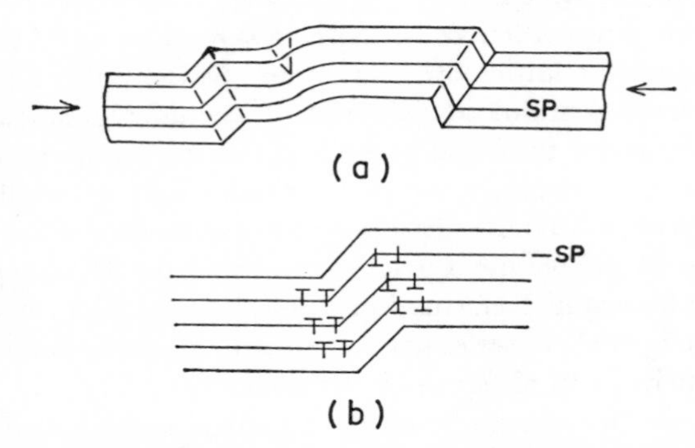

Fig. 21. (a) Kink bands in a deformed crystal with a single slip plane (SP) parallel to the length of the crystal. (b) Kinks represented as arrays of dislocations

were first described in metals but the term is now applied to similar structures which range in size from microscopic dimensions in metals and minerals up to mesoscopic size in foliated rocks. The fundamental mechanism appears to be the same for all kink bands although kinking in metals and perhaps some minerals is facilitated by the motion of dislocations.

It is not simple to decide whether some bands should be called deformations or kink bands. Siefert (1965, p. 1470) figured "deformation bands" in albite which have the same appearance as structures called kink bands in metals and minerals. He (ibid., p. 1471) considered that kink bands are a special class of deformation bands on the grounds that, by definition, kinks are simple bands in which deformation can be described by an axis of external rotation coinciding with the line of intersection of the kink band and the slip plane perpendicular to the slip direction. Siefert's structures could not be defined by simple rotation of the host crystal and thus he named them deformation bands rather than kink bands. For normal petrographic purpose such structures should be termed deformation bands until it is proved that the geometry is appropriate for kinks.

Kink bands are common in the mica, kyanite, pyroxene and amphibole of dynamically altered rocks. They have also been described in epidote and various salts. The bands may occur as a set of parallel layers, but most are somewhat lenticular and some form conjugate pairs.

Kink bands in metals are attributed to strong curvature of glide planes caused by the pile-up of large numbers of dislocations of opposite sign on neighbouring slip planes. The dislocations do not move past each other but are impeded by barriers composed of dislocations of the same kind (perhaps dislocations on an intersecting glide plane, Van Bueren, 1960, p. 161). They have been produced experimentally in minerals (Griggs *et al.*, 1960; Raleigh, 1965b). Christie *et al.* (1964) and Raleigh (1965b) gave the orientation relations used to determine slip mechanisms and a summary of their orientation patterns and relation to stress has been given by Friedman (1964).

Deformation of olivine in ultrabasic igneous rocks produces undulatory extinction, deformation lamellae and deformation bands ("banded olivines" of Hamilton, 1957). As in quartz, there appears to be a gradation (with increasing deformation) from parallel undulatory zones into sharply defined deformation bands (Ragan, 1963). Partial recrystallization may change the boundaries of the bands from straight to sutured (Ragan, 1963, plate 1D).

An analogy has been drawn between the formation of sharply bounded deformation bands and the process of *polygonization* (Bailey *et al.*, 1958; Griggs *et al.*, 1960, p. 24; Harris and Rast, 1960b; Carter *et al.*, 1964). Bending may be accompanied by polygonization and an undulose crystal may have a submicroscopic polygonized structure (Cahn, 1949).

Polygonization is the process whereby a curved or otherwise strained lattice readjusts into unstrained segments (*sub-grains*) which are separated from each other by coherent, low-angle, sub-grain boundaries. Polygonization is a *recovery* mechanism and has been studied in connection with the annealing of metals (also sapphire, May, 1959). It plays an important part in recrystallization but as the simplest model is based on dislocation theory, it may be conveniently discussed here in connection with bending.

Dislocations are initially distributed rather irregularly throughout a deformed crystal and during annealing, the first step for polygonization is for dislocations of the same sign to come together to form planar arrays (Fig. 22). The dislocations climb so that the distance between them becomes less than the original glide plane spacing. If the bend utilizes only one glide

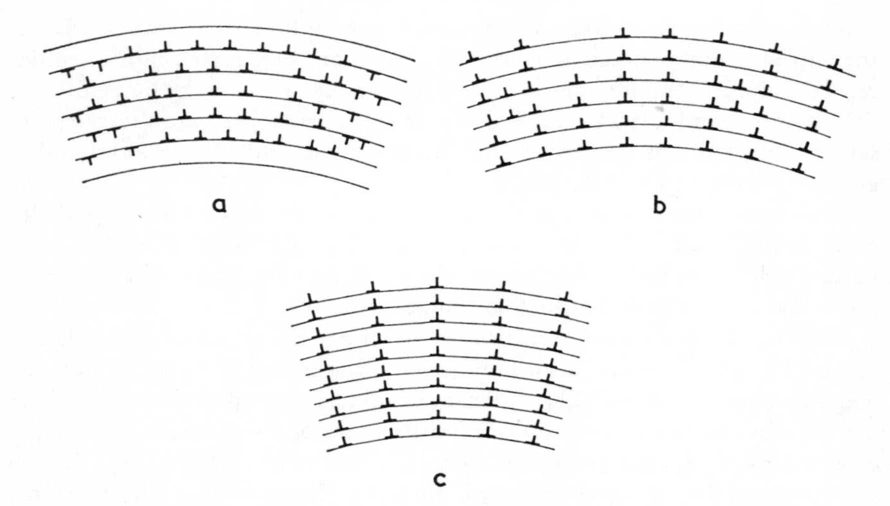

FIG. 22. Polygonization in a slipped crystal. (a) The crystal contains dislocations of both signs on the slip planes. (b) Dislocations of opposite sign cancel out. (c) Climb into vertical arrays gives sub-grain boundaries with small tilts separating unstrained segments of lattice

system then the sub-grain boundaries are perpendicular to the glide plane but if dislocations of different sign are present, the sub-boundaries will be inclined. The polygonized structure thus consists of strain-free segments of lattice of different orientations separated by walls of dislocations in stable arrays (Amelinckx, 1964, p. 35).

The initial grain size produced by polygonization during annealing of a deformed crystal is inversely proportional to the density of edge dislocations of one sign and proportional to the radius of bending of the lattice (Cahn, 1949). Polygonization takes place preferentially in the bent portions of kink bands (Towner and Berger, 1960).

Polygonization (*sensu stricto*) is a thermally activated process which may be regarded as recovery or incipient recrystallization. The formation of deformation and kink bands is, however, an athermal transformation caused by deformation alone. The two processes must be differentiated in principle although they may produce similar intragranular textures and may act together in a geological process analogous to "hot-working". If strain takes place at an elevated temperature, e.g. protoclasis in igneous crystal aggregates or low-grade regional metamorphism, then lattice strain may be closely followed, or even accompanied by, polygonization. An elevated temperature promotes dislocation movement and slip and the processes of polygonization and deformation band formation may become the same.

TWINNING

A twinned crystal may be defined as a rational symmetrical intergrowth of two individual crystals of the same species (Buerger, 1945, p. 471) or alternatively as a polycrystalline unit composed of two or more homogeneous portions of the same crystal species in contact and mutually oriented according to certain simple laws (Cahn, 1954, p. 363). Twinning is said to be *simple* if there are two units and *multiple* or *polysynthetic* if there are many.

Deformation twins are produced by stress-activated transformations, and although growth twins are thermally activated, it is convenient to consider all types of twins together at this stage.

A twin is composed of two differently oriented structures in which there is a collinear common atomic plane called the *twin plane*, and a common line called the *twin axis*. The junction is the *composition plane* and this may also be the twin plane, another rational plane or a non-rational interface. The orientations of the two parts are related by a rotation around the twin axis and many twins are reflected as mirror images of each other.

Differences in orientation of the parts of a twinned crystal render them visible by differences in reflectivity or cleavage in hand specimen or in interference colour, cleavage direction, refractive index, body colour, optical orientation, etc., under the microscope. Some twins show re-entrant angles at their boundaries.

The morphology of twins and the crystallographic relations involved have been discussed at length by Bell (1941), Clark and Craig (1952), Cahn (1954) and Pabst (1955). Recent reviews of the geometry (Christian, 1965), of twinning in metamorphic crystals (Vernon, 1965) and in sulphides (Richards, 1966) are particularly valuable.

Twinning can be regarded geometrically as consisting of a single homogeneous displacement (simple shear) of all atoms plus a small amount of minor readjustment of the individual atoms ("atomic shuffling") in their new positions. In Fig. 23a, the ions or atoms, etc., above the *AB* layer may be either in the *X* position (in which case the crystal is untwinned) or in the *Y* position (twinned).

Twinning may be produced in a number of ways. The twins are *primary* if they formed simultaneously with the formation of the crystal, and *secondary* if they were formed after the crystal. Twinning in a plagioclase crystal in an igneous rock is normally *primary* due to growth as it was precipitated from the magma; twinning in calcite in a deformed marble is normally *secondary* due to deformation. However, the distinction is not always clear in metamorphic rocks. Seven somewhat distinct genetic types are:

(1) Primary twins
 (a) Growth (twinned nucleii, radial or layer growth).
 (b) Annealing (recrystallization).
 (c) Combination (synneusis).
(2) Secondary twins
 (d) Inversion.
 (e) Deformation.
 (*f*) Interference.
 (g) Exsolution.

PRIMARY TWINS

(a) Growth Twins

Elementary growth twins form in many crystals which grow unencumbered by surrounding liquid or solid matrix. Twins may begin life as twinned nucleii, or form during layer growth of a large crystal.

TWINNED NUCLEII

Isolated, radially symmetrical twins such as cruciform twins in andalusite, radiating or sector twins in cordierite or in garnet (Harker, 1939, p. 229) may develop from twinned nucleii, i.e. those in which the atomic configuration is initiated in a twinned state. It is known that twinned nucleii can be precipitated from solution, that the proportion of twinned to untwinned nucleii increases with supersaturation and that twinned nucleii grow more quickly than those not twinned.

The process of formation of twinned nucleii may in part be related to the *stimulated twinning* of Burgers *et al.* (1953) whereby an advancing boundary releases a nucleus with the twin orientation (a kind of catalysed nucleation).

The process of formation of twinned nucleii may pass into the "multiple nucleation" of Hogan (1965) in which nucleii of a given orientation act as "seeds" and promote the formation of further nucleii with slightly different orientations. This leads to the formation of radiating and spherulitic intergrowths which may resemble twins.

Sector or *radial* twins (as in cordierite and aragonite) or *cruciform* twins (staurolite) are characterized by a number of composition planes radiating from a centre. The composition planes are thus approximately *parallel* to the direction of growth whereas they are approximately *normal* to the direction of growth in layer growth twins and variable in orientation in annealing twins. Hartmann (1956) gave reasons why sector, radial and cruciform twins may be geometrical intergrowths rather than true twins although this distinction does not appear to have any genetic significance.

LAYER GROWTH TWINS

Addition of units to any advancing crystal boundary which is a possible twin plane may be in either the normal or the twin position. If a new layer is initiated in the twin position the whole layer may take on this orientation, as will following layers until the twin process switches the orientation back to the normal position (Fig. 23). The boundary may develop one nucleation site

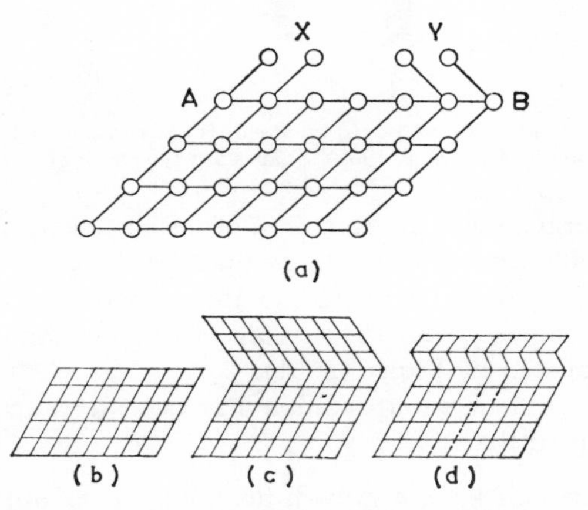

FIG. 23. Growth twins. (a) Atoms added to a crystal above the *AB* plane may be either in the normal position *X*, or the twinned position *Y*. Growth of a rhombic crystal (b) may continue in the twin position (c) to give a simple twin, then revert to the original orientation to give polysynthetic twins (d)

in the normal orientation and another in the twin orientation. Both sections may advance together and then one may revert to the alternative orientation so that a layer of uniform orientation extends across the crystal again. Growth twins may thus have an abrupt or stepped termination and be classified as (a) continuous, (b) enclosed, (c) partial, (d) edge (Irving, Miodownik and Towner, 1965; and Fig. 24).

Most interpretations of growth twinning (Buerger, Hartmann, etc.) have a thermodynamic basis: the atoms in the "normal" position have a slight thermodynamic advantage over those in the "twinned" position because the twin interface has a higher energy than that between normal lattice layers. The width of the twins depends on the thickness deposited between successive changes in orientation and this may be related to ease of nucleation of twins, i.e. the easier the nucleation of a new twin the thinner the twin lamellae. Ease of nucleation will depend on the energy difference between the normal and twinned condition, on the activation energy of twin nucleation and the

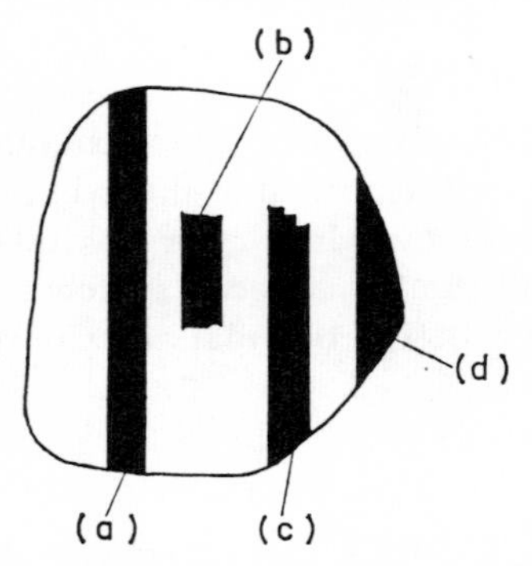

FIG. 24. Twins are (a) continuous, (b) enclosed, (c) partial, (d) edge (after Irving, Miodownik and Towner, 1965). Shapes are typical of growth twins

stacking-fault energy. These are functions of the structure of the crystal and there is strong evidence from metallurgy that the thickness of twin lamellae or "*twin frequency*" should be related to the composition as suggested for plagioclase by Donnay (1940, 1943), Gay (1956) and Smith (1958). This view has been opposed by Vance (1961).

An alternative, kinetic interpretation has been advanced by Donnelly (1967) who observed that:

(1) Twinned crystals have a growth advantage over untwinned in that, in a given assemblage, the twinned crystals are bigger than the untwinned.
(2) Some species (aragonite, feldspars) are virtually always twinned.
(3) Twinning is commonly associated with rapid growth or a viscous medium.

He concluded that twinning is not accidental and that the twin configuration may be more favourable for growth in a system in which the supply of ions to the crystal is relatively slow in comparison with the rate of incorporation.

For a given crystal species, rapid growth favours primary twin formation (Buerger, 1945; Cahn, 1954, p. 394) and twinned crystals in metamorphic rocks may show additional evidence of rapid growth (large size, poikiloblastic texture). The relation, however, is not simple and many porphyroblasts are untwinned.

(b) Annealing (Recrystallization) Twins

Grain growth in a polycrystalline aggregate such as a metamorphic rock, annealing metal or ceramic is more complicated than the elementary case above because of the mutual interference of the crystals. Growth during recrystallization involves the movement of grain boundaries such that some crystals increase in size at the expense of others. Twin formation takes place when the advancing front of a crystal growing through matrix meets a lattice domain in an adjacent crystal with an orientation closer to the twin orientation than to the normal orientation. The twins originate at grain boundaries, are nucleated by stacking faults and tend to form along slip planes (Dash and Brown, 1963).

Annealing twins may form parallel to an advancing grain boundary or obliquely to the boundary at a triple-point. Burke (1950), Fullman and Fisher (1951b), Burke and Turnbull (1952) and Aust and Rutter (1960) have advanced the theory that where a grain A grows into adjacent grains B and C (Fig. 25), then a twin A_1 of A may form if

$$\gamma_{A_1B} \cdot \text{Area}_{A_1B} + \gamma_{A_1C} \cdot \text{Area}_{A_1C} < \gamma_{AB} \cdot \text{Area}_{AB} + \gamma_{AC} \cdot \text{Area}_{AC}$$

where γ is the interfacial energy of the boundary indicated by the subscript.

Normally the areas of the AB and AC boundaries will be much greater than A_1B and A_1C but once A_1 exceeds a certain size the twin will grow at

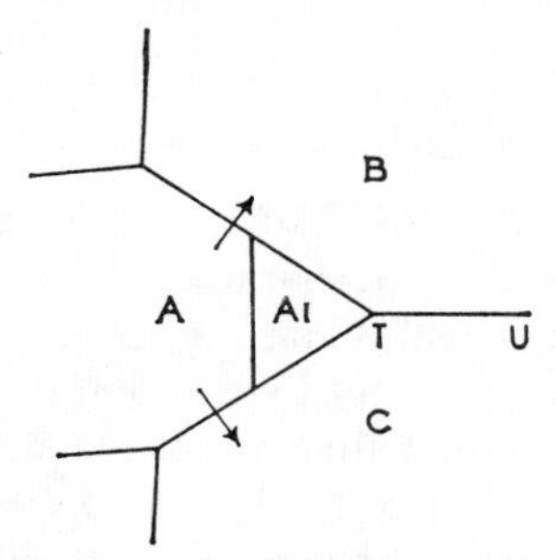

FIG. 25. Annealing twins. Growth of crystal A into two grains B and C is accompanied by movement of the triple-point T towards U (see text)

the expense of the normal orientation if surface energies due to misfit are sufficiently great (Burgers, Meijs and Tiedema, 1953; Cahn, 1954). Direct observations of the formation of annealing twins at moving grain boundaries were made by Aust and Rutter (1960b) who found that in every case a large-angle random boundary was replaced by a twin, near twin, or high-angle coincidence boundary.

Hu and Smith (1956) pointed out that this hypothesis implied that the total number of twins should be related to the grain size because twins are formed at certain boundary "encounters" and the number of encounters increases as the grains grow. There is some experimental confirmation for this.

Annealing and similar recrystallization is driven first by strain energies in the original strained aggregate and then by interfacial surface energies.

Annealing twinning is a thermally activated diffusionless transformation which takes place during grain growth in order to reduce grain boundary energy. A twin boundary is a large-angle coincidence boundary, hence its energy is lower than that of a random or misfit high-angle grain boundary (5% less in lead, 21% less in aluminium; Aust, 1961) and therefore there is a decrease in total free energy when a random boundary is replaced by a twin boundary.

The stability differs in the four kinds of twins (continuous, partial, enclosed and edge, Fig. 24). Enclosed and partial twins have a higher interfacial energy at the ends, thus they tend to shrink and disappear during annealing. The blunt ends in plagioclase, however, approximate to rational crystallographic directions (Vernon, 1965, p. 499) and may be stable, but round-ended enclosed twins may be due to shrinkage during recrystallization after twinning.

Annealing twins are common in sulphides and much of the twinning in the plagioclase of granulites and the calcite of marbles may be attributed to this mechanism.

(c) Synneusis and Combination Twins

Twins produced by the growth of adjacent crystals which happen to have a twin relation, are intermediate in character between primary and secondary types. If two crystals in a magma float into contact with each other so that the interface happens to have the twin orientation, it would be strongly adherent and the doublet could then grow as a single crystal with a simple twin (Ross, 1957; Vance, 1961, p. 1107; Van Diver, 1967, Fig. 4A).

Somewhat analogous circumstances can be visualized in metamorphic rocks where two nucleii form in close proximity and grow towards each other. If they happen to have a twin orientation then a simple *combination twin* could grow out from the centre. Vance (1961, p. 1107) has suggested that the abundance of Carlsbad twins in igneous plagioclase and the contrasted infrequency in metamorphic plagioclase is due to the liquid medium favouring synneusis. Combination twins are presumbly rare in metamorphic rocks.

SECONDARY TWINS

(d) Inversion Twins

Small-scale atomic rearrangements taking place during a polymorphic transformation or inversion may allow some parts of a crystal to adopt the twin orientation and some parts the normal orientation and produce secondary, inversion twins (e.g. high to low quartz giving Dauphiné twins).

Pairs of crystal connected by high–low transformations possess related symmetries (Buerger, 1945, p. 477; 1951, p. 195) such that the symmetry of the low form is a sub-group of that of the high form. From a purely geometrical viewpoint, the structure can be regarded as constant but a suppression of some symmetry operations occurs in passing from the high to low form. Thus there are alternative structural orientations for the low form, and transformation nucleii may form either in the normal or in the twin position.

The "cross-hatched" twinning of microcline is secondary, but it is not clear whether it is due simply to inversion from higher temperature monoclinic sanidine or orthoclase to the lower temperature triclinic microcline (Laves, 1950; Taylor, 1962) or to external deformation (Marmo, 1955a, b; Binns, 1966). The preferred occurrence of microcline rather than orthoclase in deformed rocks suggests that directed stress is the major control in that it causes the inversion from metastable orthoclase to microcline. All intermediate stages are possible between orthoclase and microcline and the position of intermediate members is indicated by their "triclinicity" as measured from X-ray photographs. The work of Binns (1966, p. 30) suggests that the triclinicity and degree of twinning both increase with increase of deformation but that cross-hatching is recognizable in orthoclase with negligible triclinicity.

An electron microscope study of the inversion of sanidine through the metastable orthoclase to triclinic andularia (or microcline) shows that the original lattice is distorted differentially (McConnell, 1965) during the transformation. Nucleii which form in the most distorted regions will tend to be in either of two orientations so that a regular cross-hatched pattern of inversion twins is initiated.

Inversion twins on the Dauphiné law produced during polymorphic transformations in silica (Frondel, 1945) are geometrically equivalent to a 180° rotation about a twin axis parallel to the *C*-axis. The positions of the silica tetrahedra in the normal and the twin position are very similar and only a slight displacement results in twinning or untwinning. Dauphiné twins in quartz are displacive; they are produced by growth in natural quartz crystals and have been produced experimentally by both inversion and deformation (Klassen-Neklyudova, 1964).

Twins produced during a chemical transformation are basically due to growth but the process is similar in many respects to inversion twinning. Growth twins are formed in MgO produced by the experimental dehydration of brucite ($Mg(OH)_2$) (Garrido, 1951). The orientations of the MgO crystals are related to that of the original brucite because the MgO bonds have a similar orientation in the two structures.

$Mg(OH)_2$		MgO
{0001}	parallel	{111}
$\{10\bar{1}0\}$	,,	{011}
$\{1\bar{1}01\}$	,,	{001}
or $\{10\bar{1}1\}$		

The two orientations, i.e. $\{1\bar{1}01\}$ or $\{10\bar{1}1\}$ parallel to {001}, have a twin relationship to each other so that brucite is converted to a mixture of crystals with the two orientations in twin relation. This kind of twinning could be expected in many minerals formed as the result of topotactic changes (see p. 89).

(e) Deformation Twins

Most intersecting, wedge-shaped, and curved polysynthetic twins in metamorphic plagioclase, microcline, cordierite, calcite and dolomite (Plate XI) can be attributed to deformation. The deformation-twinning mechanism is similar in many ways to slip; the requirements for twinning and gliding are somewhat similar, and they may occur together on the same lattice plane, except that gliding can take place on a lattice plane which is parallel to a symmetry plane of the crystal but twinning cannot. Also the twin glide direction cannot be parallel to an axis of two-fold rotation.

Translation gliding involves displacement of one layer over another along the translation plane; the Burgers vector can be any multiple of the lattice vector, glide may take place in either sense of the Burgers vector, total deformation of the crystal may be large but no change of optical or lattice orientation is involved in the homogeneous simple shear of an *unconstrained* crystal. Twinning involves a limited slip along the twin plane and the Burgers vector is a fraction of the lattice vector, twinning can only take place in one sense along the Burgers vector, total deformation of the crystal is limited but there may be a marked change in optical or lattice orientation accompanying twinning. The displacement parallel to the slip direction is directly proportional to the distance from the composition plane thus twinning is equivalent to a homogeneous simple shear.

Deformation of a rhomb (Fig. 26) may produce a twin pattern if simple shear is in the *twinning direction BA* (not *AB*) of Fig. 23 and may form a simple twin (b), multiple or polysynthetic twins (c), or a completely twinned unit (d).

The restriction in the sense of twinning (i.e. the direction of shear) depends on achieving the shortest major atom movements, a minimum amount of shuffling, and the avoidance of bringing ions of opposite sign in close proximity. Deformation twinning in rather low-symmetry lattices (e.g. cordierite or plagioclase), however, may involve complex almost reconstruc-

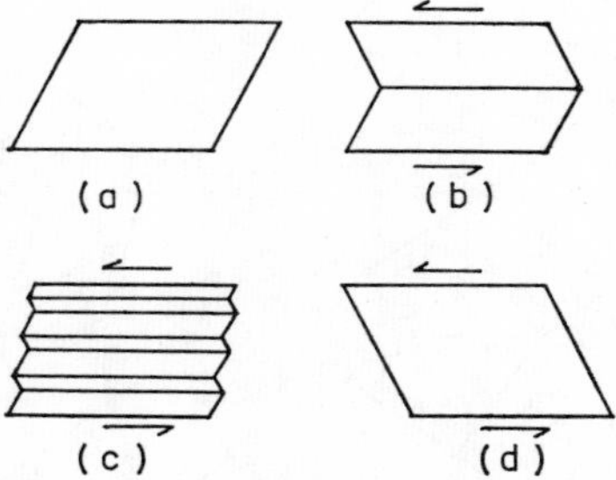

FIG. 26. Deformation twins. (a) Hypothetical rhombic crystal. (b) Deformation produces a simple twin only under exceptional conditions. (c) Polysynthetic twins are most commonly formed. (d) A completely twinned crystal represents the maximum strain

tive movements. The symmetry of calcite is sufficiently low that there is evidence (Handin and Higgs, 1959, p. 274) that even *slip* may be easier in one sense than the other.

The concept of mechanical twinning by simple shear displacements parallel to the slip plane is an over simplification because (1) the twin position can only be achieved geometrically in this way in body-centred cubic lattices; twinning in lattices of lower symmetry involves additional (but generally small) displacements ("shuffles", Fig. 27). (2) Small oscillatory movements

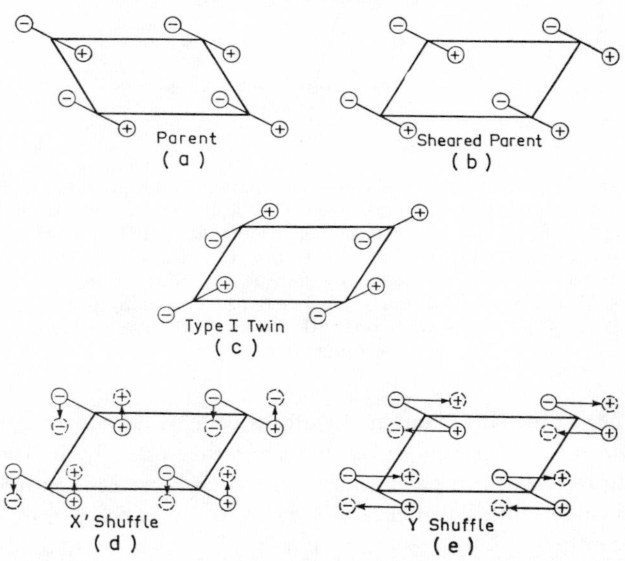

FIG. 27. "Shuffles" in deformation twinning. (a) Hypothetical rhombic lattice; atoms above and below the plane of the page shown as + and − respectively. (b) First stage of twinning shown as shear of the lattice without relative atom movements. (c) Final twin orientation after atom movements. (d) and (e) Alternative atom movements (shuffles) to obtain the twin orientation (c) from (b) (after Christian, 1965)

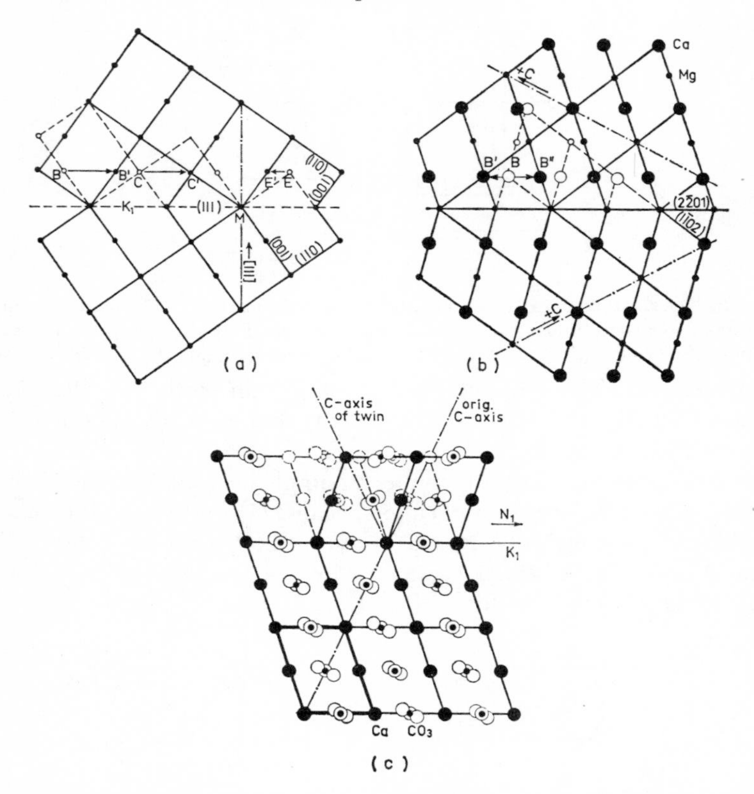

FIG. 28. Atomic rearrangements during deformation twinning (after Pabst, 1955; Handin and Higgs, 1959). (a) Face-centred cubic lattice, section parallel to $(\bar{1}10)$ with twin plane $K_1 = (111)$. The atomic movement associated with twinning is most likely E to E'; other possibilities B to B' or C to C' are less likely. (b) Twinning in dolomite on $K_1 = (2\bar{2}01)$. Cations only shown. Experiment supports atom movements B to B' (i.e. in a *negative sense*) not B to B''. (c) Twinning in calcite on $(01\bar{1}2)$. The C axis and CO_3 groups must rotate through 52° 30′. The twinning shear has a *positive* sense

occur normal to the slip plane as the atoms slide over each other. (3) Twinning may be aided by partial dislocations and take place by indirect paths.

Twinning on (0001) in calcite and related carbonates is simple in that it involves shear-like displacements of Ca atoms and CO_3 groups. However, the more common $(2\bar{2}01)$ twinning is more complex in that although the movements of Ca and C atoms are simple, the oxygen occurs as triangular, planar, O_3 groups normal to the C axis and twinning is accompanied by a rotation of O_3 groups (Fig. 28c) as this axis changes through 52° 30′ (Fig. 29).

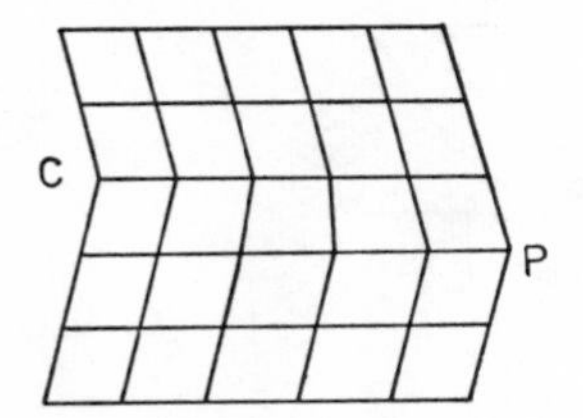

FIG. 29. A twin dislocation allows discordance between the composition plane (CP) and the appropriate lattice plane

A *twin dislocation* is formed where the composition plane (twin boundary) passes from one lattice plane to another (Friedel, 1964). This is a variety of partial dislocation which allows the formation of a non-rational composition plane. As twinning produces a new configuration of atoms, the Burgers vector cannot equal the lattice vector and therefore motion will be assisted by imperfect or partial dislocations (Hornstra, 1960; Azaroff, 1960, p. 443; Venables, 1964a, b). Deformation twinning in alpha-iron (Hornbogen, 1961) is preceded by the movement of partial dislocations that change the stacking along the twin plane. Some twinning in metals is a diffusionless, essentially instantaneous transformation and is not aided by dislocations, but twinning in metamorphic minerals involves minor to major adjustments and is semi-reconstructive or reconstructive (e.g. Raleigh and Talbot, 1967). It occurs by the formation of very small twinned regions which behave as nucleii and extend or grow. A twinned region within a crystal has a *coherent* (rational or coincidence) boundary where the composition plane is a lattice plane but meets the host in *non-coherent* (irrational) regions of misfit at the ends of the twins (if they do not extend to the crystal boundary) or where the twin changes thickness.

Deformation twinning has been produced experimentally in calcite (Turner, Griggs and Heard, 1954; Griggs, Turner and Heard, 1960) dolomite (Griggs, Turner and Heard, 1960), corundum (Kronberg and Wilson, 1962) spinels (Hornstra, 1960), magnetite, plagioclase (Mugge and Heide, 1931), diopside (Raleigh, 1965b) and galena (Lyall, 1966; Lyall and Paterson, 1966).

The size, shape and number of twins depends on the relative ease of *nucleation* and *propagation* and these are affected by the stacking-fault energy of the material, the temperature, resolved shear stress, presence of appropriate dislocations, and the strain rate. Following nucleation, the twin tends to grow *along* the twin plane to form a thin, continuous twinned layer. There are then two alternatives: other twins may nucleate to give a series of thin polysynthetic twins, or the twin may thicken by advancing the twin plane laterally (Fig. 30).

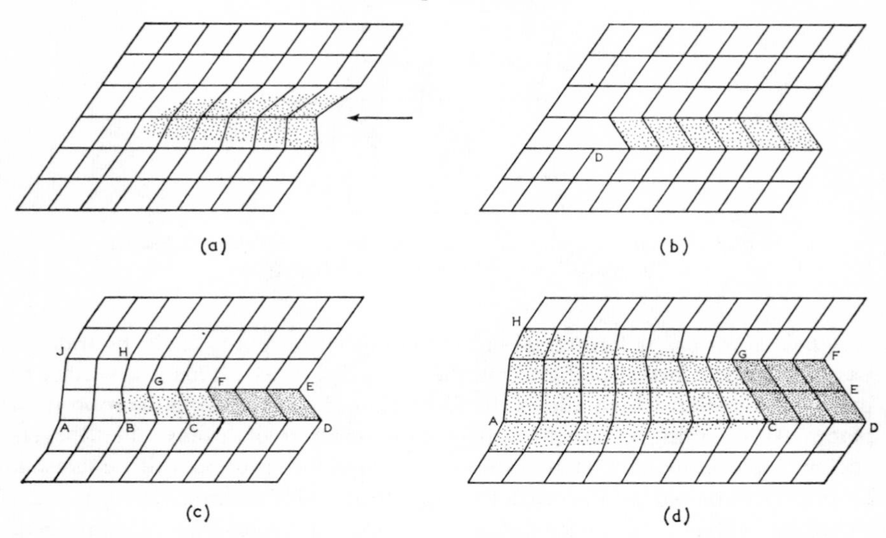

FIG. 30. (a) Deformation of a crystal with a potential twin plane causes elastic bending and strain over the stippled area. (b) Nucleation of a deformation twin takes place at the position of greatest lattice strain and a partial twin extends into the crystal. It may terminate at a partial dislocation shown diagrammatically at *D*. A twin normally ***propagates*** or extends along the twin plane *EF*. Propagation to the left is aided by movement of the dislocation. (c) The twin may terminate in a strained region without a dislocation, e.g. the crystal is twinned in the region *DEF*, is imperfectly twinned between *F* and *C* and is strained between *B*, *C* and *G*. Propagation of the twin from *E* towards *A* is preceded by and aided by, the twin dislocation region *ABHJ*. (d) *Thickening* of the twin takes place away from the twin plane, i.e. from the lattice plane *E* to plane *F*. It is easier to ***propagate*** the twin, as in (c), than *widen* it, as in (d), in most crystals. A broad region of strain *ACGH* extends ahead of the twin

It is possible on geometrical grounds that adjacent twins might grow laterally and coalesce to give larger twins and the twinning process continue to completion to give a *completely twinned crystal*—i.e. one which contains no twins but whose lattice orientation has been changed to the twinned position—but lateral growth (thickening) is difficult. There is a nucleation difficulty at each successive lattice layer in thickening a twin. The ratio of twin thickness to twin length provides a qualitative guide to the magnitude of the twinning shear, being very small when the shear is large.

Gliding and twinning may take place on the same or on different planes in a given crystal (Bell and Cahn, 1957). Deformation of a polycrystalline aggregate may cause gliding in one crystal and twinning in another depending on the magnitude and sense of the resolved shear stress. Where both twinning and gliding can take place on one plane, twinning may be precluded because it requires movement in a sense not favoured by the sense of the shear stress. The resolved shear stress to initiate twins on a given plane in metals seems to be greater than to initiate slip (100 × for basal slip in zinc) and twinning in metals is always preceded by slip and slip is enhanced by twinning. However it appears to be easier to deform calcite by twinning on $\{0\bar{1}12\}$ than slip.

There is considerable difference of opinion over whether there is a critical or minimum stress necessary to produce twinning (Fourie, Weinberg and Boswell, 1960; Bolling and Richman, 1965). It is thought that a local shear stress in the order of γ/b must be exerted to cause twinning (Bilby, Cottrell and Swindon, 1963) where γ is the stacking fault energy and b is the Burgers vector for twinning.

Nucleation takes place at positions of strain such as dislocation concentrations, grain boundaries, etc., and may be initiated at a suddenly formed fracture. The stress required to propagate twins in b.c.c. metals is less than that to nucleate them; however, other studies have shown that it is easier to produce a large number of small twins than a few large ones and that some small twins will only propagate to a large size if nucleated at high stress concentrations. In general the lower the stacking-fault energy of a crystal the lower the stress to nucleate a twin (Venables, 1964a, b) but the higher the stress to propagate it.

Twinning can be produced by very rapid deformation (shock) and may proceed in "bursts". Twins may be propagated very rapidly through undistorted crystals and on meeting an obstacle may send out shock waves which initiate other twins at a distance (Hornbogen, 1961). The rate of propagation of deformation twins in calcite is about 2 metres per second (Bowden and Cooper, 1962) and is similar to the expected rate of dislocation movement.

An increase in temperature generally requires an increase in stress to produce twinning (e.g. in metals and calcite) but dolomite is an exception in that elevated temperatures favour deformation by twinning rather than slip (Higgs and Handin, 1959, p. 276). Higher temperatures appear to result in wider twins (Bolling and Richman, 1965, p. 737). An increase of strain rate allows twinning to take place at a lower stress threshold, and high strain rates produce a large number of small twins.

Twins may form on one plane or simultaneously on conjugate planes if the resolved shear stress is sufficiently high. Twinning on the first set may be allowed by dislocations and that on the conjugate plane by jogs on these dislocations.

Priestner and Leslie (1965) showed the importance of dislocations in twinning by demonstrating that the stress required to produce twinning in undeformed (dislocation free) zinc is 25 to 50 times that for the deformed (dislocation bearing) metal. They considered that twins were nucleated in regions of stress concentration such as the ends of kink or slip bands or at the intersections of slip planes.

(f) Interference Twins

It has been suggested that deformation twins could be formed during crystallization processes by intergranular stresses caused by mutual interference (differences in the coefficient of thermal expansion of individual grains, anisotropy of thermal expansion within individual grains, volume changes during inversion, etc.). Emmons and Gates (1943) considered that all multiple twinning in plagioclase (including complex combinations such as Carlsbad-Albite, Manebach Ala-A and Manebach Acline) were due to such strains but this is not generally accepted. Cahn (1953) was able to distinguish between normal deformation twins and those due to thermal stresses in alpha uranium.

Edwards (1960, p. 36) stated that the common {111} twinning in the sphalerite of lead–zinc ores is due to stresses caused by mutual interference, and that primary or growth twins are rare in ore minerals because they rarely form as "free" crystals in the sense of intratelluric crystals growing in magma. Neither of these opinions is valid. Twins in ore minerals may be formed by growth analogous to annealing twins, and as Stanton (1960, p. 8) pointed out, twins may well be due to deformation of external origin. There seems to be very little evidence for the existence of interference twins in metamorphic rocks.

(g) Exsolution Twins

An exsolution twin is a mechanical twin produced by stresses due to volume changes during exsolution of a solid solution. The concept was advanced by Buerger and Buerger (1934) to account for twinning in chalcopyrite which contained exsolved cubanite, the twins having formed later than the exsolved bodies. This twin mechanism, if valid, is similar in many respects to the inversion and interference twin mechanisms.

There is little evidence to support the existence of the process but it is conceivable that some twins in perthitic intergrowths might be due to stresses caused by exsolution and it might be possible to find some relationship between the degree of twinning in either host or inclusion and the degree of exsolution; however, twinning and exsolution may both result from stress.

DIFFERENCES BETWEEN PRIMARY AND SECONDARY TWINS

It is important (Vance, 1961, p. 1103; Siefert, 1964; Vernon, 1965; Richards, 1966; Cannon, 1966) to attempt to decide whether the twinning in a crystal in a metamorphic rock is:

(a) Pre-metamorphic in origin.
(b) Primary and due to growth or recrystallization.
(c) Secondary and due to deformation after crystallization.

Deformation twinning of low-symmetry silicates does not involve simple displacements of ions or ion groups and the transformations necessary for some kinds of twinning are so complex that it is doubtful whether these twins could be produced other than by growth (Pabst, 1955). No detailed structural examination has been made of twinned feldspars to evaluate this factor but it has been postulated that Carlsbad, Acline, Baveno and Manebach twins can form only by growth, that Pericline twins can form only by deformation, and that Albite twins can form in either way. Growth twins appear to be more common in igneous than metamorphic rocks.

Phillips (1930), Turner (1951), Gorai (1951), Rao & Rao (1953) and Tobi (1961) have shown that there are differences in the distribution of twin laws in the plagioclase of igneous and of metamorphic rocks. This should not be taken to indicate necessarily a difference in twin laws of growth and deformation twins, because the significance of the results is not clear (Vance, 1961; Smith, 1962). It is difficult to distinguish between growth and deformation twins and thus between twins of different ages or origin and it is necessary to recognize the following genetic categories:

(1) Most twins in metamorphic rocks which have not been strained after crystallization (e.g. some hornfelses) are primary, i.e. due to growth, inversion or annealing, in the solid state.

(2) Most twins in igneous rocks which have not undergone post-consolidation strain are due to growth, combination, or inversion of crystals in a dominantly liquid environment.

(3) Twins in deformed rocks of igneous, sedimentary or metamorphic origin (schists, gneisses, granulites) may be due to growth, inversion or deformation.

(4) Twins in dynamically metamorphosed rocks such as mylonites are primarily due to deformation but they may also be relict, pre-metamorphic, growth twins.

Untwinned feldspars are common in metamorphic rocks and where twinning is present the number of individuals in a twin is small and there is generally only one twin law. Simple twins on the Albite Law in albite are common but have been frequently mistaken for Carlsbad twins (Tobi, 1959, p. 203). Growth twins in igneous crystals are commonly complex; abundant individuals and a variety of twin laws may be present in one crystal. Deformed calcite crystals may contain three sets of multiple twins but these are generally conjugate and only on one twin law. Conjugate deformation twins form in diopside on two laws (Raleigh and Talbot, 1966).

Voll (1960, p. 533) found that albite in contact with calcite in limestone was commonly twinned (simple, broad twins on the Albite Law) but albite in contact with quartz in adjacent schist was untwinned. He attributed the twinning to growth and considered that the contrast in twin frequency was due to a greater control by orientation on interfacial tension at albite–calcite than at albite–quartz interfaces.

Multiple twins in calcite and plagioclase in deformed rocks are commonly curved and twisted but it may not be clear whether the twins themselves were *due* to the deformation or existed before it and were twisted by the deformation.

Many primary and secondary twins have simple, straight, rational composition planes but variations in the nature of the composition plane are very useful in distinguishing between twins of different kinds. The complex and irregular composition plane of an interpenetrant twin and the radiating composition planes of a sector twin are probably only formed by simple growth or inversion and not by deformation.

It has been suggested (Cahn, 1954, p. 408) that the composition plane of a growth twin *must* be parallel to a lattice plane; if this is true, then pericline twins which are parallel to the irrational rhombic section could only be formed by deformation. Twin dislocations allow the composition plane of a deformation twin to be irrational although this is a high energy condition. Cahn's argument is probably not valid but is supported by Vernon (1965, p. 489) who found that all pericline twins in the plagioclase of mafic gneisses from Broken Hill had the shapes of deformation twins whereas all the Carlsbad were growth twins and the Albite law was common for both deformation and growth twins.

It is seen empirically that most *growth* twins are parallel to the more slowly growing faces of a crystal but this simply implies that the twin plane is the most closely packed lattice layer. Twinning and slip take place along similar close-packed layers and growth twins and deformation twins in some crystals obey the same twin law.

The composition planes of growth twins are straight and the twins commonly change width or terminate within the crystal abruptly; lamellae thicken and thin independently of each other, the changes being step-like and unrelated to any later bends; lamellae terminate independently and without relation to later bends or fractures. Square ends to internal or enclosed twins imply growth, and tapered ends suggest deformation although deformation twins commonly have square ends or steps where they meet cracks, oblique slip planes or other twins. Lattice distortion at the termination of a deformation twin is reduced by a gradual reduction in width and can be supported by twin dislocations, but a deformation twin is unlikely to grow with a blunt end as this would enlarge the terminal distortion. A lack of accompanying deformational features is characteristic of growth twins but the twins in deformed rocks may be either growth or deformational types. Vance (1961) and Raleigh and Talbot (1967) have suggested that primary twins in plagioclase and diopside respectively tend to be broader with fewer lamellae per grain than deformation twins, and metal studies support this; simple twins in feldspar are probably due to growth.

Mugge (1883, 1885) suggested that the two parts of a mechanical twin will generally be symmetrical about the twin plane but those of growth twins will not. To produce polysynthetic twins of constant width by growth, e.g. from a melt, implies some kind of rhythmic growth control such that nucleation of successive twin orientations take place on one side or both sides of a crystal and this has been regarded as improbable (Siefert, 1964). However, remarkable rhythmic processes in magmas produce zoning in plagioclase, compositional differences in alternate twin lamellae, layering in gabbros, etc., and it should be noted that many primary annealing twins in metals are very regular in width. Adjacent polysynthetic twins in plagioclase of granitic rocks differ in composition (Emmons and Gates, 1953; Vogel, 1964, p. 629) and this may be explained by some rhythmic growth process related to that producing oscillation zoning. To produce such twins from homogeneous crystals by deformation would involve a very difficult reconstructive transformation with considerable diffusion of various elements. De Vore (1955b) has suggested that a variation in optical properties between some adjacent twins may be due simply to a difference in the degree of ordering and not to chemical differences.

The "optical crystallographic scatter" method of Vogel (1964) may be a fruitful way of distinguishing between growth and deformation twins. The "scatter" refers to the scatter of poles of composition planes of adjacent twin lamellae relative to X, Y and Z when plotted on a stereographic projection. The pattern of scatter between twins in separate grains ("external scatter") is different for growth and for deformation twins. Laves (1965, p. 512) however has criticized some of Vogel's arguments.

The abundance of untwinned albite in metamorphic rocks and the failure by Mugge and Heide (1931) to produce deformation twins experimentally in albite (although they achieved it in plagioclase) has suggested that twinning in albite is difficult to produce mechanically. The ability to twin may depend on order–disorder relations (Laves, 1952b, c, 1965) and hence mechanical twinning might not be possible in completely ordered albite (Vogel and Siefert, 1965). This implies that a crystal of pure albite containing deformation twins must have inverted from a high-temperature disordered type. Tobi (1961) and Jones (1961) found only Albite or Carlsbad-Albite twins in low-grade rocks and Brown (1962, 1965) suggested that albite crystals in Greenschist Facies rocks show only growth and not mechanical twins, and that Carlsbad twins can only be produced by growth. Disorder may be sufficiently high in the Amphibolite Facies to allow mechanical Albite and Pericline twins. Complex twins such as Albite-Ala A in charnockite (Subramaniam, 1959) are almost certainly growth-twins.

Twinning, like other transformations, varies in structural complexity and might be regarded as either *displacive* or *reconstructive* depending on whether the transformation is simple and rapid, or difficult and sluggish, although all twinning in silicates is more complex than a true displacive process. "Displacive" twinning is achieved if the process involves only slight structural change, e.g. twinning of calcite or albite-law twinning of plagioclase. Reconstructive twinning involves the breaking down and rebuilding of the structure and cannot be effected by simple displacement. Reconstructive twins (Carlsbad and various complex laws) would be expected to be restricted to growth twins but displacive twins to be either primary or secondary. This may however be an oversimplification and experimental investigation is necessary.

Deformation twinning may obliterate original features such as the compositional zoning of plagioclase (Emmons and Mann, 1953; Cannon, 1966) but inversion from a high- to a low-temperature state apparently need not destroy pre-existing twinning (Cannon, 1966, p. 537).

Chess-board or *chequer-board* twinning in albite (Becke, 1913a) consists of short alternating Albite-Law twins which do not cross the whole crystal (Starkey, 1959, p. 142). In sections normal to (010) a chess-board pattern results when one set of twins is extinguished and the albite has a strained patchy appearance in sections parallel to (010). Starkey concluded that it results from two processes: first, conversion of potash-soda feldspar into albite by metasomatism with some residual irregular microcline-twinning persisting in the later feldspar, and second, deformation to produce the chess-board pattern. Chess-board albite has been regarded as indicative of replacement and albitization (Anderson, 1928; Niggli, 1954) although it may be primary (Battey, 1955).

Cordierite forms both deformation and growth twins (Venkatesh, 1954; Singh, 1966). Cordierite is orthorhombic with interfacial angles (110) $\wedge$ $(1\bar{1}0)$ of 60° 50′ and (130) $\wedge$ $(\bar{1}30)$ of 59° 10′. The twin composition planes are (110) and rarely (130) and twinning may be simple, polysynthetic, parallel, interpenetrating, or cyclic (sector) with radial, stellate or concentric twins forming 3, 6 or 12 segmented groups or stars. The pseudo-hexagonal symmetry of the sector twins is due to the prism interfacial angles being very close to 60°. All the twin planes of similar identity in a twin are either parallel or inclined to each other at 60° and 120°, whereas dissimilar twin planes make angles of 30° or 90°.

Cordierite resembles plagioclase in some respects; twinned crystals are very common in a magmatic or thermal metamorphic environment and untwinned crystals are abundant in regional metamorphic rocks. The symmetry of radial and sector twins implies that they are formed by growth and not deformation and this is supported by their abundance in hornfelses, buchites, synthetic glasses and contaminated igneous rocks and their comparative rarity in regional metamorphics.

Simple twins appear to be growth forms. Measurements by Singh (1966) indicated that 4% of the cordierite in acid granulites from British Guiana contained this type. The figure reached 15% in hornfelses, 22% in gneisses, 25% in xenoliths and 61% in contaminated granite.

Polysynthetic twins appear to be formed by either growth or deformation. Regular polysynthetic twins which are parallel, wedge-shaped, tapering or curved would be expected to be due to deformation even though they have been found in hornfelses. Interpenetrant polysynthetic twins are very abundant in hornfelses where they are presumably due to growth although there is no apparent reason why twins of similar appearance in other rocks could not be conjugate deformation twins.

The type of twinning in the cordierite of buchites can be related to the rate of growth (Venkatesh, 1952, p. 487). Sector and stellate twins are found in perfect, slowly formed crystals, and irregular, concentric twins (parallel to the "hexagonal" faces) are found in skeletal, rapidly formed crystals.

Calcite commonly contains polysynthetic twins on *e* $\{01\bar{1}2\}$ and one, two or three sets of conjugate twins may be visible in a single grain. There is no doubt that many such twins are secondary. The atomic rearrangements involved are simple and can easily be achieved at room temperature by pressing a crystal in the appropriate direction with a knife. Calcite single crystals and aggregates have been twinned experimentally by deformation (Turner *et al.*, 1954; Griggs *et al.*, 1960). Twinned calcite however occurs in thermally metamorphosed marbles and apparently undeformed limestones where twinning is presumably due to growth. It is not possible to decide whether most twins are primary or secondary because they are polysynthetic, continuous and non-tapering, thus lacking diagnostic characteristics.

MARTENSITIC TRANSFORMATIONS

Martensite transformations occur in a wide variety of metallic and non-metallic materials where stress or temperature changes cause rearrangement of atoms to a new structure by well-defined and correlated atomic movements. Thus, like twinning, these transformations may be either stress or thermally activated. As in twinning, the movement of each ion is somewhat less than the interatomic distance (Bilby and Christian, 1961; Wayman, 1964).

The transformation is diffusionless but the change in structure or orientation of the crystal lattice lacks the rigid symmetry requirements of twinning. The transformations are named after the iron–carbon alloy *martensite* (Troiano and Greninger, 1946; Cohen, 1951). Austenite (gamma-iron) is stable at high temperatures and changes to ferrite (alpha-iron) at low temperatures, but if austenite is quickly cooled below the alpha–gamma transition, a new phase (martensite) is formed as plates or rods. The transformation is from the face-centred cubic lattice of alpha iron to the body-centred tetragonal lattice of martensite such that {110} of austenite is parallel to {112} of martensite. The transformation can be regarded crystallographically as the superposition of a shear and an extension, e.g. in martensite there is 1/3 shear along [110] plus 6% extension parallel to [112] as though the f.c.c. alpha-iron lattice had been compressed along ⟨110⟩.

Barrett (1952, p. 573) and others use the term to cover any thermally activated diffusionless process which causes regular atom movements and gives rise to a change in structure, e.g. In-Th alloy from f.c.c. to f.c. tetragonal; In-Cd, Cu-Mn, Cr-Mn alloys from cubic to tetragonal, Zr, Li, L-Mg from b.c.c. to h.c.p.

The process referred to as *synchro-shear* by Amelinckx (1964, p. 272) in which graphite is changed from hexagonal to rhombohedral by shearing is a martensite-type transformation. The change in lattice symmetry is produced by restacking the carbon atoms, distorting the bonds and inserting stacking faults and other defects.

Inversion from one polymorph to another is martensitic where there the change only involves slight structural change, e.g. the inversion of enstatite (orthorhombic) to clinoenstatite (monoclinic) in kink bands within experimentally deformed enstatite (Turner, Heard and Griggs, 1960, p. 61). Stress may be essential for the enstatite-clinoenstatite transformation and the latter could be regarded as a "stress" mineral (ibid., p. 62).

THERMALLY ACTIVATED TRANSFORMATIONS

Most metamorphic changes are thermally activated and take place as a result of a change in temperature. This class of transformation includes a wide range of processes which may be subdivided on their degree of complexity.

I. Insignificant degree of transport
   Inversion
   (a) Displacive
      (1) change in bond angle and structure
         α–β quartz

(b) Reconstructive
  (1) change in bond angle
    quartz–tridymite–cristobalite
  (2) change in structure (symmetry or packing)
    quartz–coesite–stishovite
    orthoclase–microcline
    indialite–cordierite
    calcite–aragonite
    andalusite–sillimanite–kyanite
  (3) change in bond-type
    diamond–graphite
  (4) twinning

II. Short-range transport
  (a) Order–disorder in solid solution
  (b) Exsolution from solid solution

III. Short- to long-range transport
  (a) Crystallization and recrystallization

Inversion

A compound exhibits polymorphism when it forms crystals of the same composition but of different structure. The structural differences may be small (low and high quartz) or large (andalusite–kyanite). The transformation from one polymorph to another is called *inversion* and it takes place at an inversion temperature at a certain inversion pressure.

Some inversions are markedly dependent on temperature and others on pressure; the slope of the inversion curve is given by the Clausius–Clapeyron equation $\frac{dP}{dT} = \frac{\Delta S}{\Delta V}$ where ΔS is the entropy change and ΔV the volume change. Some transformations involve only slight volume changes, thus the inversion curve is steep; they are driven by changes in temperature and are rather insensitive to pressure, e.g. quartz–tridymite–cristobalite; orthoclase–microcline; cordierite–indialite. Recent studies on the possible mineralogical composition of the mantle and the formation of eclogites, granulites and glaucophane schists have drawn attention to a series of transformations whose inversion curves have low slopes and which involve considerable volume and density changes. The processes are thermally activated but are pressure-sensitive and consist of structural repacking to give more dense structures at higher pressures. Polymorphic changes include aragonite–calcite, quartz–coesite–stishovite and diamond–graphite. Similar transformations include the entrance of Mg into garnet, of Na into pyroxene or amphibole (jadeite, omphacite or glaucophane) and the change from an enstatite structure to a spinel or olivine plus stishovite (Ringwood and Major, 1966).

The high–low change in quartz consists of a slight rotation of the silica tetrahedra to change the second coordination slightly but retain initial Si-O first coordination. The transformation at 573°C is displacive and instantaneous so that the high temperature form is not preserved in metamorphic rocks. The change involves an increase in volume so that the low temperature form may retain the typical hexagonal bipyramids of the high temperature form as a pseudomorph but with some cracking.

The inversion from monoclinic (high temperature) orthoclase to triclinic (lower temperature) microcline may take place in response to falling temperature but may be accelerated or caused by strain at low temperatures. The inversion is generally accompanied by typical cross-hatched twinning. The two polymorphs may occur together in one crystal. It is generally accepted (Barth, 1934; Laves, 1950, 1952b) that the transformation is due to an increase in order in the lower temperature microcline although Mackenzie (1954) has suggested that the process might depend on the soda content, i.e. soda-bearing orthoclase can only invert to microcline if albite is exsolved. The indialite–cordierite transformation is similar to that of orthoclase to microcline (Eitel, 1958, p. 139).

The calcite–aragonite inversion is pressure-sensitive and throws light on the pressure conditions of the glaucophane schists (Coleman and Lee, 1962; Brown, Fyfe and Turner, 1962). The transformation is topotactic and reconstructive; under experimental conditions (Dasgupta, 1964) it takes place by reorientation of CO_3 groups along the [001] axis of aragonite together with migration of Ca within (001). Natural replacements of aragonite by calcite (Brown *et al.*, 1962) suggest an equal volume transformation (with removal of both Ca and CO_2 because of the density decrease) with a number of alternative structural relations, all involving parallelism of an *a* axis and [*f*:*f*] of calcite with two of *a*, *b* or *c* of aragonite.

The three aluminium silicates consist of chains of silica-tetrahedra and alumina-octahedra and the change from one to the other involves mainly rotation of octahedra with some additional displacements. Andalusite is orthorhombic with density 3·13 to 3·16 and unit-cell dimensions $a = 7{\cdot}78$, $b = 7{\cdot}92$, $c = 5{\cdot}57$ Å, whereas kyanite is compact with the oxygen ions in a slightly distorted cubic close-packed structure and the density is 3·53–3·65 with unit cell dimensions $a = 7{\cdot}10$, $b = 7{\cdot}74$, $c = 5{\cdot}57$ Å. Transformations from one aluminium silicate to another are reconstructive. Two or even three (Hietanen, 1956) of the group may occur together in one rock but there seems little doubt that most such rocks are polymetamorphic and that the polymorphs are not in equilibrium (Rutland, 1965; Pitcher, 1965).

Small free-energy differences and difficulty in nucleation make these transformations difficult and sluggish (Waldbaum, 1965). Kyanite replaces andalusite (Workman and Cowperthwaite, 1963) directly although the pseudomorphs are the multi-crystal type because the transformation is reconstructive. Sillimanite does not commonly replace kyanite directly (but see Francis, 1956, p. 357); early kyanite in a rock probably supplies alumina for the growth of sillimanite but this is nucleated elsewhere (generally in biotite) and does not form pseudomorphs. Replacement of the higher pressure varieties kyanite and sillimanite by the lower pressure andalusite apparently is very rare (MacKenzie, 1949) and indeed such retrograde transformations are kinetically difficult.

Transformations with Short-range Transport

Crystallization of new mineral phases (as opposed to the recrystallization of existing minerals) is an important part of metamorphism and is due to a change in composition of a single phase (e.g. dehydration) or a reaction between two or more mineral phases.

Such mineralogical transformations differ in their complexity and range of transport of constituents, hence in their rate and resultant textures although all are reconstructive in that they require nucleation of the new phase and its subsequent growth as discussed in detail later. Two special processes of textural importance are discussed here:

(a) Topotaxy (control by original structure),
(b) Order–disorder, exsolution.

TOPOTAXY

The simplest transformation retains most of the original lattice; it simply gains or loses ions with a minimum of rearrangement and is said to be *topotactic* (Grim, 1953; Johns, 1965; Wahl, 1965). Topotaxy differs from epitaxy (p. 164) in that it involves a structural relation between a primary and a secondary mineral during *replacement* whereas epitaxy refers to the structural control of an *overgrowth* by the substrate. Some mineralogical transformations are completely reconstructive in that original lattices are entirely destroyed and new ones built but this requires a considerable amount of energy and time. Other mineralogical changes take place by the shortest possible displacements of ions and the least breaking of bonds and this means that new minerals will use old lattice fragments which are as large as possible. Silica tetrahedra will almost certainly retain their identity but the "fragments" may be much larger in some cases.

The old texture may be completely destroyed if large-scale transport accompanies nucleation and growth of new phases. However, activation energies for nucleation and transport are lower in the crystallization of new phases which use existing lattice fragments. In some circumstances it may not be too gross an oversimplification to regard a rock as an essentially rigid, close-packed aggregate of oxygen atoms or silica tetrahedra with interspersed cations which are mobile during reaction. Grain boundaries can be regarded as defects.

Reaction fabrics (Lauder, 1961) result from the control of orientation of one mineral by another and examples include topotactic replacement textures, oriented intergrowths due to exsolution, twinned or layered crystals, eutectoids due to simultaneous crystallization, and parallel growths (epitaxy). The term *replacement* is used petrographically to signify that a given mineral has been changed chemically and/or structurally to a new mineral; the chemical change may be profound but at the same time the process may be a gradual dissolution such that the identity of the replaced mineral as a unit is retained.

The substance which is replaced is the *host* or *primary mineral* and the replacing mineral the *guest*, *metasome* or *secondary mineral*. The alteration process may be the result of *metasomatism* (change in bulk chemical composition accompanying metamorphism) or due to some kind of chemical rearrangement within a closed chemical system. Replacement may take place either through the agent of some solvent as an intergranular fluid or by solid diffusion processes.

PSEUDOMORPHS

A pseudomorph is a textural unit (e.g. a crystal or fossil) which has been altered or replaced so as to retain the original shape of the unit but to change the internal structure or composition.

Pseudomorphs have been classified genetically (Niggli, 1920; Glover and Sippel, 1962) as follows:

(1) *Physical pseudomorph:* where atoms are slightly displaced giving a new crystal structure.
(2) *Exchange pseudomorph:* where an ion or atom in a crystal is exchanged for one from the surrounding environment.
(3) *Pure displacement pseudomorph:* where there is complete materia exchange while maintaining the external form

The classification followed here is a simple descriptive one:

(1) *Single-crystal pseudomorph:* a single original crystal (or other textural unit, e.g. a fossil) replaced by a single crystal of another species.
(2) *Multi-crystal pseudomorph:* a single crystal (or other textural unit) replaced by a number of differently oriented crystals of a second mineral.
(3) *Aggregate-pseudomorph:* an aggregate of crystals of one mineral replaced by a single crystal of a second mineral (plate XIIb).
(4) *Multi-phase, multi-crystal pseudomorph:* a single crystal (or other textural unit) replaced by an aggregate of crystals of a variety of minerals.

Single-crystal pseudomorphs are not common and examples include alteration of plagioclase to scapolite, biotite to chlorite and garnet to chlorite. This kind of replacement involves a considerable degree of structural similarity between the primary and secondary mineral and probably also a considerable retention of original structural units (topotaxy). Changes such as pyroxene to bastite and olivine to iddingsite result in what is fundamentally a single-crystal pseudomorph (even though two secondary mineral species are present) where the major secondary mineral forms a single continuous crystal with the minor secondary constituent dispersed through it.

Such transformations follow the Concept of Structural Continuity (Bernal, Dasgupta and Mackay, 1957, and Brindley, 1959) which may be stated as follows: whenever steric conditions permit, solid state transformations (even where material is gained or lost to the system) proceed by a minimum rearrangement of the structures of single crystals. Excellent reviews are given by Dent Glasser, Glasser and Taylor (1962) and Brindley (1963).

A simple example is the experimental transformation from alpha to gamma polymorphs of simple oxides where there is a restacking of existing units from the hexagonal close-packing of the alpha form to the cubic (spinel) structure in the gamma form. The transformation of hematite (alpha) to magnetite (gamma) is an oxidation–reduction reaction which may be regarded as a transformation from a hexagonal close-packed array of oxygen atoms with interstitial Fe^{3+} ions to a spinel-type packing with interstitial Fe^{3+} and Fe^{2+} ions (Bernal *et al.*, 1957). Hematite and magnetite pseudomorph each other so that (0001) of hematite becomes (111) of magnetite. Similar changes include ilmenite to hematite plus rutile; groutite, MnOOH, to pyrolusite, MnO_2 (Lime-de-Faria and Lopes-Vierira, 1964); goethite, alpha-FeOOH, to hematite; diaspore, alpha-AlOOH, to gamma-alumina.

Experimental decarbonation of a carbonate such as calcite, dolomite, magnesite, siderite, ankerite, etc., to its oxide takes place by a homogeneous, topotactic, semi-reconstructive process to give a pseudomorph (Bernal *et al.*, 1957; Dasgupta, 1964, 1965). The carbonate structure consists of alternate layers of CO_3 and cations normal to [111]; loss of CO_2 is accompanied by simply closing up the cation layers. A single crystal of dolomite can be converted to a single-crystal pseudomorph of calcite containing reoriented MgO. Ankerite, $Ca(Mg,Fe)(CO_3)_2$, can be transformed into calcite with the same orientation plus MgO and vaterite. Every (Mg,Fe) layer plus an adjoining CO_3 layer is destroyed and the remaining Ca and CO_3 layers close up to give calcite plus MgO, FeO and CO_2.

A considerable degree of structural retention may be seen in the dehydration processes which are common in metamorphism. The basic sheet structure is retained in the change kaolinite–illite–sericite–muscovite, and the clay particles act as nucleii for metamorphic micas which take on a mimetic fabric. The structural controls which are exemplified in single-crystal pseudomorphs are displayed to a lesser degree in various transformations showing structural retention leading to oriented intergrowths and complex multi-crystal pseudomorphs. The conversion of muscovite to mullite (Sundius and Bystrom, 1953) can lead to a 120° triangular network of mullite as a result of oriented nucleation of the mullite in the mica structure. This control has also been found in natural specimens of sillimanite in biotite (Chinner, 1961; Binns, 1964). The general process of mullite formation in layer silicates may be deduced from the work of Brindley and Nakahira (1959) on the kaolinite–mullite transformation.

Serpentine and chlorite dehydrate topotactically to forsterite or enstatite (Brindley and Zussman, 1957; Brindley and Hiyami, 1965) under experimental conditions and the reverse process appears to be common in nature. This process may be regarded either as an example of *homogeneous* transformation (leading to single-crystal pseudomorphs) or *heterogeneous* transformation (leading to multi-crystal pseudomorphs).

(a) **Homogeneous Transformation** (Brindley and Zussman, 1957)

Four unit-cells of serpentine become three unit cells of forsterite because of the following similarities in unit-cell dimensions:

$$2a \text{ (serp)} \frown b \text{ (forst)} \frown 2b \text{ (chlor)}$$
$$2b \text{ (serp)} \frown 3c \text{ (forst)} \frown 2b \text{ (serp)}$$

and the following atomic structural similarities:

$$\begin{aligned} a \text{ (serp)} &= b \text{ (forst) and [013] forst} = a \text{ (chlor)} \\ b \text{ (serp)} &= c \text{ (forst) and [011] forst} = b \text{ (chlor)} \\ c \text{ (serp)} &= a \text{ (forst)} \qquad\qquad\qquad\quad\; = c \text{ (chlor).} \end{aligned}$$

Serpentine or chlorite could thus be changed to forsterite with a minor amount of reorganization of silica tetrahedra and breaking of Si–O bonds with the rejection of H_2O and SiO_2. The orientation would change so that *a*, *b* and *c* of serpentine or chlorite become *b*, *c* and *a* respectively of forsterite. The orientation relationships between olivine, hematite and chlorite produced by the conversion of olivine to "iddingsite" have been given by Smith (1959, 1961) and Brown and Stephen (1959).

(b) Heterogeneous Transformation (Brindley, 1959, 1963; Brindley and Hiyami, 1965)

It is suggested that the reaction is initiated at certain favourable sites within a crystal (dislocations, external surfaces, twins, etc.) and is maintained by diffusion of appropriate cations to or from the reaction front. The serpentine ($Mg_3Si_2O_5(OH)_4$) behaves in two ways; one part becomes a "donor region" and the other an "acceptor region". The donor region accepts H^+ which it liberates as water and donates Mg^{2+} and Si^{4+} to the acceptor region which is changed into structurally disordered forsterite (Mg_2SiO_4) thence to organized forsterite, or with extra Mg^{2+} and Si^{4+} from the donor region, into enstatite (according to Brindley and Hiyami). It had been thought earlier that the acceptor region became an amorphous serpentine anhydride with breakdown into a magnesia-rich region which ultimately became forsterite and a silica-rich region which ultimately became enstatite.

Other examples of appreciable retention of structure include the alteration of talc to anthophyllite (Greenwood, 1963), the dehydration of apophyllite (Lacey, 1965a, b) and the reversible reaction, pyroxene to amphibole. An amphibole could be expected to convert to a pyroxene by retaining the chain structure, expelling the weakly bonded water, and rotating alternate silica tetrahedra to change Si_4O_{12} groups into Si_4O_{11}. Experiments on the thermal decomposition of crocidolite (Patterson, 1965; Hodgson, Freeman and Taylor, 1965) confirm this hypothesis. The process takes place by dehydroxylation, then oxidation of the iron, then disintegration of the amphibole anhydride to form pyroxene with the simultaneous production of hematite, spinel and cristobalite. The products form pseudomorphs of the original amphibole fibres with parallelism of *c* (amphib)//*c* (pyrox)//*a* (hem)// [211] spinel//*a* (cristob).

A structural control of the orientation of a secondary mineral by a primary one has been demonstrated even in extremely complex replacements such as sphalerite after dolomite (Robertson, 1951). The preservation of an original hour-glass inclusion pattern in a pseudomorph of garnet after actinolite (Sturt and Harris, 1961) despite considerable changes in composition and structure implies a slow, gradual replacement process.

A multi-crystal pseudomorph is the common variety in which a single original crystal has been replaced by many smaller crystals, e.g. garnet by chlorite, pyroxene by amphibole, andalusite or kyanite by mica, a calcitic fossil by wollastonite (Plate XXIVb and Harker, 1939, Fig. 106B). Considerable compositional and structural changes occur, the transformation is reconstructive, and the process involves nucleation at a number of sites in the old crystal (or at its borders) and growth of the new mineral phase.

The most common example is the serpentine network which has been called *mesh serpentine* (Deer, Howie and Zussman, 1962, Vol 3, p. 183), replacing pyroxene or olivine. Lauder (1965) has classified serpentine networks as follows:

(1) *Net* serpentine: a reticulate (i.e. like a net) series of serpentine cords enclosing areas of different crystallization and/or composition and possibly some unaltered olivine.
(2) *Cord* serpentine or vein bands. The straight to curved layers forming the network. These can be subdivided into *primary cords* which are the earliest and best developed, together with *secondary* and *tertiary cords* which form short lengths between the primary cords.
(3) *Mesh* serpentine: that which occurs in the areas between the cords. It may be called *primary*, *secondary* or *tertiary mesh* depending on whether it is enclosed by primary, secondary or tertiary cords.

An *aggregate-pseudomorph* is one in which a group of small, differently oriented crystals of one species has been replaced by a single crystal of a secondary mineral. The pseudomorph is skeletal and mirrors the original shapes of the primary crystals but is structurally continuous and uniform. The transformation is completely reconstructive but only one nucleus is formed so that the second phase grows as a single crystal. Examples include garnet after mica (Rast, 1958; Harker, 1939, Figs. 82, 83) or staurolite after mica (Plate XIIb and Harker, 1939, Fig. 101A).

A *multi-phase, multi-crystal pseudomorph* contains a number of crystals of different minerals but still retains the outline of the original mineral. The structure may be regular, zonal, completely irregular, complex or symplectic e.g. garnet pseudomorphed by a mixture of cordierite, orthoclase and biotite (Chinner, 1962).

VOLUME CHANGE DURING REPLACEMENT

Experimental examples of topotactic transformations almost invariably involve a loss or gain of certain atoms and a consequent increase or decrease in volume while still retaining the identity of the crystal. Petrographic and structural evidence, however (Turner and Verhoogen, 1960, p. 318; Thayer,

1966), seems to favour serpentinization and similar alteration processes at constant volume although there is no wide agreement on this point. Constant volume transformation of olivine or pyroxene to serpentine requires the removal of about 30% by weight of bivalent oxides and silica (Thayer, 1966, p. 707) but a topotactic relationship may be retained. High-energy changes will show only limited topotactic control and will be characterized by veining and non-pseudomorphic replacement or complex multi-crystal pseudomorphs.

Volume change is indicated by expansion cracks radiating from altered crystals and by dilation veinlets which show displacement of grain, twin, kink, or inclusion boundaries. An absence of volume change is indicated by the preservation of such original boundaries without displacement or distortion.

ORDER AND DISORDER IN COMPOUNDS AND SOLID SOLUTIONS

Atoms, ions and ionic groups of one kind can associate together to give homogeneous structures such as the close-packing in metals and a second species of atom or ion can be fitted into this structure if it has the appropriate size and valence properties to give an alloy or a compound. The structure may be sufficiently flexible that widely varying amounts of the second species can fit within the first (as in alloys), or it may be accepted only in fixed stoichiometric proportions (as in compounds), or there may be a continuous series of intermediate solid solutions between two end members.

A crystalline solid contains constituent units (atoms, ions, ionic groups etc.) arranged in a certain pattern. At absolute zero the position of every unit would be fixed with respect to all others, i.e. the crystal is completely *ordered*. *Short-ranged order* refers to the relation of a certain unit to the immediately adjacent units, and *long-range order* refers to its relation to units at some distance. Increase of temperature above absolute zero causes thermal vibration and the positions of the units become less fixed relative to the others. The units move further apart until just above the melting point there may be only limited short-range order and the atoms some distance apart are *disordered* with respect to each other. Disorder and entropy both increase with increase of temperature.

A solid solution consists of a solid phase containing two or more kinds of atom whose relative proportions may be varied within limits. The most abundant is called the solvent and the least abundant the solute. Most solid solutions are of the *substitutional* type in which the different atoms are distributed between equivalent sites; an *interstitial* solid solution is possible when small atoms of solute can fit within the solvent structure. If the solute occupies random positions in the solvent lattice it is said to be disordered but if it occupies certain rationally related positions it may show various degrees of order. Differences between solute and solvent atoms are reduced at high temperatures and the solute positions become disordered. Reduction of temperature causes an increase in order and reduces the freedom of choice of position of solute atoms (Fig. 31). A solid solution is stable under a given set of physical conditions ("in equilibrium with its environment") when it has the lowest possible free energy and highest possible disorder and entropy.

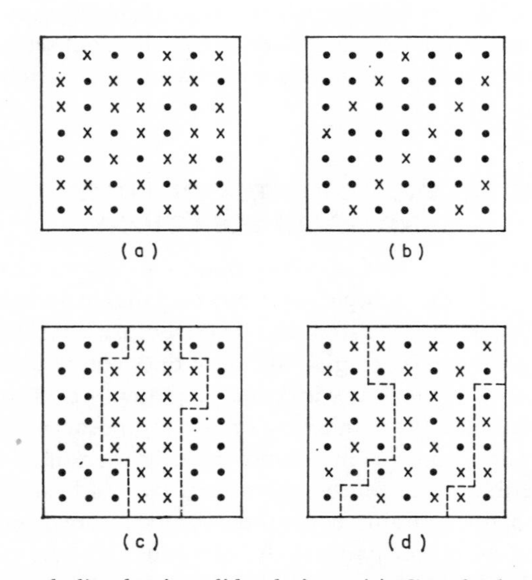

FIG. 31. Order and disorder in solid solutions. (a) Completely disordered solid solution with two components. (b) Superlattice produced by partial ordering. (c) An intergrowth of two separate phases produced by exsolution. (d) Metastable antiphase domain outlined by dotted boundary where partial ordering has placed like atoms adjacent to each other

Changes in the *amount* of disorder must be distinguished from changes in the *kind* of disorder (Flinn, 1960; Megaw, 1962). In *substitution disorder*, crystallographically equivalent positions in the lattice can be occupied at random by chemically different atoms (e.g. Si and Al in silicates); this is the common kind of substitution solid solution in alloys.

In *position disorder*, there are two or more energetically equivalent sites in the structure for a particular kind of atom which may thus occupy these sites at random.

Stacking disorder implies irregularity in an ordered sequence of stacked units; it is common in metals and also occurs in layered silicates such as micas where the total structure depends on the way in which sheets are stacked.

A decrease in temperature causes a change from disorder to partial order and possibly the formation of a superlattice which is an atomic structural pattern *within* the general solid solution atomic pattern, produced by one kind of unit occupying related sites (Fig. 31b). It is not clear whether superlattice regions are formed by a process of nucleation and growth, beginning at discrete centres and continuing outwards until the whole crystal is consumed, or whether the reaction is one of continuous atomic interchange which proceeds simultaneously in all parts of the lattice.

Increase in order may also lead to the formation of metastable *anti-phase* regions which are bounded by surfaces across which like-atoms are adjacent (Fig. 31d). An *anti-phase domain* is 180° out of phase of periodic variation with its neighbour, e.g. a sequence *ABABAB*, shows a regular periodic variation but a sequence *ABABBABAABA* contains an anti-phase region *BABA* bounded by two places where two *B*'s and two *A*'s occur together. The boundary of an anti-phase region in a crystal has high energy.

Exsolution from Solid Solution

On decrease of temperature, a solid solution may display immiscibility between solute and solvent and the strain between adjacent regions of differing composition may become so great that an interphase boundary is produced (Fig. 31c), and the solute becomes recognizable as a separate, inclusion phase (Geisler, 1951; Brett, 1964; Christian, 1965). The boundary may be coherent or non-coherent, depending on the degree of similarity of the structure of host and inclusion.

A rapid lowering of temperature of a solid solution reduces the ionic mobility of the solute atoms and "freezes" it in the structure so that an apparently homogeneous but distorted structure results. If the reduction of temperature is slow and allows ionic mobility, the solute ions will become increasingly organized and will finally separate as discrete bodies to give an *exsolution texture* with the impurity phase in bodies of various sizes and shapes dispersed through or around the host material (Fig. 32 and Plate XXVII). The relative sizes and shapes of the inclusions will depend partly on the relative surface energies and structures of host and inclusions and on the temperature at which exsolution takes place, the rate of cooling and the relative proportions of the two incompatible phases.

If a process of homogeneous nucleation (p. 116) is considered, then the size of the exsolved particles will depend on the total amount precipitated and on the number of nucleii. The number of nucleii will increase for greater amounts of the foreign phase in solution or for greater amounts of under-

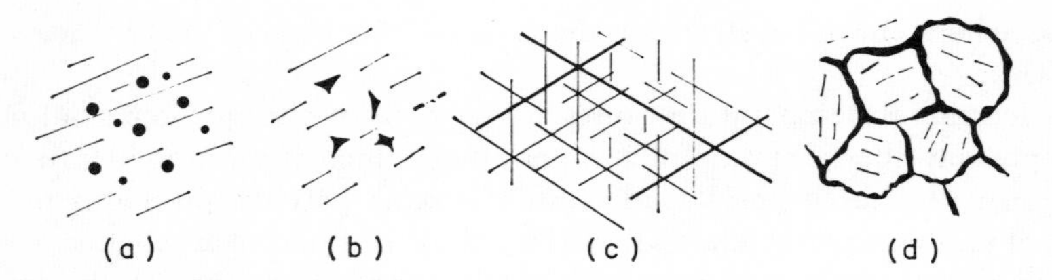

FIG. 32. Exsolution textures. Non-oriented; (a) globular inclusions (γ greater than host?), (b) cuspate inclusions (γ less than host?). Oriented: (c) Widmanstatten type with strong structural correspondence between host and inclusions (γ similar?). Segregation: (d) exsolved phase has become completely separated and forms an intergranular phase

cooling before precipitation. A greater supersaturation will give a greater driving energy for precipitation and hence a smaller critical radius for the nucleii. Figure 33 shows the classical relationship for homogeneous nucleation in a solid solution; greater undercooling or greater concentration of the solute results in a smaller critical radius of the nucleus, hence easier nucleation, more nucleii and smaller exsolved bodies.

Concentration of the exsolved phase at the grain boundaries might be due to precipitation within the host followed by migration to the boundaries (Toney and Aaronson, 1962; Westbrook, 1964) but is best interpreted in terms of heterogeneous nucleation at grain boundaries (Embury and Nicholson, 1965). Nucleation of the solute will take place preferentially at defects in the structure of the solute, but if the solvent is structurally good, nucleation may take place only at the grain boundaries.

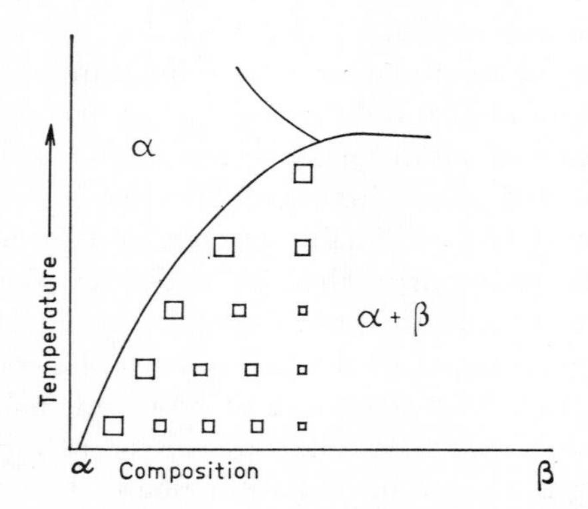

FIG. 33. Possible relationship between composition, temperature and critical nucleus size (after Smallman, 1963, Fig. 126). Cooling of the solid solution to the $\alpha + \beta$ field may cause exsolution of β from α. The size of the critical nucleus (shown by the squares) decreases with concentration of β and with the degree of undercooling. A smaller critical radius means easier nucleation, more nucleii hence smaller inclusions

If cooling is very slow, the exsolved phase may form independent large crystals outside of the host. This is particularly important in sulphide minerals and Stanton (1960, p. 5) considered that rapid cooling (hundreds of degrees in a matter of hours or days) is necessary to preserve exsolution intergrowths of sphalerite in chalcopyrite and to prevent complete segregation. It has been shown, however (Brett, 1964, p. 1266), that oriented intergrowths can be produced experimentally in sulphides by slow cooling.

Two main modes of precipitation from solid solutions have been recognized (Christian, 1965, p. 60) in exsolution intergrowths:

(a) Continuous (general).
(b) Discontinuous (cellular).

Continuous precipitation occurs where the supersaturated phase deposits the precipitate in all parts of its lattice by a continuous draining away of the excess solute. A steady state is obtained and the growth law of the precipitate is linear. The host retains its identity (shape and size), there is an orientation relationship between host and inclusion, the distribution of solute is uniform and thus nucleation sites are evenly distributed. The shape of the exsolved phase may depend on interfacial energies if equilibrium is approached; however many exsolution textures are metastable and have been controlled by kinetics.

Discontinuous precipitation occurs where the supersaturated phase (*A* with dissolved *B*) breaks down into a mixture of two phases (*A* and *B*) which precipitate in a small unit or "cell" which grows at the expense of the host. The orientation of *A* in the cell differs from *A* as the host. The duplex (two-mineral) cells are not uniformly distributed, there is no orientation relationship between host and cell, the boundaries are not coherent and the growth law is parabolic (i.e. second-order kinetics). Examples include lead–tin alloys.

All common mineral exsolution textures, particularly oriented intergrowths such as perthites, are formed by *continuous precipitation*. No clearly defined analogues of alloys formed by discontinuous precipitation occur among mineral textures but symplectic intergrowths, eutectoids, and some multiphase multicrystal pseudomorphs may be formed by an analogous process.

If there is a volume change during exsolution, the intergrowth pattern may be controlled by the distribution of strain energy which will then be much greater than the surface energy. The configuration of inclusions in exsolution intergrowths has been attributed to the interaction of strain energy and space-filling requirements (Geisler, 1951, p. 423; Christian, 1965, p. 13). The shapes of inclusions range from the globules of emulsion texture on the one hand (where the inclusions have lowest surface area but greatest interfacial energy) to oriented plates, rods and needles (with large surface area but low interfacial energy).

The total free energy is reduced when the strain in the host lattice is reduced. This occurs when host and inclusion share lattice features because the energy of the interface is lowered. The activation energy of transformation is lower if the inclusion can use structural units from the host so that nucleation is easier.

Inclusions are oriented and geometrically regular if there are structural similarities between host and inclusion such that low energy (but high area) mutual boundaries are formed and this implies that host and inclusion have similar surface energies. Inclusions are irregular in shape if they cannot achieve structural continuity with the host because host and inclusion have differing structures and different surface energies. The mutual boundaries will tend to have a high energy. Brett (1964, p. 1246) interpreted a spherical shape of inclusions as implying similar structures in host and inclusion but this conflicts with other evidence.

Most non-oriented inclusions tend to have curved boundaries and it has been suggested (p. 167) that an inclusion will be globular if its specific surface energy is greater than that of the host but will form cuspate wisps with concave boundaries if the surface energy is less than that of the host.

Some experimental evidence supports this concept. The mutual textures of a cooled mixture of periclase, chromite and silicate (Scheerer, Mikami and Tauber, 1964) can be related to differences in surface energies (γ chromite greater than γ periclase greater than γ silicate). Silicate forms cuspate crystals in periclase, and chromite forms globular crystals in periclase; chromite and periclase are idioblastic towards silicate but chromite-periclase boundaries are irregular.

Desch (1934) pointed out that in certain eutectic alloys, metals with a high surface tension (gold at 1200°C has $\gamma = 1120$ dynes/cm) form globular clusters but those with a low surface tension (antimony at 80°C has $\gamma = 350$ dynes/cm) form angular grains. Graphite forms lamellae in iron with γ below 1100 dynes/cm, but the addition of magnesium raises γ to 1400 dynes whereupon graphite forms spherical rosettes. However, there must be some doubt as to the validity of this concept in respect to inclusions within a silicate host crystal. The energy of the host–inclusion interface depends entirely on the interface itself and need bear no simple relation to the relative energies or tensions of specific solid/gas or solid/vacuum surfaces. As a general rule, curved and cuspate interfaces are probably not in equilibrium.

Stanton (1964) determined relative surface energies of self-boundaries using interfacial angles at triple-points and found that the grain boundary energy (γ) of galena is about 0·6 of that of sphalerite and γ of chalcopyrite is about 0·75 of that of sphalerite. Galena has a globular form when enclosed by sphalerite and sphalerite is cuspate when enclosed by galena (Stanton, 1964, Figs. 7 and 8), thus the evidence from interfacial angles is compatible with that from emulsion textures. The relationships of chalcopyrite and sphalerite are not clear, however; chalcopyrite commonly has convex boundaries towards sphalerite, but Stanton's measurements suggest that chalcopyrite has a lower grain boundary energy than sphalerite. Discrepancies may be due either to emulsion textures not being stable low-energy intergrowths or to γ anisotropy. Estimates (Gilman, 1959, p. 204) suggest values for γ_{110} sphalerite of 360 ergs/cm, γ_{110} and γ_{100} of galena of 440 and 625 ergs/cm respectively. The anisotropy of both minerals suggests a considerable overlap in values of surface energy. Regularity in the shapes of exsolved phases (Brett, 1964) can be interpreted to indicate that γ digenite $>$ γ bornite $>$ γ chalcocite and chalcopyrite.

Exsolution processes take place by diffusion of cations to appropriate positions leaving the host framework generally undisturbed. The more abundant constituent generally becomes the host and the less abundant forms the inclusions. The shape (Fig. 32) and size (Fig. 33) of the intergrowth pattern is determined by:

(1) The structure of the host and thus whether there is a structural plane which may be shared by host and inclusion.

(2) The atomic structure of the inclusion and hence its tendency to form plates or rods of a rational shape.

(3) The degree of structural similarity of host and inclusion; hence the coherence, energy and stability of the interface. This controls whether there is a tendency for the inclusion to persist.

(4) Diffusion power of the inclusion. Sufficient high mobility of the ions in the inclusions due to temperature, defects, diffusivity, etc., may allow a reduction of surface energy by moving inclusions to the external grain boundaries of the host. The impurity will form an intergranular film if it "wets" the grain boundary but will break up into "beads" or discrete grains if there are appreciable differences in surface tensions. Grain boundary segregation is ubiquitous in metals and ceramics (Westbrook, 1964; Scheerer *et al.*, 1964) but is much less common in silicates where difficulties of diffusion do not allow migration to the boundaries. However, there is some evidence (Dromsky, Lord and Ansell, 1962) that diffusion is not the rate-controlling factor in segregation but that it depends on the rate of solution and exsolution in the solid solvent.

(5) Partial ordering in the host. Although interfacial energy relations explain why inclusions are oriented in certain lattice directions it does not explain why the exsolved phase forms continuous layers. The lamellar exsolution structure in pyroxene or feldspar must involve either a partial original high-temperature ordering in the host with the impurity atoms concentrated in certain layers which become more ordered and finally

separate, or very restricted nucleation followed by a very strong lattice control of the growth of the new phase along certain lattice planes.

(6) Exsolution of one phase within another is basically a nucleation and growth problem. The inclusion phase must be initiated as a separate mineral phase in the form of a tiny nucleus which then grows, i.e. it expands its interface. The activation energy (and hence the ease of nucleation) depends on the surface energy of the interface and hence on the degree of similarity of structure of host and inclusion. Nucleation is favoured by small differences in structure but it must be remembered that the driving force for the exsolution process itself is a difference in structure which prevents the solid solution remaining stable. The relative size of host and inclusion will depend on their relative proportions and in general the more abundant the impurity, the larger the size of the inclusion, however for a given composition slight atomic structural differences between host and inclusion will result in a very fine intergrowth because nucleation of the exsolved phase is easy and abundant nucleii will be formed.

Short- to Long-range Transport

DISCONTINUOUS REACTION TEXTURES

Two phases may crystallize together as separate crystals or in various intimate intergrowths ranging from regular oriented intergrowths to irregular, duplex mixtures in eutectoid and symplectic textures. Such textures may be produced by the breakdown of a single crystal of one mineral to a mixture of minerals, or by the combination of dispersed reactants to form units of intergrown minerals.

All crystallization involves nucleation and growth but duplex crystallization in symplectites, myrmekite and eutectoids involves co-nucleation and mutual growth of two phases. Eutectoid textures are strongly dependent on the nucleation process and may be classified (following Mondolfo, 1965) as:

(1) *Normal eutectoid texture*, in which one phase nucleates on the other phase and both grow together.
(2) *Anomalous eutectoid texture*, in which both phases are nucleated by foreign impurities.
(3) *Degenerate eutectoid texture*, in which the second phase is not nucleated until the first phase has crystallized.

Most intergrowths formed by simultaneous crystallization of two phases are normal eutectoid types. The crystals generally have an irregular shape but some are oriented intergrowths in which each mineral is structurally related to the other and the shapes of the crystals are controlled by epitaxy.

Complex layered kyanite–staurolite crystals form as the result of the structural correspondence of (100) of kyanite and (010) of staurolite (Deer, Howie and Zussman, 1962). Complex layered crystals of clays, chlorite, micas and chondrodite can form because of the close correspondence of the silica layers. Differences in cell dimensions can be compensated by periodic dislocations which allow formation of regions of predominantly good fit with intervening regions of bad fit.

Irregular fine-grained mineral intergrowths are called *symplectites* (symplektite) or *symplectic intergrowths*. Special symplectites containing vermicular (worm-like) quartz plus plagioclase are called *myrmekite* and those consisting of quartz plus orthoclase have been called *micrographic*, *microgranophyric* or *dactylitic* (finger-like). The first symplectite was named *kelyphite* by Schrauf (1882) in a reaction rim around garnet in serpentine; however, it was not a single mineral but a mixture of amphibole and spinel. There is some tendency to restrict the term "kelyphite" to mineral mixtures around olivine or to mixtures of ferromagnesians (Grout, 1932, p. 361), but this is not general.

Symplectic intergrowths occur between quartz and feldspar, amphibole and spinel, plagioclase and magnetite, garnet and quartz, bytownite and pyroxene, biotite and quartz, diopside and spinel, epidote and quartz, scapolite and orthoclase, scapolite and quartz, and rare examples such as chondrodite and calcite, and orthoclase and sodalite. Most symplectites consist of pairs of minerals intergrown in a manner similar to *eutectoids*. They may occur at boundaries between reacting minerals or may be formed in a mass by the breakdown of a crystal of primary mineral into a pair of secondary minerals. Most normal symplectites result from simultaneous co-nucleation and interdependent growth of the two phases by discontinuous precipitation but similar textures can be produced by partial replacement (Augustithis, 1962). Conditions of growth are such that crystallization takes place rapidly under strong chemical or energy gradients and dendritic growth results. The geometry of the forms produced depends on the branching of the original duplex nucleus; the spacing between branches or fingers depends on the kinetics of diffusion, the rate of release of free energy of transformation, and on interface energy requirements.

Where the process operates under a high-energy gradient, there is little or no tendency for the lattices of the original and the secondary crystals to be related. The irregular texture of symplectites suggests that, as is the case with all dendritic growths, surface energies are not minimal and the form of the intergrowth is not controlled by surface energies. Most symplectites consist of two minerals but a few contain three; in some of the three-mineral intergrowths a slight control of form by relative surface energies is seen, e.g. vermicular hornblende and garnet with granular spinel.

Myrmekite is common in high-grade regional or plutonic rocks such as gneisses and granulites. It forms (Binns, 1966, p. 30) by:

(1) Simultaneous crystallization of crushed quartz and feldspar following marginal cataclasis (Sederholm, 1916; Shelley, 1964). Myrmekites are commonly associated with cataclastic textures although this is not universally true (Watt, 1965).
(2) Replacement of orthoclase by quartz plus acid plagioclase (Sederholm, 1916; Binns, 1966). Lobate or "cauliflower" growths of myrmekite projecting from plagioclase–K feldspar boundaries into the K feldspar have a clear replacive aspect. This is probably the most common origin.
(3) Exsolution of albite from K feldspar. The irregular distribution and proportion of quartz does not favour this explanation.

Coronas, Reaction Rims

Synantetic minerals (Sederholm, 1916, from the Greek "meet") are those which occur at the contact between two minerals. Rims around minerals have come to be known as *kelyphitic rims*, *coronas*, *corrosion mantles* and *reaction rims* (Shand, 1945). A rock containing abundant coronas is a *coronite.* The terminology is rather vague and complications in the nomenclature have arisen because terms have been used by some authors in very specific ways. "Corona" is non-genetic but "reaction rim" has a genetic ring and it has been suggested that the term "corona" should be restricted to primary depositional rims in igneous rocks and "reaction rim" used for secondary rims in metamorphic rocks. This distinction is not an easy one to make and the term corona will be used here to refer to simple or complex structures consisting of concentric layers about a central crystal (Plate XVIII). The layers are generally regular in thickness and range from one to five in number (Alling, 1936; Shand, 1945). Each layer may consist of one mineral as a granular or radiating fibrous aggregate, or may consist of a complex intergrowth or symplectite. Synantetic coronas occur at contacts between felsic and femic minerals (in basic rocks), between iron oxides and various minerals, and between plagioclase and potash feldspar (in acid rocks, charnockites, etc.).

Layered coronas are particularly well developed around olivine in metabasic rocks, garnet in eclogites, and feldspar in charnockites and granulites. They were first described in basic (Törnebohm, 1877; Holland, 1896) and ultrabasic (Schrauf, 1882) igneous rocks. They are very well developed in troctolites (Huang and Merritt, 1952) and in metagabbros or metadolerites (Sederholm, 1916; Brögger, 1934; Buddington, 1939, 1952; Shand, 1945;

Huang and Merritt, 1952; Gjelsvik, 1952; Poldervaart and Gilkey, 1954; Murthy, 1958; Bose, 1961).

Perfect, *complex layered coronas* are restricted to altered basic igneous rocks; poorly developed examples occur in some gneisses, charnockites, granulites (Lovering and White, 1964) and acid to intermediate plutonic rocks, but even poor ones are noticeably rare in hornfelses and schists. The lack of reaction rims in a rock is no indication that the minerals are in equilibrium as many polymetamorphic schists contain minerals from as many as three metamorphic phases coexisting without coronas.

Coronas are abundant in inclusions in kimberlite (a brecciated, serpentinized peridotitic rock which contains xenoliths of country rock, garnetiferous ultrabasic rocks, eclogite, etc.). Kelyphitic borders on chrome-pyrope consist of a narrow, almost opaque brownish-black intergrowth of magnetite and chlorite, or distinct hornblende and chlorite crystals, or intergrowths of chrome-spinel plus enstatite and/or hornblende, mica and/or chlorite (Nixon, Knorring and Rooke, 1963).

A *coronite metagabbro* (Buddington, 1952) is a gabbroic rock which retains its ophitic texture and contains relict olivine, augite and hypersthene, but with abundant multi-layer coronas. The rock occurs as less-altered cores which grade outwards into sheaths of pyroxene-gabbro gneiss or garnet amphibolite.

The classic multi-layered corona occurs at the junction of olivine and plagioclase in gabbroic rocks. The zone nearest the olivine is generally of hypersthene as tiny parallel prisms perpendicular to the olivine surface, followed by a symplectite of actinolite plus spinel, and finally a layer of garnet next to the plagioclase.

A great many varieties of compound layered coronas have been described. They include (inner mineral first and outer last):

(1) Iron oxide-biotite or hornblende-garnet, pyroxene or spinel.
(2) Clinopyroxene-hypersthene, hornblende or biotite-garnet.
(3) Orthopyroxene-biotite-hornblende or garnet.
(4) Olivine-hypersthene-hornblende, biotite, augite-garnet, spinel or plagioclase.
(5) Garnet-amphibole, biotite or chlorite.

The following different geneses have been suggested (Buddington, 1939, p. 295). Coronas may be:

(1) Primary magmatic and zoned according to the order of crystallization.
(2) Late magmatic and related to discontinuous reaction of early formed crystals with a melt.
(3) Due to reaction in the solid or semi-solid state in the presence of deuteric intergranular fluid.

(4) Due to growth in the solid state during thermal metamorphism.
(5) Due to growth in the solid state during regional metamorphism.

There seems little doubt that coronas have been produced by all five of the processes above. Shand (1945) compared coronas with compositional zones and considered that zoning was a continuous reaction series whereas a corona was a discontinuous reaction series.

It seems possible to divide coronas into two main types such as Sederholm (1916, p. 34) has done:

(1) *Simple coronas* with one, two, or rarely three layers of minerals such as pyroxene, amphibole and biotite. These occur in non-metamorphosed igneous rocks and form by the first three processes listed above.
(2) *Complex coronas* consisting generally of a number of layers (commonly two or three and even four) and containing garnet and minerals such as spinel-symplectite, pyroxene, amphibole, mica, scapolite, feldspar, sphene and chlorite. These occur in metamorphosed igneous rocks and form by the last two processes listed above. However, some simple metamorphic coronas may consist of igneous-type minerals and be indistinguishable from the igneous kind. It is also useful to divide metamorphic coronas into those containing entirely anhydrous mineral assemblages and those with some hydrous minerals.

The mechanism of formation of multi-layer, complex coronas is not understood and it is at present not possible to distinguish between the five different genetic types of coronas using the characteristics of the corona alone without considering additional information such as field relations. The considerable debate which has arisen over the coronas in Adirondack metagabbros hinges largely on whether the coronas belong to the igneous history of the gabbro or are a metamorphic phenomenon (Shand, 1945; Shaub, 1949; Jaffe, 1946; Buddington, 1952). The apparent resistance of "dry" basic igneous rocks to recrystallization during metamorphism, even high-grade regional metamorphism, may mean that the only evidence of the metamorphic process within a gabbroic body is the existence of coronas and clouded plagioclase.

The corona occupies space which once belonged to the primary minerals if the rocks were solid when metamorphosed. The layering follows the outline of the primary minerals and its form is controlled closely by the intergranular boundaries. The types of minerals in the corona and their order is not random but occurs repeatedly in different rocks. The compositions of the corona minerals are not related in a *simple* way to those of the primary minerals. It would appear that anhydrous metamorphic coronas have formed by reaction between the two bounding primary minerals essentially in the solid state. The rims replace a pre-existing mineral without reflecting its structure and the boundaries of the rim are controlled by the distance from the original interface. It seems likely that a succession of rims represents successive mineral phases in equilibrium at one instant within a compositional gradient produced by solid diffusion.

The approximate compositions of successive rims in the classic compound corona, olivine-hypersthene-hornblende-garnet-plagioclase are plotted in Fig. 34. It can be seen that the composition of each rim depends mainly

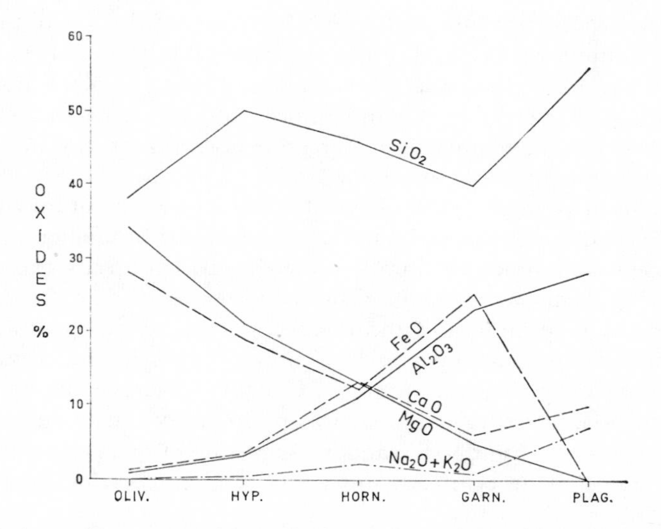

FIG. 34. Approximate compositions of minerals in successive layers of a compound layered corona containing olivine, hypersthene, hornblende, garnet and plagioclase

on its position between the end members olivine and plagioclase and represents the mineral (or in symplectites, the mixture of minerals) appropriate to the composition at that point in a chemical gradient produced mainly by migration of FeO and MgO outwards from olivine and CaO and Al_2O_3 inwards from plagioclase.

Some coronas represent an arrested stage in the alteration of a mineral during progressive metamorphism, and others (kelyphitization and uralitization of pyroxene in eclogite) are retrograde. The presence of minerals whose compositions do not lie directly between the end members (especially hydrous phases) implies that the system was not closed chemically and it would appear that chemical movements, probably with an intergranular hydrous phase, result in the addition of amphibole, etc., by alteration of early pyroxene, garnet, etc. Thus in the example quoted above, the igneous pair olivine and plagioclase reacted under anhydrous metamorphic conditions to give hypersthene and garnet, one (or both) of which later changed to hornblende, etc., under hydrous metamorphic conditions.

Transformations Controlled by Heat Transport

MELTING AND METAMORPHISM

Metamorphism normally takes place in the solid state and the temperatures are not normally sufficiently high to cause melting. A disperse fluid phase (the "intergranular fluid") may be present but it can only constitute a very small part of the total rock. Partial melting does take place, however, in the Sanidinite Facies at high temperatures and low pressures and under high-grade regional metamorphism at high temperatures and water pressures to produce migmatites and anatectic melts.

The atoms in an ideal crystal above absolute zero temperature vibrate but are held in position by mutual bonding. An increase in temperature causes an increase in amplitude of vibration and the solid generally expands. The melting of an ideal crystal can be regarded as taking place when the amplitude becomes very large and the frequency zero, or as the result of the co-operative calamitous appearance of imperfections in the crystal as the temperature reaches the melting point. Complete disorder may be attained with respect to lattice sites and interstitial positions. A silicate melt, however, is not a completely disordered, random aggregate of separate ions; it has a considerable degree of *short-range* order and consists of various molecular or ionic groupings.

A liquid is partway between a solid and a gas, but because its density is nearer to that of the solid (particularly just above the melting point), the atoms in a liquid must be about the same distance apart as they are in a crystal. However, the atomic arrangement is different in that there is no long-range order in liquids. The high viscosity of silicate melts is correlated with the presence of organized (*cybotactic*) groups.

Lowering of the temperature of the liquid below the melting (freezing) point may result in solidification in three different ways:

(1) Immediate but slow change from melt to crystals at the melting point.
(2) Supercooling with rapid change from melt to crystals below the melting point.
(3) A gradual increase in viscosity and change from liquid to glass.

It is convenient for many purposes to regard melting and solidification as identical processes merely acting in opposite directions although this is not strictly true (Buckley, 1951, chap. 9).

The *velocity* of growth is much less than that of solution or melting not only because it is easier to disrupt a complex crystalline solid than it is to rebuild it but also because the presence of impurities on crystal surfaces slows growth but does not affect melting.

The reorganization of ions into some complex oxide and silicate structures is so difficult that any but extremely slow cooling of the melt causes only partial organization and the liquid configuration does not change into the crystal configuration but gives a glass which is a highly polymerized solid with a moderate to high degree of short-range order (less than a true crystal) but little long-range order (like a liquid).

Glass or glass-like material is rare in metamorphic rocks but occurs in those in which melting has taken place, e.g. buchites and possibly some hyalomylonites (pseudotachylytes). Glasses have been reviewed by Morey (1938), Weyl (1951), Douglas (1965) and recently in great detail by Eitel (1965).

Morey (1938)defined glass as an inorganic substance which is continuous with, and analogous to, the liquid state of that substance but which as the result of having been cooled from a fused condition has attained so high a degree of viscosity as to be for all practical purposes rigid. Organic glasses are now recognized.

The necessary condition for glass formation (Douglas, 1965) is that the material should be cooled through the temperature range where crystals can grow, so quickly as to prevent any appreciable growth. Silicate glasses are composed of SiO_4 tetrahedra linked in a random, three-dimensional network in which the formation of chains or sheets is not excluded but is not a requirement. Glasses represent the most generalized form for an array of ions in the solid state. They are electrostatically neutral but do not have ions in simple proportions (stoichiometry). Crystalline silicates have simple ratios of Si:O but glasses do not. The internal distortion and lack of symmetry means that glass has a greater energy than the appropriate crystalline state so that crystallization results in a decrease of internal energy. The process whereby a glass crystallizes to an extremely fine-grained crystalline aggregate is called *devitrification*. The rate of devitrification depends on the rate of growth and possibly also of nucleation of crystallites which in turn depend largely on the viscosity of the glass. Devitrification must take place below the "melting point" (although this is not exactly definable) but at a sufficiently high temperature that the rate of growth is appreciable. A complex, incompletely understood, relation exists between the viscosity (hence temperature and composition) and the rate of devitrification. Theory and experiment on nucleation and growth in the recrystallization of glasses have been reviewed by Marshall (1961), Stookey and Maurer (1962) and Knapp (1965).

Devitrification is an order–disorder phenomenon which can be regarded as a reconstructive transformation. In glasses, each SiO_4 tetrahedron is a little imperfect and the tetrahedra are strung together in loose chains and networks. Crystallization involves changing the shape of the tetrahedra, rotating them to form appropriate lattice patterns, and producing grain boundaries, i.e. crystal–glass surfaces. Spherulitic textures result where the rate of growth exceeds that of nucleation, and a very fine-grained granular, fibrous, dendritic, graphic or eutectoid texture results where the rate of nucleation exceeds that of growth.

Experiments on the sintering of ceramics give a great deal of information relevant to the textures of partially melted hornfelses. Sintering (Herring, 1951; Gray, 1957; and various papers in Kingery, 1959) is a complex process leading to densification of a finely granular aggregate below the melting point without the formation of a liquid phase but it passes into melting with elevation of temperature. It involves an initial interaction between the surface atoms or ions of the individual particles or any material adsorbed on them, followed by the reconstitution of normal bonds within this zone. Material is transferred along and through boundary regions. It is to be expected that processes similar to sintering take place during high grade thermal metamorphism and are promoted by a high temperature and the presence of a gas phase.

Increase in temperature in an aggregate of a single species of pure crystalline solid results in complete and abrupt melting at the melting point. Increase in temperature in a multiphase polycrystalline aggregate such as an impure sandstone causes melting in the lower melting point minerals, increases diffusion and sintering processes in higher melting-point minerals and partially converts them into liquids at temperatures far below their melting points by dissolving them in an interstitial melt.

Grains are pulled together by the surface tension of the intergranular melt. Centres of grains approach each other because of solution along contacts. Melting or solution of solids converts solid into liquid which fills pores, and the presence of an intergranular liquid allows repacking of grains to a more closely packed aggregate.

There is an internal rearrangement during sintering when shrinkage of the aggregate is proportional to time, but after about 35% of liquid has been produced, shrinkage is proportional to the cube root of time. During the initial stages the rate of shrinkage is inversely proportional to the radius of the particles, whereas in the later stages it becomes inversely proportional to $\frac{4}{3}r$, and solution takes place at positions of greatest pressure rather than greatest curvature (Kingery, 1959, p. 190).

A rise of temperature of a multicomponent, polycrystalline aggregate causes melting in order of increasing melting points. Minerals may melt one at a time, in pairs or in groups.

MELTING UNDER PLUTONIC CONDITIONS

The melting point of a solid crystal with a positive coefficient of expansion increases with increasing hydrostatic pressure. The melting point of a crystalline solid is reduced when placed in contact with a liquid (i.e. a phase which has a much lower melting point) in which it is soluble. If the solid has a positive coefficient of solubility with pressure, then its melting point will be progressively reduced by increase in pressure. Pressure does not have a great deal of effect on solubility in a solid–liquid system because the liquid is essentially incompressible and the solid is slightly elastic. Solubility in a gas–solid system is, however, very pressure sensitive because of the high compressibility of the gas. The melting points of common silicate minerals fall by about 1°C for every 1 kg/cm^2 (or atmosphere) of H_2O pressure for the first hundred kg/cm^2 but the decrease becomes progressively less at higher pressures.

Melting of quartz-rich sedimentary, metamorphic or igneous rocks under contact metamorphic conditions begins at approximately 950°C. Melting takes place at lower temperatures under regional metamorphic conditions and migmatites, etc., can be expected to form by partial melting. The low melting point minerals quartz and feldspar melt before the ferromagnesians to give a mixture of crystals and liquid (migma) but crystallization of the melt is complete and the textures are igneous. They differ from normal metamorphic textures, and hence will not be discussed here.

Melting is not a common metamorphic process within the upper parts of the earth's crust because of the high melting points of the common rock-forming minerals. However, three crystalline solids of geological significance (calcite, salts and ice) have low melting points. It is not intended to deal with these materials in any detail because each is a field of study in itself and because each has been reviewed recently in some detail; the metamorphism of salt deposits by Borchert and Muir (1964), the metamorphism of ice by Shumskii (1964), and the melting of calcite by Wyllie and Haas (1966).

Ice and the various salts (and to a lesser extent, calcite) are crystalline solids with rather high crystallographic symmetry and with considerable powers of translation gliding; they are deformed and recrystallized under natural conditions either at temperatures just below their melting points, or just above their melting points in contact with a melt (or solution) phase. Ice and salts undergo normal geological processes and although they are clearly special cases, their textures should be considered together with those of metamorphic rocks and ore deposits. Salt deposits consist mainly of gypsum, anhydrite, halite, dolomite and magnesite. The rocks are initially formed in sedimentary beds as precipitates from solution and thus their textures are controlled by the same processes as those of igneous rocks. There is a recognizable order of crystallization with early formed euhedral and porphyritic crystals and later anhedral minerals filling cavities.

Authigenic changes take place readily and the movement of interstitial water allows recrystallization by solution and deposition. The textures produced resemble those of metamorphic rocks except that cavities are preserved because the confining pressure is low. The low strength of salt aggregates, however, allows compaction under moderate loads of superincumbent rocks and "load metamorphism" occurs at shallow depths. The behaviour of salts in association with a solvent phase is like melting in rocks and the textures are similar to those of high-grade thermal metamorphics except that melting does not result in a glass phase.

Folding of salt beds and movement of salt domes causes deformation and the production of textures related to those of high-grade regional metamorphic rocks. Ease of recrystallization even at low temperatures tends to prevent metamorphism strictly analogous to dynamic metamorphism although rapid deformation, particularly near faults, causes brecciation, bending of crystals (Goldman, 1952, plates 20 no. 1, 35 no. 1), shearing and a kind of mylonitization, etc. The similarity between crystalline schists and salts was the subject of comment by Koenen as early as 1905 and the concept was further enlarged by Rinne (1926).

It is regrettable that the terminology used by Borchert and Muir (1964, pp. 93, 94, 108) differs from normal metamorphic usage; their nomenclature is as follows:

"Dynamic metamorphism" relates to continuously changing compositions of the liquor in the salt masses. "Progressive metamorphism" is that due to the normal rise of temperature during simple burial of the salt. "Normal metamorphism" gives minerals which are successively richer in $MgCl_2$; "retrograde metamorphism" acts in the opposite sense and consists of changes caused by dilute solutions percolating through the salt, accompanied by solution of salts, and changes of composition of salts and liquors.

A wealth of textural data is given for the Sulphur Salt Dome in Louisiana (Goldman, 1952) and for the metamorphism of the Stassfurt beds in the Hildesia Mine (Kokorsch, 1960). Braitsch (1962, p. 117) listed the minerals of salt deposits in order of increasing tendency to form porphyroblasts and idioblasts (analogous to Becke's "crystallo blastic series" in metamorphic rocks) and this is no doubt related to relative solubilities and surface energies.

The textures of many "salt rocks" are complex and difficult to interpret because of the lack of equilibrium. The minerals and textures appear to respond readily to slight changes in conditions such as small stresses, slight heating or cooling, slight changes in the chemistry of intergranular solutions, etc.

Syntectonic crystallization produces crystalline textures with both lattice and dimensional preferred orientations. Gypsum recrystallizes readily and forms granular aggregates with polygonal texture but also non-equilibrium feathery aggregates with ragged or sutured margins. Anhydrite forms idioblastic porphyroblasts when surrounded by a matrix of gypsum or calcite. Monomineralic aggregates of anhydrite (both in sedimentary beds and regenerated crystals in cap rock) have strongly developed idiotopic ("pile of bricks") texture consisting of close-packed idioblastic crystals (Goldman, 1952, plate 30 nos. 3, 36). Coalescence of gypsum to give large crystals appears to be common. Le Comte (1965) discussed experimental creep in rock salt and found that deformation reduced the grainsize although no quantitative correlation was found.

Ice is another geological material which normally exists under conditions close to its melting point. Its behaviour then may range from that analogous to cold-working in metals at low temperatures, hot-working of metals at higher temperatures, recrystallization, and finally pressure- and temperature-dependent melting (or even sublimation) at higher temperatures. Unfortunately work on glacial flow and ice deformation has not yet shed much light on the processes of rock metamorphism.

The various kinds of ice are analogous to sedimentary (snow, firn), igneous (surface, bottom, pack ice) and metamorphic (glacial) rocks. Glacial flow is in some ways analogous to the deformation of metamorphic rocks and ice textures are similar to some metamorphic rock textures. Cataclasis in ice produces undulose extinction, biaxiality, mortar texture, brecciation, thrusts, shear zones, fissures and a preferred orientation. Deformation and recrystallization together produce folded and foliated ice with a preferred orientation. Melting and solidification (regelation) forms banded ice with a single layer or a number of intersecting layers of tectonic origin (Shumskii, 1964, p. 362). Static recrystallization produces coarse, evenly grained ice with a hornfelsic, "annealed" texture from snow or from glacier ice.

CHAPTER 6

Crystallization and Recrystallization

THE term *crystallization* as applied to metamorphic rocks refers to those processes which involve the formation of crystals which are essentially new. It is the most important of the thermally activated transformations. Changes in size and shape of crystals such as rotation, gliding, crushing, etc., which take place during metamorphic reconstitution resulting from mechanical processes do not affect the identity of the lattices and are not considered as part of crystallization.

The terminology used in metamorphic literature is confused, therefore metallurgical practice is followed here. *Crystallization* refers to the formation of crystals of a new phase and *recrystallization* refers to the reconstitution of an existing phase. Crystallization (e.g. the formation of garnet in pelitic hornfels) involves the formation of tiny new units (*nucleii*) and their subsequent *growth*. Recrystallization (e.g. the metamorphism of chert to quartzite) may involve nucleation and growth but as crystals of the mineral already exist, the process may consist merely of an increase in grain size (*grain growth*) or change in grain shape by grain boundary movement or by coalescence without nucleation.

Rocks may undergo slight metamorphic changes:

(1) Without recrystallization, e.g. grain growth of quartz in sandstone to produce quartzite without the destruction of the original clastic grains.

(2) With partial recrystallization, e.g. recrystallization of pelitic matrix but retention of large clastic grains in metagreywacke.

(3) With complete recrystallization of the phases present without new minerals, e.g. sandstone to quartzite.

(4) With complete reconstitution and the crystallization of metamorphic mineral phases, e.g. clay-shale to andalusite-mica hornfels.

The texture of a rock is largely defined by the three-dimensional grain boundary network. Recrystallization is a time-dependent process and as the texture evolves from the original to a final equilibrium texture, various kinds of grain boundary can be recognized:

(1) *Original boundaries.* These are normally destroyed quickly during metamorphism but may be preserved if they mark a strong chemical discontinuity.

(2) *Arrested boundaries.* If equilibrium is not achieved, a grain boundary may not have been completely adjusted to its most stable form. Decreased boundary mobility due to falling temperature may "freeze" a moving boundary and leave it irregular, sharply curved, dentate or sutured.

(3) *Equilibrium boundaries.* These belong to a low free-energy configuration and tend to be straight or slightly curved and regular in disposition. Rational, coincidence and twin boundaries are special cases.

Any texture depends on what existed before it, on the laws of nucleation and growth, and on the cessation of growth by the disappearance of the driving energies and the mutual impingement and interference of the crystals. The geometry of every part is predestined by the position and orientation of the first stable nucleii and this effect may persist in some degree through every subsequent generation of change that does not involve complete disorder. An equilibrium texture is approached as the historical accidents of nucleation are obliterated. The final texture depends on the interaction of two factors; the local geometrical requirements resulting from a need to minimize the interfacial energy at junctions, and the overall requirements of space filling which limits the possible connections between the junctions (Smith, 1964).

Much of our knowledge of the process of recrystallization depends on a study of annealing (Burke and Turnbull, 1952; Beck, 1954; Barrett, 1952; Burgers, 1963). A cold-worked metal or compressed powder is in a fine-grained condition and the total free energy consists of the normal lattice free energy plus the strain energy (that due to bending of the lattice, dislocation, twins, stacking faults, etc.), plus a considerable surface energy due to the large surface area of the fine-grained material. The behaviour of ionic and silicate minerals during annealing is similar to that of metals (Griggs *et al.*, 1960).

Annealing results in a decrease in dislocation content, e.g. a reduction from about $10^{11}/cm^2$ to $10^6/cm^2$ in cold-rolled polycrystalline copper. Dislocations can be removed by migration to boundaries, cancellation of defects of opposite sign, etc.

Recrystallization also allows a reduction in grain boundary energy by decreasing the grain boundary area. The decrease of grain boundary area in increasing the grain size of a chert (grain diameter 0·001 mm) to a quartzite (grain diameter 3 mm) is about $10^5 cm^2/cm^3$. If the grain boundary energy is about 400 $ergs/cm^2$ this gives an energy release of about 4×10^7 $ergs/cm^3$, i.e. 10^9 ergs/g mole or 25 cal/g mole. A similar calculation for recrystallization of limestone to marble indicates about 5 cal/g mole.

This is a very small quantity and is much less than the strain energy released during recrystallization which is from 20–200 cal/mole for most metals and possibly 500 cal/mole for calcite (Paterson, 1959; Gross, 1965). A cold-worked metal retains between 1% and 10% of the expended energy as strain energy in the lattice but much of this is released during the recovery stage prior to recrystallization proper and is probably not available to help overcome the activation energy of nucleation.

Heating to the *annealing* or *Tamman* temperature, which is about half the melting point in degrees Kelvin, allows sufficient ionic mobility for recrystallization. The energy losses described above constitute the driving force for recrystallization.

NUCLEATION

Nucleation may be classified as follows:

(I) *Homogeneous:* where nucleii appear within a homogeneous isotropic medium. A much studied and useful but perhaps over-simplified concept referred to also as "transient" nucleation or the Volmer–Becker–Flood–Doring hypothesis. Nucleii are formed spontaneously by statistical thermal fluctuations.

(II) *Heterogeneous* (inhomogeneous): nucleation occurs at favoured sites.

- (a) Nucleation at sites of high strain concentration or on "pre-formed" nucleii and strain-free sub-grains in strained solids.
- (b) *Surface, secondary or catalysed nucleation:* clusters form on a foreign surface of some kind. Nucleation is catalysed by the presence of "seeds" which form a substrate and a structural relation may exist between substrate and nucleus.
- (c) *Auxiliary nucleation:* a large number of nucleii are suddenly produced following the introduction of a few "seeds".

A new phase of any kind does not normally appear instantaneously as crystals of a finite size, the nearest approach being in diffusionless structural transformations and even here growth takes place. A new mineral, crystal, twin, etc., normally is formed as a very small region which then grows. A group of a few atoms or ions in an unstable configuration is known as an *embryo* or *cluster*. This may grow until it reaches a stable size when it is called a *nucleus*. The minimum size for stability is called the *critical radius* and calculations suggest a figure of 10Å for a range of materials. Homogeneous nucleation may require a nucleus containing between 0 and 100 atoms but heterogeneous nucleation may require less than 10 atoms (Walton, 1965b, p. 607).

Homogeneous Nucleation

The classical picture of nucleation originally considered the formation of a spherical drop of liquid condensed from a vapour, but has been extended to cover nucleation in metals and other solids. The energy of the interior of the nucleus and that of the host material are considered to be comparable but the total energy of the nucleus is greater because of the energy of the interface. Thus a nucleus will tend to disappear as soon as it is formed because it has a higher energy than its surroundings.

A new phase will tend to form, i.e. a nucleus will tend to be produced, if a reduction in total free energy is achieved, thus the free energy will fall by an amount equal to the product of the new volume and the loss in free energy ($\frac{4}{3}\pi r^3 \Delta F$). However, production of this new volume and its interface involves an increase of free energy equal to the product of its surface area and unit surface free energy ($4\pi r^2 \gamma$). If new material (i.e. the embryo) has a radius less than the *critical radius*, it has a greater total free energy than the material it has replaced, thus is unstable and tends to disappear. Once the critical radius is exceeded then the nucleus is stable and continues to grow.

The fundamental reason (Smoluchowski, 1951) for the appearance of nucleii in a homogeneous substance is the existence of thermal fluctuations, i.e. transient local deviations from the normal state. These deviations can occur in any part of the substance as fluctuations of local energy, density or chemical activity. They occur all the time, but only under certain conditions can they become stable and form nucleii. The fluctuations may occur within the stability field of the phase, i.e. are *homophase fluctuations*, but some fluctuations may form configurations approximating to the structure of a second phase, i.e. are *heterophase fluctuations*.

The activation energy barrier (ΔE_A) which must be overcome to allow nucleation where γ is the surface energy is given as:

$$\Delta E_A = \frac{16\pi\gamma^3}{3\Delta F^2}$$

The critical radius (r_c) of a nucleus can be considered as the size at which the decrease in volume free energy equals the increase in surface free energy ($4\pi r^2 \gamma = \frac{4}{3}\pi r^3 \Delta F$) or as the smallest stable size for which the free energy of formation is a maximum.

$$r_c = \frac{2\gamma}{\Delta F}$$

If it is accepted (Turnbull, 1948; Smoluchowski, 1951) that nucleii of critical size are formed by statistical fluctuations, the probability of a fluctuation of sufficient magnitude to overcome the activation energy of nucleation E_A depends on the number of atoms of sufficiently high energy. If the energy distribution of the atoms is regarded as normal, i.e. if the number with energy sufficiently great to overcome an energy barrier E is given by the Arrhenius factor $\exp(-E/RT)$ where R is the gas constant (and depends on the Boltzmann constant k) and T is the temperature in degrees Kelvin, then the probability of nucleation is proportion to $\exp(-E_A/kT)$ where E_A is the activation energy for nucleation and the rate of nucleation N,

$$N = D \exp(-E_A/kT)$$

where D is a rate constant.

The homogeneous nucleation theory has been widely used in chemistry and metallurgy because reasonable agreement with experiment has been obtained but it can be applied to geological problems only in the most general way. McLean (1965, p. 105) has pointed out that the equation for nucleation rate above draws attention to the immense influence of surface energy and volume free energy on the rate of nucleation but when the equation is applied quantitatively it can be seen to be invalid. Taking what seem reasonable figures, a difference in interfacial free-energy of only 3 ergs/cm^2 (and the interfacial energies of minerals must have far larger differences than this) makes the time for one nucleation change from a fraction of a second to all geological time (McLean, 1965, Fig. 2).

The classical theory is of doubtful validity even when restricted to the formation of a drop of liquid from a vapour and is less valid when applied to a crystal forming from a solution (Walton, 1965b) or a crystal forming in a solid matrix (Christian, 1965, p. 720).

Nucleii of crystals deposited from solution (Walton, 1965) have been accepted as more soluble than large crystals, thus nucleii tend to dissolve as they are formed and deposition is opposed. This concept is probably based on the so-called Gibbs–Kelvin equation which relates the vapour pressure of small drops to the radius of curvature of the drop, hence the solubility of a tiny crystal in solution or stability of a crystal surrounded by solid, to the diameter of the crystal nucleus. Walton (1965b) considered that the analogy might not be applicable, that the surface energy of a small cluster is very little greater than that for an infinite surface and that nucleii are not as unstable as has been considered.

One major weakness of the classical theory is that it ignores the fact that the radius of curvature of the nucleus is of the same order as the width of the transition zone between nucleus and host. When a nucleus is about 20 atoms across, and its disturbed boundary is two or three atoms wide and is transitional into the host, the internal free energy and surface energy of the nucleus become rather approximate concepts. There is no good reason to assume that the surface energy of such an interface is equal to that of the same species in large crystals.

Heterogeneous Nucleation

Real single crystals are not ideal but imperfect, and a polycrystalline aggregate (particularly a strained one) is not even remotely homogeneous thus there is not an equal probability of nucleation at any point as considered in the homogeneous nucleation hypothesis. The activation energy of nucleation is markedly reduced by various imperfections such as dislocations and grain boundaries (Christian, 1965). Cahn (1950, p. 324) considered that "the evidence is overwhelming that nucleii are formed in regions where the damaging effect of deformation is most marked". He considered that there is a strong inhomogeneity in the lattice with a sharp shear gradient at places of maximum strain and that small units polygonize to strain-free "blocks" which act as nucleii.

Crystallization is regarded by many in terms of a nucleation model involving a rearrangement of dislocations and a growth model involving movement of grain boundaries. Heterogeneous nucleation can be interpreted in two ways. The "block" theories propose that crystallization begins from pre-existing units or "blocks" in which the strain is either much larger than the average ("high-energy" or strain-assisted nucleation) or much smaller ("low-energy" nucleation). It is suggested that both are valid and that one may predominate over the other depending on the conditions of recrystallization (e.g. amount of strain prior to crystallization).

(1) STRAIN-ASSISTED ("HIGH-ENERGY") NUCLEATION

The presence of impurity particles, grain boundaries, dislocations and strained areas, enable nucleii to be formed with much lower activation energies than in homogeneous crystals. The formation of a nucleus at a boundary or dislocation involves the destruction of part of an existing surface or dislocation, so that extra free energy is released to aid formation of the nucleus interface.

The energy contribution towards nucleation by grain boundaries or other defects depends on their nature and hence on their energy. The energies of grain boundaries increase with decreasing coherence so that nucleation is favoured at corners, triple points or high angle boundaries. Dislocations can be treated similarly.

Grain boundary nucleation is easier (compared with nucleation within a perfect crystal) on grain boundaries which have a high energy compared with that of the nucleus boundary. Nucleation may be spontaneous at positions of very high energy and the nucleus will grow steadily from a zero size.

Nucleation in the solid state is further complicated by small volume changes and resultant strains which accompany any kind of transformation. Nucleation accompanied by slight volume change at grain boundaries will be similar to that without volume change; however, nucleation within a crystal is not only controlled by considerations of minimum surface area or energy but involves a shape which will reduce the elastic strains. A certain combination of size, shape and orientation constitutes a critical nucleus (Christian, 1965, p. 418). The nucleus may have a coherent interface with its host and the strains may be taken up elastically by either host, nucleus or both. The interface may, however, be incoherent and as its energy will be greater than that of a coherent one, nucleii tend to form coherently at first and become incoherent later. Thus there is a distinct tendency for the nucleus to be formed so that the orientation of its lattice and that of its interface is so related to its host as to reduce elastic strains due to volume changes. The significance of this control is clear in the nucleation of material exsolved from solid solutions giving oriented intergrowths or in topotactic transformations. It does not seem to be important in the recrystallization of single-phase aggregates or the crystallization of new phases in a complex mixture.

(2) "LOW-ENERGY" THEORY OF NUCLEATION

Work on deformed metals suggests that dislocations which are rather evenly dispersed during cold-working, congregate during the early part of annealing (mainly during "recovery") to form networks or tangles (Plate V). The three-dimensional network of closely packed dislocation tangles surrounds very small regions which are unstrained and free of dislocations. These regions constitute "blocks" or "pre-formed nucleii" which readily grow outwards into more stressed crystals which are less stable. The "incubation period" in annealing was considered by Turnbull (1948, p. 779) to be due to comparatively slow rates of adjustment of the blocks which are bent and must polygonize and become free of dislocations (Cahn, 1950; Cottrell, 1953, p. 188). Turnbull considered that there is a slow approach to an initial steady state nucleation but that the nucleation rate increases during recrystallization.

NUCLEATION AND GRAIN BOUNDARY MOVEMENT

Crystallization in metamorphic rocks involves nucleation and growth of several kinds. A new metamorphic phase must be nucleated *ab initio* but the nucleation will probably take place at high-energy regions such as grain boundaries or triple-points in undeformed rocks or highly strained regions in deformed rocks. Recrystallization of existing stable phases may take place at high-energy positions or from pre-formed nucleii. Recrystallization involves competition between the process of nucleation plus growth on the one hand and of grain growth of pre-formed nucleii on the other (Burke and Shiau, 1948). Recrystallization of Ti–Mn alloy following *small* deformations shows nucleation in regions of high strain but sub-grain growth in less deformed regions; recrystallization following *large* deformations shows nucleation at many points (i.e. nucleation is easier) and nucleii form continuously with time (Schofield and Bacon, 1961).

The stability of a nucleus depends on its internal energy and its surface energy. The latter depends on the specific surface energy, the surface area and the orientation of the nucleus boundary. The mobility of a boundary, hence the ability of the crystal to grow, depends on the relative orientations of the lattice of the nucleus, its boundary and the lattices of adjacent crystals. Beck (1954, p. 300) regarded nucleation as being a kind of boundary migration as a nucleus block absorbed surrounding strained material by moving the boundary out from the sub-grain. The mobility of a boundary is related to the degree of misorientation across it and low-angle boundaries are immobile and tend to become fixed, but high-angle, high-energy boundaries move more quickly. Strained materials adjust initially by the formation of a sub-grain structure in which the grains are composed of regions of slightly different orientation.

The energy of the nucleus must be low in the early stages or it would disappear spontaneously; however, the mobility of a grain boundary increases with increase of energy, thus the interface must have a minimum energy or the nucleus will not grow. The higher the energy, i.e. the greater the discordance with the matrix, the faster the growth. There is a conflict between the favoured *formation* of nucleii in orientations relative to the host which give a low interfacial energy and the favoured *growth* of nucleii which are so oriented as to have a high interfacial energy.

CRYSTAL GROWTH

Growth of crystals in metamorphic rocks takes place for a variety of reasons. The driving forces for recrystallization are lattice strain energy and grain boundary energy, both of which are small. The driving forces for *de novo* crystallization are chemical and may be large. It is necessary to consider separately the nucleation of a new mineral and the recrystallization of existing minerals. It is also essential to distinguish between the growth of a crystal surrounded by its own kind and the growth of a new foreign species.

One of the simplest examples of the growth of crystals is the annealing of a fine-grained aggregate of strained grains to give a coarse-grained aggregate of unstrained grains (Fig. 35). The driving force is the lattice strain energy

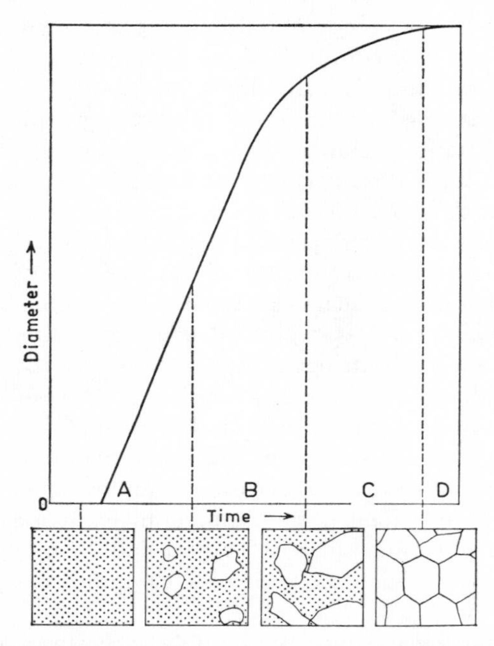

Fig. 35. Isothermal growth during annealing. *OA* is the period of incubation, *AB* of steady-state growth from isolated nucleii, *BC* of mutual impingement and interference, *D* of slow adjustment of grain boundaries (Cutler, 1959) in *Kinetics of High Temperature Processes*, ed. W. D. Kingery. By permission of M.I.T. Press, Cambridge, Mass.

plus the grain boundary energy. On heating a deformed polycrystalline mass (e.g. annealing a cold-worked metal) there is an initial *recovery* period when it changes shape slightly without textural change. This is followed by the most important part of annealing which is a period of steady grain growth and coarsening (*primary* recrystallization) to produce a granular aggregate. Under certain circumstances, continued heating causes rapid enlargement of a few crystals by *secondary* or abnormal recrystallization.

In addition to a great deal of detailed experimental information about ecrystallization of metals (well summarized by Byrne, 1965, p. 60), some significant experiments have been carried out on non-metallic materials such as alumina (Cutler, 1959), magnesia (Spriggs, Brissette and Vasilos,1964), fluorite, periclase and anhydrite (Buerger and Washken, 1947), carbonates, oxalates and hydrates (Norton, 1959; Griggs *et al.*, 1960).

Some relationships have been found between the grain size of a recrystallized aggregate and the original grain size, amount of strain, temperature, time, and various specific parameters such as interfacial surface energies and activation energies (Burke and Turnbull, 1952; Byrne, 1965, p. 91):

(1) A mineral will not recrystallize in the absence of solvents unless it is heated to a critical temperature, this temperature being a characteristic of the mineral.

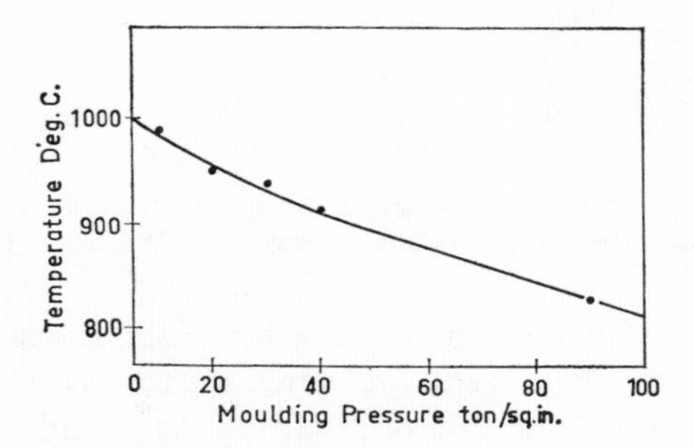

FIG. 36. Reduction of critical temperature for annealing of fluorite powder with increasing precrystallization strain, i.e. moulding pressure (Buerger and Washken, 1947, Fig. 7)

(2) The amount of recrystallization is directly proportional to the annealing time in the early stages when isolated crystals grow in the matrix, but the growth law differs once the new grains impinge on each other and the rate of recrystallization decreases markedly (Fig. 35).
(3) The temperature at which crystallization commences is less for increasing amounts of pre-crystallization strain (Fig. 36).
(4) In the early stages of recrystallization, before a steady state is reached, the grain size rises sharply with annealing temperature (Fig. 37). Continued heating causes a slow increase in grain size.
(5) The time required for a given amount of recrystallization decreases logarithmically with the annealing temperature (Fig. 38).

Recrystallization begins with the formation of nucleii which may or may not be abundant. At an early stage there may be only limited nucleation and new minerals will grow at isolated points leaving the original texture essentially unaltered. Pseudomorphs and porphyroblasts develop. As time

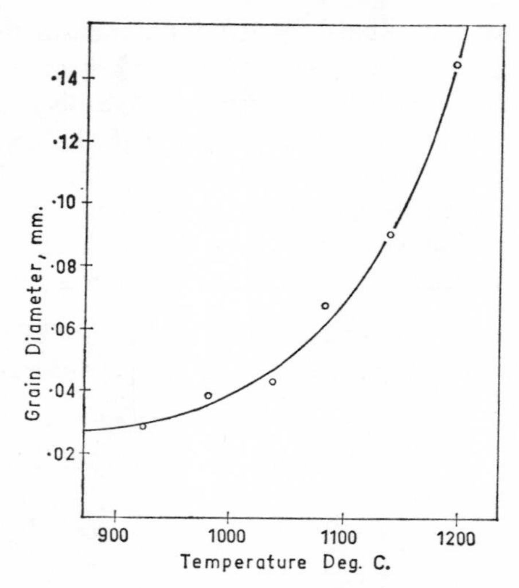

FIG. 37. Grain size as a function of annealing temperature for compressed fluorite powder for short-term experiments (Buerger and Washken, 1947, Fig. 5)

goes on, and possibly as the temperature increases, the sizes of the crystals increase until a second stage is reached when the rock is composed essentially of new crystals. At this stage the original texture is indistinct and possibly will have been obliterated but the new crystals are small. Low-energy textures begin to develop. As the time-dependent grain growth process continues, the grain size increases, there is competition between adjacent crystals

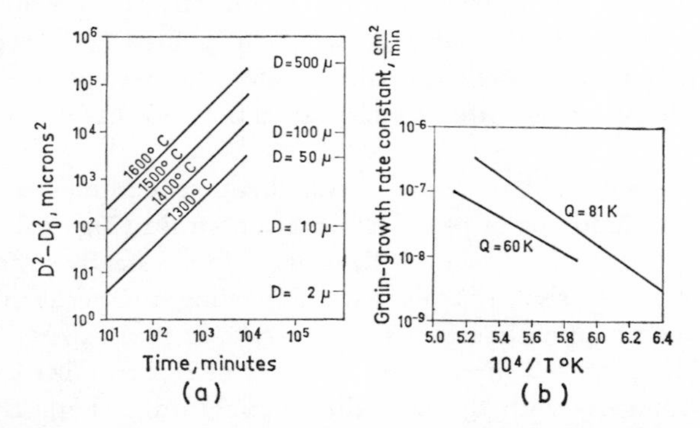

FIG. 38. (a) Isothermal growth of MgO at various temperatures shows that increase of diameter is proportional to the logarithm of the time, and that the grain size is larger for successively higher temperatures (Spriggs, Brissette and Vasilos, 1964, Fig. 2). (b) Temperature dependence of rate of grain growth in MgO. Two values of activation energy of recrystallization ($Q = 60$ and 81 kcal/mol) are given (Spriggs *et al.*, 1964, Fig. 3)

so that less stable crystals or less stable grain boundary configurations are destroyed until finally a comparatively coarse-grained aggregate is formed. The ultimate grain size will depend primarily on growth factors such as time and the diffusivity of material, and hence on temperature, the presence of solvent phases and the nature of the material itself.

The Size of Crystals

Two aspects of size must be considered: the grain size of a rock is an *absolute* feature. The mean or average grain size can be measured but is only significant if the range of grain sizes is small. If the rock consists of crystals of different sizes, the *relative* grain sizes are significant.

The average size of the crystals in a given volume of an equigranular, monomineralic, crystalline rock depends on the number of crystals present. This depends first on the number of original nucleii formed and thus on the ease of nucleation. If the energy level (i.e. temperature) is kept sufficiently high, crystals will continue to grow until they interfere with each other. The total number per given volume may then decrease as less favourable crystals disappear and more favourable ones grow.

The homogeneous recrystallization of an aggregate of crystals of a single mineral does not involve long-range transport of material. True nucleation may take place but grain growth by movement of boundaries may be easier, thus it may be reasonable to assume that the grain size depends on the growth process (mutual boundary interference) rather than on the rate of nucleation. An increase in average grainsize results from the disappearance of less stable crystals.

The grain size of metamorphic rocks of some thermal aureoles decreases approximately exponentially with distance from the contact (Fig. 39a; Edwards and Baker, 1944; Grigorev, 1965, p. 179) and it is commonly held that the grain size of a rock is an approximate index of its grade of metamorphism. The maximum temperature reached in the aureole and the grain size show a similar relation to distance from the contact (Fig. 39b; Jaeger, 1957, p. 314) and experiments show a correlation of grain size and temperature under certain conditions. The size of the crystals produced by annealing crushed fluorite for a given length of time increases with increase of annealing temperature in Fig. 37 (Buerger and Washken, 1947; Norton, 1959, p. 117). However, it is an oversimplification to assume that the grain size of a metamorphic rock depends on the temperature alone. The ultimate grain size may depend on the amount of pre-crystallization strain (Fig. 36) or on time relations (Fig. 38). Transformations are time-dependent, i.e. at any temperature the amount of transformation increases with time until a state of minimum free energy is reached (Christian, 1965, p. 11). The amount

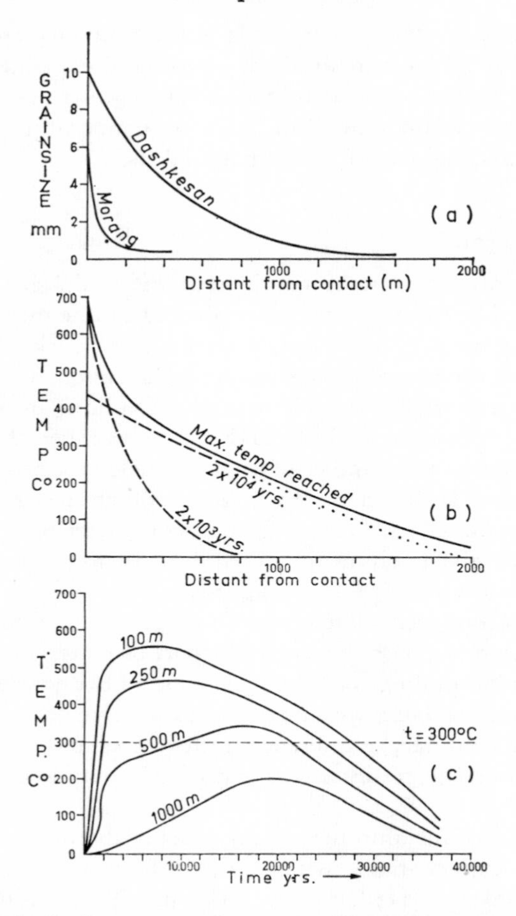

FIG. 39. (a) Grain size in contact rocks as a function of distance from the contact. Values for calcite in marbles (Dashkesan: Grigorev, 1965) and quartz in hornfels (Morang Hills: Edwards and Baker 1944). (b) Temperature distribution in aureole adjacent to an igneous sheet 1000 metres thick (after Jaeger, 1957) for 2×10^3 and 2×10^4 years and an approximate curve for the maximum temperature reached. (c) Variation of temperature with time for various distances (in metres) from the contact in (b)

of transformation does not depend on the temperature except in so far as the rate is temperature-dependent. Thus the variation in grain size in an aureole can only be attributed to the maximum temperature reached if the textures are regarded as non-equilibrium and the grain size attributed to a process which is in an advanced stage close to the contact but has not proceeded far in the outer parts. Many thermal aureoles do not show a correlation between grain size and distance from the contact and presumably approach textural equilibrium more closely than those which show this correlation.

The time-dependent (i.e. kinetic) control on the grain size is illustrated in Fig. 39c which shows the change of temperature with time in the inner and outer part of an aureole (after Jaeger, 1957). If it is accepted that there is a minimum temperature below which recrystallization does not take place or takes place exeedingly slowly (taken on the diagram as 300°C for some imaginary process), then the time available for recrystallization in the inner part is very much greater than in the outer.

The relationship between temperature, time and grain size is complex:

(1) The rate of nucleation first increases then *decreases* exponentially with temperature, i.e. fewer nucleii are produced at very high temperatures and thus the number of crystals will be small and the grain size greater for crystallization which begins and ends at the higher temperatures.
(2) Crystals increase in size with time provided a certain "critical" temperature has been exceeded.
(3) The rate of crystal growth increases with temperature once this "critical" temperature is exceeded.

It seems to be commonly supposed that the number of nucleii formed is greater at high temperatures but experimental work of Norton (1959) and others has shown that this is not necessarily so. As the temperature is increased, the free energy of formation of a critically sized nucleus increases much more rapidly than does the available thermal energy, thus the probability of nucleation decreases rapidly with increased temperature (Fig. 40a, and Christian, 1965, p. 12).

The effect of initial grain size is not clear. One might expect that a coarse aggregate of unstrained crystals of a stable phase (e.g. a quartz sandstone or quartzose conglomerate) would be reluctant to recrystallize, but that an aggregate of very small crystals with consequent large total surface (chert) would recrystallize readily. However, many meta-cherts are fine-grained and many meta-sandstones are coarse-grained. Fragments of different sizes and composition in a heterogeneous rock (greywacke) react at different rates and have a complex behaviour although the finer clastic material alters first.

The presence of foreign materials can affect the grain size considerably. Nucleation of a mineral may be enhanced by the presence of "seeds" which need not be related in composition to the mineral. Minor amounts of impurities can act as catalysts in reactions, but on the other hand, some minerals (carbon appears to be the main offender) depress grain growth, possibly by impeding grain boundary movement.

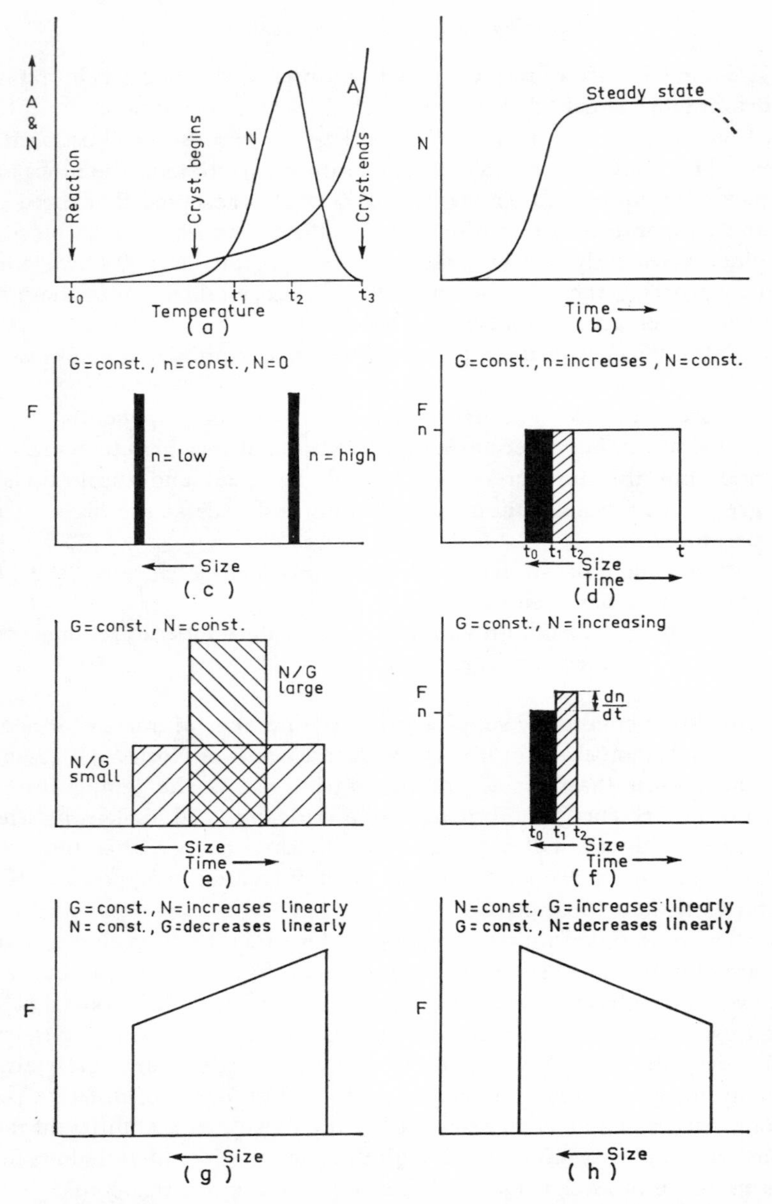

FIG. 40. (a) A solid state transformation takes place at t_0. Nucleation begins at a temperature above this and the nucleation rate $\mathcal{N}$ increases with temperature until t_2 when it decreases because of the exponential increase of the activation energy of nucleation A. (b) At constant temperature the nucleation rate $\mathcal{N}$ increases with time to become constant. (c), (d), (e), (f), (g) and (h) Grain size distribution frequency (F) for crystallization with various values of nucleation rate $\mathcal{N}$ and growth rate G (see text)

It has been suggested repeatedly that minerals have differing abilities of growth and that some minerals with a high "power of crystallization" tend to grow into large crystals whereas others only reach comparatively small sizes. However, the reason that cordierite commonly forms isolated large crystals in knotted hornfels is probably more a function of difficulty of nucleation than ease of growth. The "power" has been said to be associated also with the ability to develop crystal faces although the two characteristics (size and shape) are distinct.

The sizes of garnets from Glen Clova have been correlated with the composition (Chinner, 1960); garnets which are low in manganese are large and those with progressively more manganese are smaller. Specimens with a ratio $\left(\frac{\text{MnO}}{\text{MnO} + \text{FeO}} \times 100\right)$ of 32·7 have diameters near 0·1 mm and range up to 5 mm with a ratio of 3·1. An increase in manganese content may produce easier nucleation or slower growth.

The spacing of nucleation sites is important but not very predictable. Crystal growth involves a change from a less stable to a more stable condition (decrease in free energy). As each small unit of material is added to a nucleus, there is a decrease in free energy (strain and surface) which is liberated as heat energy, thus there will be a tendency for an increase in temperature in the vicinity of the nucleus and this might lower the probability of nucleus forming in close proximity. In addition, a "large" (old) nucleus would be more stable than a "small" (new) one, and should tend to grow at the expense of potential new nucleii.

The grain size of metamorphic rocks is controlled by the relations of two rate processes:

(1) The rate of nucleation $N = \frac{dn}{dt}$ (the number of new crystals produced per unit time).

(2) The rate of growth $G = \frac{dr}{dt}$ (the increase in size per unit time).

A high ratio of N to G means many nucleii plus slow growth and hence a small ultimate size. A small ratio means a large grain size. The ultimate size reached by a crystal, however, depends on a large number of variables and it is not possible to determine the kinetics of crystallization of a metamorphic rock from a study of grain sizes alone (Galwey and Jones, 1963; Kretz, 1966b).

The total number of nucleii present depends on the rate N and the length of time nucleation has operated. N is temperature-dependent and the effects of temperature and time on the number of nucleii (n) are similar and easily confused. Figure 40a shows the classical (Becker) view of the variation of the rate $\mathcal{N}$ with temperature for a process by which a material A (stable below t_0) is transformed to B above t_0. The free-energy driving force is zero at the transformation temperature and increases with increase of temperature until it overcomes the activation energy of nucleation (E_A) and nucleation begins. Nucleation is slow at first because of the low rate of diffusion at lower temperatures but $\mathcal{N}$ increases with temperature. The rate of change in the rate of nucleation $\mathcal{N}$ depends on the properties of the system but the classical case assumes that $\mathcal{N}$ increases to a maximum a certain temperature above that of the transformation but then decreases to zero because of the exponential increase of activation energy of nucleation with temperature. The curve for $\mathcal{N}$ is represented by:

$$\mathcal{N} = C\left(\exp - \frac{(Q + E_A)}{kT}\right)$$

where
$\mathcal{N}$ = nucleation rate,
C = constant,
Q = activation energy of diffusion,
E_A = activation energy of nucleation,
k = Boltzmann's constant,
T = temperature in degrees Kelvin.

This interpretation is not entirely satisfactory even for simple examples (Smoluchowski, 1951, p. 162) but is a useful beginning to an understanding of solid systems. It should be noted that $\mathcal{N}$ increases approximately linearly between t_1 and t_2 (Fig. 40a) and most metamorphic reactions might take place in this temperature range.

At a given temperature the nucleation rate will increase from zero at zero time to a temperature-dependent constant in the steady state condition (Fig. 40b and Christian, 1965, p. 399). However, after a certain period, growth dominates over nucleation and $\mathcal{N}$ decreases to zero. Thus the graph of $\mathcal{N}$ against either time or temperature rises from zero to a maximum then decreases to zero.

The Becker interpretation assumes that the rate of transformation is so much faster than the rate of heating that the latter can be ignored; however, diffusion and recrystallization are so slow in metamorphic systems that it is by no means certain that equilibrium is commonly achieved or that the rate of nucleation $\mathcal{N}$ is very much greater than the rate of change of $\mathcal{N}$. Time is more important in a discussion of growth than nucleation as the size of a crystal depends on the length of time that a certain growth rate has operated.

Once a stable nucleus has been formed, the crystal will grow to an ultimate size which depends on the rate of growth G, the time available for growth, and the continued supply of nutrient. Activation energies for growth are generally less than those for nucleation, thus growth is generally easier than nucleation. Growth depends on rate processes in the body of the rock (diffusion control) and at the crystal surface (interface control). The diffusion-controlled processes depend on temperature, diffusivity, time, and distance of travel of required materials. The interface-controlled processes involve addition or removal of atoms at the crystal surface and mutual rearrangement of crystal boundaries.

Some degree of understanding of the significance of the sizes of crystals in rocks can be obtained by considering the grain size distributions predicted by simple models with various values of N and G. The following discussion is general but is most applicable to separate crystals such as porphyroblasts. Grain growth and recrystallization in mutually impinging crystals such as quartz in a quartzite are a matter of grain boundary movement rather than of nucleation followed by growth.

(1) The most simple model (Fig. 40c) considers an initial fixed number of nucleii with a constant growth rate (n = constant, G = constant, nucleation rate $N = 0$). All crystals begin (and end) crystallizing together, thus all are of the same size; they will be relatively large if the number of crystals (n) is low, or small if n is high.

This model fits any system in which the new units are of the same size, but the implication of simultaneous nucleation of all units depends on the assumption that their subsequent growth was equal and that growth ceased in all units simultaneously. The model might fit some exsolution perthites and approximates to the simple grain growth (without nucleation) of a monomineralic stable aggregate, e.g. quartzite or marble.

A group of porphyroblasts of the same size may, however, result from other circumstances than all crystals being nucleated simultaneously and growing at the same rate. Some crystals may be nucleated earlier but grow more slowly and others may be nucleated later but grow more quickly due to some inhomogeneity in the system. In this case the rate of growth is a function of the time of nucleation. A more likely reason is that there is an upper limit to the possible crystal size and each crystal ultimately reaches this size (governed perhaps by the supply of nutrient material) no matter what its time of nucleation or rate of growth. However, Kretz (1966, p. 158) measured the distribution of crystals of various sizes with respect to the distance to their nearest neighbours but could find no correlation. This implies that the size of a porphyroblast is independent of its position within the rock and of the local supply of nutrient.

(2) A slightly more complex model assumes a constant growth rate G and an increase in the number of crystals, n, at a constant rate N. Crystallization begins at t_0 and dn crystals are produced in the first instant from t_0 to t_1; these will ultimately grow to a radius r_0 (Fig. 40d). If growth takes place

from t_0 to t_1 at a constant rate G, the final size range of these crystals is from Gt_0 to Gt_1. In the next (equal) instant from t_1 to t_2, a further *dn* crystals are nucleated; these will grow ultimately to a group ranging in size from Gt_1 to Gt_2 and thus to a smaller size than those nucleated between t_0 and t_1. Thus a histogram of size distribution consists of blocks of equal height (*dn*) and equal width ranging from r_0 to r, where r is the size of the smallest and latest crystals. The rectangular histogram will be tall if N/G is large and short if N/G is small (Fig. 40e).

This model approximates to recrystallization (with nucleation) of a monomineralic aggregate, e.g. quartzite, marble or biotite-hornfels.

(3) If G is constant, and N increases at a constant rate (Fig. 40f), *dn* crystals are formed in the first instant (t_0 to t_1), but more crystals are produced in the next instant because N increases. The first crystals formed will ultimately be the largest and they will be followed by increasing numbers of smaller crystals as in Fig. 40g and the variation diagram will be a straight line with positive slope. The line will be variously curved if N increases non-linearly. If N decreases while G is constant, then the line will slope downwards.

(4) Changes in G and N affect the slope of the frequency curve or histogram in an opposite manner. The explanation of (3) above where G is constant and N increases is contrasted with that where N is constant and G increases (Fig. 40g). In the first unit of time t_0 to t_1, a number of crystals *dn* is formed, these range in size from r_0 to r_1; thus the first block in the histogram has a height *dn* and a width $r_1 - r_0$. In the second equal unit of time (t_1 to t_2), the same number (*dn*) is formed because N is constant, and these will range from r_1 to r_2 in size. However, as the rate of growth G has increased, the range r_1 to r_2 will be greater than r_0 to r_1. A fundamental requirement of a histogram (or of a frequency distribution curve derived from a histogram) is that the width of the blocks (*dr*) must remain constant, thus a size range equal to that in the first unit of time is achieved in a shorter time during the second unit because of the greater growth rate. With a constant nucleation rate, a smaller number of nucleii will thus be formed in the second unit. The second histogram block will be the same width as the first but will be shorter. The frequency curve will thus slope downwards as G increases with N constant (Fig. 40h).

The simplest explanation of the most simple size distribution (Fig. 41a) is that the growth rate was constant and the distribution is due entirely to variations in nucleation rate (Fig. 41b, c) i.e. nucleation increases to a maximum then decreases (Jones and Galwey, 1964; Kretz, 1966b).

It must be emphasized that the grain size distribution contains only one variable (diameter) and it is not possible to make absolute predictions about two (or more) independent processes leading to this distribution. It is

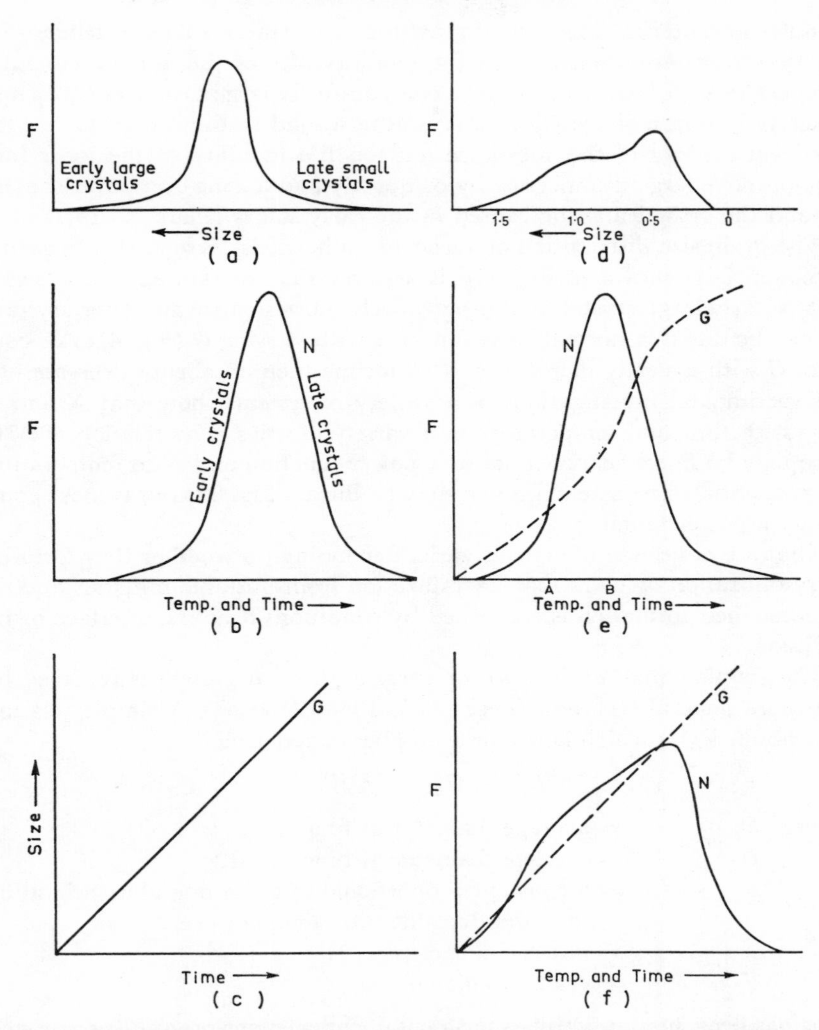

FIG. 41. (a) Grain size variation curve (frequency against size) for normal nucleation (b), and constant growth (c). (d) Size distribution of garnets in a hornfels (Kretz, 1966b) can be explained by (e) a normal variation in nucleation rate plus varying growth rate or (f) by irregular nucleation combined with a steady growth rate

possible to postulate for each diagram an almost infinite combination of changes in nucleation rate and growth rate, in turn dependent on diffusion rate, supply of reactants, change in H_2O content, catalytic or depressive effects of impurities, etc.

The growth of any crystal depends on the rate of supply and incorporation of nutrient material. The rate of growth of a crystal in solution depends on the degree of saturation and on the mobility of ions (i.e. on the viscosity, temperature, etc.) which can move comparatively large distances easily and quickly. The rate of growth of a crystal in a solid medium depends on the chemical activity of the substance and on the mobility of the ions. Ions cannot move large distances easily or quickly and a zone ofimpoverishment around the crystal may be caused in the early stages (Plate XVIb).

The grain size distribution of garnets in a hornfels (Kretz, 1966b, garnet specimen 2–1) shown in Fig. 41d is asymmetrical or skewed and shows a lack of early large crystals and progressively more smaller and later crystals. It may be due to a normal variation in N with varying G (Fig. 41e) or constant G with a steady increase in N with time then an abrupt decrease.

Experimental investigations of a variety of systems show that N and G vary with time and temperature in a variety of ways. The relation of N to time may be linear, exponential or a power function and N to temperature is exponential. The rate of growth may be linear (Fig. 35) but is more commonly a power function.

The rate of growth of crystals varies depending on whether they form by recrystallization or by *de novo* crystallization from unstable minerals and on whether their formation is controlled by conditions near the interface or by diffusion.

The simple equation $D = Kt$ for early isothermal growth is replaced by the more general equation (Beck *et al.*, 1948) $D = Kt^n$. A simple relation (Turnbull, 1951) which holds for a number of systems is:

$$D_t^2 - D_0^2 = K\gamma Vt$$

where D_t = average diameter at time t,
D_0 = average diameter at time $t = 0$,
K = a constant proportional to grain boundary migration rate and dependent on temperature,
γ = interfacial energy,
V = gram atomic volume.

This has been supported by experimental evidence of Spriggs, Brissette and Vasilos (1964) shown in Fig. 38a.

The reaction rate of a homogeneous isothermal reaction can be taken as a function of the various reactants, and the change in concentration (C) of a component with time (t) is

$$\frac{dC}{dt} = KC^n$$

The index n is zero for a reaction with *zero-order kinetics* where the rate is independent of concentration, e.g. the phase change calcite $\rightleftharpoons$ aragonite

(Brown, Fyfe and Turner, 1962). $n = 1$ for a reaction with *first-order kinetics* in which the rate is proportional to the amount of unreacted material, e.g. common metamorphic dehydration such as amphibole to pyroxene. $n = 2$ for a reaction with *second-order kinetics* where reactants must diffuse through a layer of product.

The general equation for the amount of material transformed isothermally (Christian, 1965, p. 488) is:

$$\delta = 1 - \exp(-kt^n)$$

where δ = amount transformed,
k = Boltzmann's constant,
t = time.

Values of n range from $\frac{1}{2}$ for precipitation on dislocations to 1 for growth of needles and plates, $1\frac{1}{2}$ to $2\frac{1}{2}$ for most crystal shapes growing under a diffusion control, 2 for grain boundary nucleation after saturation, to 4 for a constant or increasing nucleation rate in processes not involving long-range diffusion.

A simple grain size distribution such as that in Fig. 41a could be explained as being simply due to a normal variation of N (Fig. 41b) coupled with a constant growth rate (41c) under approximately isothermal conditions. Asymmetry or skewness could be explained by variations either in growth rate or in nucleation. It is not possible to distinguish between these.

It is unlikely that most metamorphic crystallization is even approximately isothermal and it probably takes place under rising then falling temperatures. Nucleation might take place in the region between t_1 and t_2 on Fig. 40a as the temperature rises and then between t_2 and t_1 as it falls. This would give a symmetrical curve like the normal one in Fig. 41a for N with time. It is not possible to distinguish between the effects of time and temperature on a grain size distribution curve and the two are combined in Fig. 41.

The problems involved in explaining the various sizes of crystals in rocks are illustrated by reference to two examples of "graded bedding" in metasediments. A regular gradation in grain size has been described in granoblastic polygonal scapolite-diopside rocks from the regional metamorphic province near Cloncurry, Queensland (Edwards and Baker, 1954, p. 8). The metamorphic minerals are larger at the bottom of each bed and smaller near the top, followed by a sharp reversion to a coarse grain at the base of the next layer. The orderly variation is repeated through a considerable thickness of beds and was interpreted as a reflection of graded bedding in the original sediment. The crystals showing the range of grain size are entirely metamorphic and the rock is thought to be originally a calcareous shale and as this would not normally contain graded bedding, any reflection of original sedimentary differences must be a subtle one. The lack of detailed data such as the range of sizes and relative proportions of the various minerals prevents reaching an explanation. It may be that the grain size reflects an original compositional difference and is due to scapolite (which forms the largest crystals) being more abundant towards the base. It is possible that some minor impurity was differentially dispersed throughout the original sediment and could

(1) inhibit nucleation and cause large crystals,
(2) inhibit grain boundary movement and cause smaller crystals,
(3) affect the rate of ionic diffusion, etc.

It has been suggested (Read, 1936; McCall, 1954, pp. 162; Read and Watson, 1962, p. 502) that the size of andalusite porphyroblasts in some Dalradian schists in Banff and Donegal reflect original compositional differences in that the andalusites are largest in the fine-grained, clay-rich tops of graded units and smaller in the more siliceous bases. It is not clear, however, whether the higher alumina results in a larger *amount* of andalusite, or in larger *sizes*, or whether it affects the nucleation process and produces a larger *number* of crystals.

The grain size of a monomineralic metamorphic rock depends on the relative roles of crystallization and deformation, and on time, temperature and the original grain size. Cataclasis tends to reduce the grain size but this will be discussed later and at present only the effects of crystallization are considered. Rast (1958) suggested that an increase in the rate of deformation at high temperature may result in enhanced growth and an increase in grain size, but work on ice (Macgregor, 1950) and on metals implies that an increase in the rate of deformation during recrystallization results in a smaller grain size.

Although recrystallization generally involves an increase in grain size there appears to be a preliminary reduction even in the thermal metamorphism of unstrained crystal aggregates. Thomas and Campbell Smith (1932) produced evidence that xenoliths of medium-grained gabbro with flakes of biotite of 8–10 mm across were reduced to hornfels with an average grain size of 0·25 mm, and Joplin (1935) observed that many basic xenoliths in granites are very fine-grained.

Polygonization of an aggregate of strained crystals or mechanical granulation are the most common processes leading to a reduction in grain size but neither seems to be applicable to the thermal metamorphism of basic xenoliths. The process is probably due to the replacement of an original coarse aggregate by many small new metamorphic crystals so that the fine grain is due to easy nucleation and a limited amount of growth.

Porphyroblasts

A metamorphic rock which contains large crystals in a matrix of smaller ones (the ratios of diameters should be about 5 or 10 to 1) is said to be *porphyroblastic* by analogy with the igneous texture *porphyritic* (Fig. 53). The large crystals are called *porphyroblasts* (cf. phenocrysts). The term *blasto-porphyritic* means a relict texture after an original porphyritic texture in which the range in sizes is a reflection of a previous igneous process not a metamorphic one. The relict textures of poorly sorted sediments may give rise to large crystals in a fine-grained matrix but again, as the large size is a relict feature, such crystals are not properly called porphyroblasts.

PORPHYROBLASTS IN HOMOGENEOUS ROCKS

Metamorphism of a rock of uniform composition and grain size ultimately tends towards an even-grained equilibrium texture, but crystals of contrasted sizes may be produced in the very early or the very late stages. In dynamic metamorphism, reduction in grain size may leave large relict crystals (which are not *porphyroblasts* but *porphyroclasts*) in a crushed matrix. In the early stages of diagenesis or metamorphism of a very fine-grained sediment such as chert, limestone or clay, certain crystals may begin to grow before others resulting in a patchy or clotted texture with large (new) crystals surrounded by small (old) crystals. This *grumous* texture (Bonet, 1952; Pettijohn, 1957, p. 412) is an interrupted, non-equilibrium texture.

Normal or primary recrystallization during the annealing of metals generally results in an even-grained or granular texture. There is competition between grains during the process as some grains grow at the expense of others but the final equilibrium texture of primary recrystallization is even-grained (granular-polygonal). This may be followed by *secondary* recrystallization or *abnormal grain growth* when certain crystals suddenly grow very rapidly and engulf a number of the adjacent crystals to become many times larger than the average grain size. Excellent illustrations of this process in ceramics (alumina, spinels) are given by Burke (1959). This process is not clearly understood but it seems likely that certain crystals which have special orientation relations with adjacent crystals (i.e. have very mobile boundaries) expand to absorb these surrounding crystals. Having reached a larger size, the crystals now have a lower surface energy per unit volume than the smaller ones and are thus more stable and can continue to eat up the smaller crystals in a manner somewhat similar to that by which a large soap bubble connected to a small bubble will expand. The lack of porphyroblasts in monomineralic metamorphic rocks implies that this process is not important in nature although it might apply to some large crystals in calcite or salt aggregates.

PORPHYROBLASTS IN HETEROGENEOUS ROCKS

Some metamorphic minerals such as garnets, feldspars, aluminium silicates and cordierite have a tendency to form large crystals when surrounded by other minerals such as quartz, calcite, micas, etc. (the "matrix-formers" of Rast, 1965, p. 82). Porphyroblasts of one mineral in a foreign matrix may be considered as being due to the growth of separate nucleii. If the composition of the porphyroblast is not markedly different from that of the matrix, then growth will involve only short-range diffusion of reactants and will be interface-controlled. If the porphyroblast differs appreciably in composition from the matrix then growth requires movement of constituents over appreciable distances and growth is diffusion-controlled. The size of porphyroblasts thus depends on the rate and duration of nucleation (N) and growth (G) as discussed on p. 128.

If the porphyroblasts use up all the necessary reactants in the matrix they will cease growing. If all porphyroblasts are of the same size and all the material available for the porphyroblast species has been used up on a certain number (n) of crystals, then the size of each crystal is given by

$$V = \tfrac{4}{3}\pi r^3 n \text{ or } r = \sqrt[3]{\frac{3V}{4n\pi}}$$

where r = radius of each crystal,
V = total volume of the species,
n = number of crystals present.

V is a function of composition and is a constant for a given rock and thus the size of each crystal depends on the number of crystals.

The number of crystals present is equal to the total number of nucleii less those which have disappeared. If it is assumed, first, that no crystals have disappeared during the growth stage then the size is a function of bulk composition and ease of nucleation. The ability to form stable nucleii will depend on:

1. The mineral itself:
   (a) The higher the surface energy of the mineral the more difficult is its nucleation.
   (b) The more complex the structure, the more difficult the nucleation (rate of growth also probably depends on complexity of crystal structure).

2. The presence of foreign material acting as seeds or as a substrate for nucleation, e.g. quartz is nucleated by hematite, sphene grows around ilmenite, etc.

3. The presence of foreign impurities acting as depressants and hindering grain boundary movement, e.g. graphite in schists.

If the porphyroblasts cease growing because of depletion of constituents, then their absolute sizes depend on the amount of constituents available and their relative sizes on the amount available to each nucleus. The porphyroblasts will be of equal size if the matrix is homogeneous and the nucleii are equally spaced. Variations will result from chemical inhomogeneity of the matrix or from depressed or enhanced nucleation in certain places. Incomplete diffusion of reactants to the nucleus may result in the porphyroblast being surrounded by a halo (Plate XVIIIb).

It is possible that the number of porphyroblasts is less than the original number of nucleii but any loss of nucleii must have happened at a very early stage if all the porphyroblasts are of the same size. It is unlikely that the number of porphyroblasts should be reduced with an increase in size of some and the consequent disappearance of others. The slight difference in surface energy between widely separated crystals would certainly not be sufficient driving force to cause this.

There is an important relationship between the chemical composition, the size and the number of the crystals. A sediment with layers alternately richer and poorer in clay may be metamorphosed to a banded rock in which the more pelitic layers are richer in mica. The total amount of mica in these layers is greater because of the initial composition; the rate of diffusion is slow and the chemical gradient will not be flattened out. A given quantity of mica may crystallize as a small number of large crystals or a large number of small crystals and Rast (1965, p. 80) suggested that mica was preferentially nucleated in layers richer in iron and magnesia because the high concentration favoured nucleation and thus many small crystals. The large number of small crystals of mica in lowgrade rocks appears to be due to easy nucleation. Subsequent elevation of the temperature may increase the size of the crystals by destroying some, but the temperature relation could be interpreted as indicating the appearance of a smaller number of nucleii at the higher temperature. It is not yet clear that a high concentration of constituents in a certain layer results in *more* crystals rather than *larger* crystals.

The amount of garnet present as porphyroblasts in a given specimen of pelitic hornfels depends on the amount of ferrous iron available in the rock as would be expected from a consideration of the *equilibrium chemistry* (Jones and Galwey, 1964). However there is some tendency for a complex relation between the size (the numbers of garnets per unit volume) and the bulk chemical composition (the relative proportions of lime, magnesia and alumina to alkalies). In simple terms, an increase in the relative proportions of oxides required by garnet causes an increase in the number of garnet nucleii. This implies that the bulk composition of a rock may control the *kinetics* of nucleation as well as the thermodynamics of metamorphism processes.

In an attempt to clarify the processes involved in the nucleation of garnet in calc-silicate hornfels, Galwey and Jones (1963) and Jones and Galwey (1966) made the assumption (1963, p. 5684) that "the rate of growth of a crystal is determined by temperature and concentration of reactants, but is independent of the size of the crystal. It thus follows, for crystals growing in an environment where diffusion of reactants is fast compared with the rate of consumption of reactant, that the final size of the crystal is a measure of the time in which it grew". Thus the large crystals were nucleated before the smaller ones and as they found relatively more smaller crystals than larger, then nucleation was deduced to be slower in the early stages than later. They considered that if the amount of reactant is considered to be constant, then accelerating nucleation is due to *increase* of temperature during the nucleation process. This is not in agreement with conclusions given earlier.

Nucleation and growth involve a large number of variables: chemical activity of different components, temperature, possible catalytic effects by water, etc., rates of nucleation, of diffusion and of growth, etc. There are so many debatable assumptions involved in the model of Galwey and Jones that their conclusions are open to question. Nevertheless, this kind of textural analysis is producing some useful information.

CRYSTAL SHAPE

The shapes of individual crystals are due to the interaction of five main processes: free growth, impingement, corrosion, mechanical disruption, and plastic flow. The first two processes control crystallization fabrics in metamorphic rocks and such crystals may be classified genetically as:

I. Free growth forms (e.g. single crystals grown in liquid or matrix)
   (1) *Xenoblastic;* irregular, rounded, or spherical shapes due to growth of a crystal with negligible growth anisotropy or rapid growth of an anisotropic crystal. The segregation of liquid globules in immiscible glasses is analogous.
   (2) Regular (*idioblastic*) crystals of relatively anisotropic materials under ideal conditions of growth (slow, unimpeded).
   (3) Regular branching (*dendritic*) growths from a single nucleus.
   (4) Amoeboid or skeletal growths.

(5) Radial linear growths from a single or complex nucleus (split crystals, spherulites).

II. Mutual growth forms (growth in a polycrystalline aggregate with mutual interference). Mutual boundary textures.

(1) Single-phase aggregates.

(a) Individual columnar crystals (fibre or columnar impingement textures) due to limited nucleation (generally at a surface) with mutual impingement to produce columnar or acicular crystals, fascicular bundles and spherulites. The individual members have a form due to mutual growth but the group may have a free growth form.

(b) Individuals in a texture controlled by mutual impingement and interfacial energies, e.g. granoblastic-polygonal and decussate textures.

(2) Two-phase aggregates

(a) Individuals in polygonal texture whose shapes are controlled by impingement, relative growth and interfacial energies.

(b) Individuals in duplex (two mineral) intergrowths:

(i) Oriented intergrowths due to exsolution, epitaxial growth or partial replacement; the form is controlled by interfacial relations.

(ii) Non-oriented intergrowths where the forms are controlled by growth and nucleation factors rather than interfacial relations, e.g. eutectoids, symplectites.

(iii) Individual elements in duplex crystals formed by partial twinning or other structural transformations.

(c) Skeletal, reticulate or fish net crystals of a second phase along grain boundaries. The form is favoured by grain boundary diffusion (or an intergranular fluid) coupled with difficulty of nucleation.

The rather confusing nomenclature of crystal shapes in metamorphic as well as sedimentary and igneous rocks has been recently reviewed by Friedman (1965). A crystal is *idioblastic* (or is an idioblast) if it is bounded entirely, or almost entirely, by crystal faces formed by growth *in situ*. A crystal is *xenoblastic* (or is a xenoblast) if it has no crystal faces. There is no commonly used term for crystals with a few crystal faces but *sub-idioblastic* is used.

The most general practice at present is to apply the terms *euhedral*, *subhedral* and *anhedral* to igneous crystals with many, a few, or no crystal faces respectively. Relict, pre-metamorphic *euhdral* crystals are not *idioblastic*. Those constituents whose shapes have not been changed by metamorphism have been called *allothimorphs*. The various textural terms have come to have a genetic association and most metamorphic petrologists restrict euhedral, subhedral and anhedral to igneous crystals and use idioblastic, sub-idioblastic and xenoblastic when describing metamorphic crystals. Similarly porphyroblastic and poikiloblastic are metamorphic terms analogous to porphyritic and poikilitic in igneous terminology. As with any genetic terms, difficulty is experienced in selection of the best term when it is not known whether the crystal is igneous or metamorphic.

Free Growth Forms

The shape of a crystal growing in matrix or fluid depends on certain specific properties of the crystal species and on the growth process. Crystals may be bounded by few crystal faces, by a complex array of faces, by simply curved interfaces or complex irrational boundaries.

The significance of crystal faces in solids may be seen by first comparing them with liquids. An isolated drop of liquid, unaffected by the distorting effects of gravity (e.g. a globule of oil suspended within an immiscible liquid of the same density) has a spherical shape because the surface tension reduces the bounding surface to the least area. In this form, the surface energy is least because the surface area is least. By contrast, however, a sphere cut from a single crystal does not have the least surface energy for that volume because of the anisotropy of the atomic structure. The least energy shape is controlled by the internal structure as well as the intrinsic surface energy (Wells, 1946). If a crystal is ground into a sphere and placed in a supersaturated solution, it grows by the addition of crystal faces indicating that these are lower energy surfaces than the lower area spherical surface.

A number of empirical relationships for crystal faces (the law of Hauy, Wulff's theorem, Barker's principle of simplest indices, Goldschmidt's law of complication, and the "rules" or "laws" of Bravais, Federov, Donnay and Harker) have been discussed by Buerger (1947) and Christian (1965) who showed that the following relations are most significant.

(1) If the surface energies at all possible faces were equal, large numbers of faces would appear in equilibrium and the shape would be approximately spherical. If, however, a particular face has a much lower surface energy than any other, only faces of this form would appear in the equilibrium shape even though the surface area per unit volume is greater than that for a sphere.
(2) Crystal faces with the highest density of atoms (or the least interplanar spacing) tend to have the lowest surface energy.
(3) In a crystal in equilibrium (with respect to shape) the distance of any

face from the origin (centre) is proportional to the surface free energy of that face.

(4) Faces with higher energy advance more quickly than those with lower energy and thus tend to be removed from the crystal.

(5) A crystal face has a lower energy than a random surface and the crystal approaches thermodynamic equilibrium with a lowest total surface energy by developing lowest surface energy crystal faces.

Kretz (1966a) showed that (3) above could not be applied to metamorphic minerals. Phlogopite (ibid., p. 84) and pyroxene (ibid., p. 86) in hornfels showed such a variation in shape that $\frac{\gamma(hk0)}{\gamma(001)}$ could not be estimated. However, he pointed out that each species of crystal in an equilibrium texture should approach a constant ratio of length to width.

The development of a crystal is a competition between the extension laterally and growth radially of the various faces and if a face is considered to extend outwards by addition of layers, initiation of a new layer is easiest on a face with lowest density of atoms, i.e. this is the fastest growing face (Buerger, 1947). The relationships are shown in Fig. 42 which represents

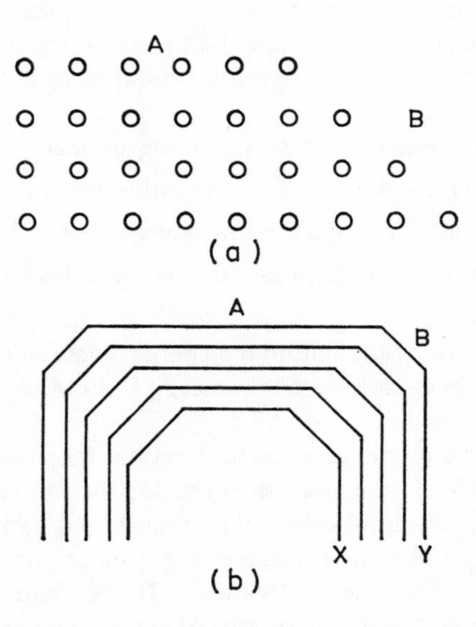

FIG. 42. (a) Structure of a hypothetical crystal with faces *A* and *B*. Atomic packing is closer on *A* which thus has lower energy. (b) Growth of a crystal from *X* to *Y* shows that the lower energy face *A* grows more slowly than *B*. *B* grows more quickly but decreases in area relative to *A*

two faces (A and B) of a growing crystal. The atoms on face A are more closely packed than on B, thus there is more mutual coordination of unsatisfied bonds at the surface and the free energy of A is less than that of B. As the faces grow by adding new units in layers, the rate of permanent attachment to B will be higher than that to A thus B will grow more quickly. For geometrical reasons, the more slowly advancing face A will increase its area with respect to B. Thus even though the total surface area and hence the surface energy of the crystal increases as the crystal grows, the increase is lessened by favouring the lowest-energy faces. In terms of kinetics, the slowest growing faces predominate and in terms of thermodynamics, the lowest-energy faces predominate.

A flat surface has a lower energy than an irregular one because it has less disturbed bonds, but alternative atomic rearrangements are possible in some structures. Thus a low-angle facet or *vicinal face* (face with large Miller indices almost parallel to a face with low indices) may form if the bonding and surface energy is similar to the flat face (Nicholas, 1963). This is the probable explanation for curved crystal faces on dolomite and siderite. The topography of crystal faces has been studied in considerable detail by means of interference microscopy by Tolansky (1960) who determined the form and size of vicinal faces, growth "steps" and "islands".

Crystal faces will tend to form if the surface energies along these rational lattice directions are substantially less than those along irrational directions. A metamorphic crystal is able to develop crystal faces only if :

(1) It can advance against the surrounding matrix.
(2) It grows under conditions of considerable ionic mobility.
(3) There is sufficient time for the low-energy face to be maintained.
(4) Its grain boundary energy is sufficiently different from that of the matrix grains.
(5) Its anisotropy of grain boundary energy is sufficiently high to allow certain surfaces to have a lower energy than others.

The surface tension model of a straight crystal face has been shown to be unsatisfactory (p. 46). The surface *energy* model of the crystal face explains its stability in terms of immobility rather than of strength. The crystal face, being rational, has a very low surface-energy; its orientation, relative to the lattice, places it in a low-energy "valley". If conditions tend to rotate it, by even a few degrees, from this orientation with respect to the lattice, the surface energy would rise very markedly and a large amount of work would have to be done. Thus low-energy crystal faces, twin planes, etc., tend to be immobile and straight.

The role of surface energy in forming crystal faces is visualized as follows: A crystal being deposited from a solution or melt has a surface energy whose magnitude depends on there being a difference between the liquid and the solid. If the melt is well ordered and has a structure and degree of packing similar to that of the solid (e.g. copper or zinc) then the interfacial energies are low. There is little energy difference between surfaces of various orientations thus there is no great tendency for crystal faces to form. Where the solid and liquid differ greatly in structure (a difference in packing will be shown as a difference in density) then the solid–liquid surface energy will be high. A crystal does not develop faces because it has a greater surface energy than the surrounding medium; it may be greater or less. The important thing seems to be that there should be a considerable difference, i.e. a potentially high energy interface so that external surfaces with least energy (rational) allow a substantial saving of energy. A crystal tends not to develop crystal faces when surrounded by a aggregate of the same species with no fluid present although some faces are developed in decussate aggregates of biotite, wollastonite, amphibole, etc., and the unusual idiotopic texture of some dolomite and gypsum aggregates may be an exception to this rule.

The formation of an ideal crystal from a liquid (solution or melt) is difficult because of the necessity of nucleating each surface layer in turn. Nucleation is avoided and growth is easier in real crystals because of the presence of imperfections, steps, ramps and spiral dislocations on the surface (Fig. 43). Growth of synthetic garnets from melts (Lefever, Chase and Torphy, 1961; Lefever and Chase, 1962) illustrates a number of characteristic features. High supersaturations or undercooling of the melt induces dendritic growth and new layers are nucleated particularly at corners and edges to give "Hopper" crystals because the supersaturation is higher there than in the centres of faces. Reduction of supersaturation results in symmetrical or layered growth parallel to crystal faces. Precipitation takes place at growth spirals; one spiral may occupy a face at low supersaturations but higher supersaturations result in the development of a number of spirals.

Simple growth spirals appear to be very rare on crystals grown in a solid medium presumably because of interference by neighbouring crystals. Dendritic crystallization is rare because of the necessity for high supersaturations and an imperfect layer growth probably takes place independently at a large number of sites across the advancing face.

The presence of various kinds of defects (mosaic structure, lineage, dislocations, vacancies, etc.) may change the surface energy of a crystal face. The presence of an emergent screw dislocation may cause a low-energy face (a slowly growing face) to grow much more quickly. "Books" of mica in some

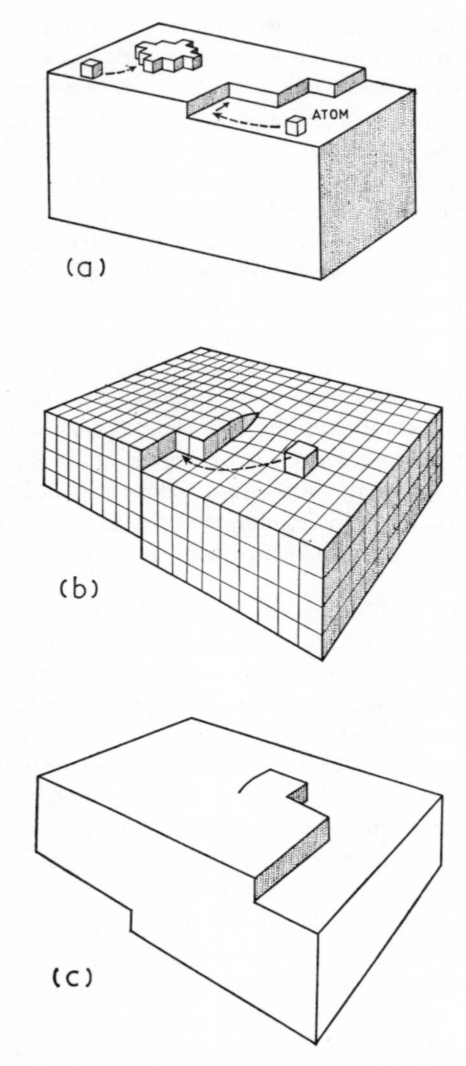

Fig. 43. (a) Layer growth of a crystal involves addition of atoms to a step or an island. Growth is controlled by the difficulty of beginning new layers. (b) and (c) Growth of a crystal with an emergent spiral dislocation is facilitated because new units are continuously added without nucleating new layers (after Read, 1952)

pegmatites are somewhat elongate along [001] instead of (001) possibly because of a spiral dislocation aiding growth. The effect of surface defects on shape may be so strong as to obscure the energy–shape relation.

Herring (1951) has used the Wulff construction to relate the lowest energy equilibrium shape for free growth crystals to the anisotropy of surface energy. The limiting cases are the sharply edged polyhedron for high anisotropy and the sphere for zero anisotropy but between are various shapes bounded by combinations of flat and curved surfaces, sharp and rounded corners. The shapes of individual crystals in matrix or even of inclusions within a host crystal can be explained in this way by considering grain boundary energies.

The relation of structure to shape is illustrated by the crystallization of mica. The layer structure gives a very great surface energy anisotropy and the (001) faces tend to be large, regular and stable. The sheet structure consists of separate layers of silica-tetrahedra arranged in rings, of Al octahedra, and cations, etc., parallel to (001). A mica nucleus is bounded by low-energy (001) faces in which the ions are close together and by higher-energy faces approximating to (*hk*0). Growth by attachment of appropriate units is easier around the edge of the (001) face both because a variety of units can be added and because the weaker mutual bonding extends a "force field" to attract ions. Thus most mica grows into sheets as the low-energy, slowly growing (001) faces dominate over the higher-energy, more quickly growing (*hk*0) faces. Where a crystal is bounded by non-perpendicular faces with different rates of growth, the more-quickly growing face will be "grown out". In the case of muscovite, (001) and (*hk*0) are perpendicular and both persist, but (001) predominates.

The particular crystal faces which develop on a given species depend in the first instance on relative surface energies and thus on the structure, but they are also influenced by the physical and chemical environment. The Donnay-Harker "law" predicts from the crystal structure that the forms {211} and {110} should dominate in garnet and this has been found to be so (Pabst, 1943). Chinner (1960) showed a correlation between the composition and habit of garnets from Glen Clova. Garnets which are low in manganese are xenoblastic with abundant inclusions and those with progressively more manganese become better formed. The correlation of shape with manganese content may be due to manganese increasing the surface energy anisotropy or reducing the rate of growth.

Minerals display differing degrees of a tendency towards crystal shape and a given mineral may possess crystal faces when in contact with one mineral but not another; e.g. calcite may be idioblastic towards quartz but when calcite and garnet are in contact it is the garnet which is idioblastic towards the calcite.

Burke (1959) illustrated the not uncommon occurrence in ceramics of a single-phase aggregate containing large idioblastic porphyroblasts of spinel in a finer-grained polygonal matrix of spinel. The common association of a large size and development of crystal faces has been noted also in metamorphic rocks. The explanation of this texture is that the low-energy polygonal texture is changing to a coarser, even lower energy texture, probably by abnormal or secondary recrystallization.

TABLE 6

Decreasing importance of crystal faces	Rutile, pleonaste, sphene, pyrite; garnet, sillimanite, tourmaline, staurolite; epidote, zoisite, wollastonite, arsenopyrite, magnetite, ilmenite; andalusite, pyroxene, amphibole, forsterite, pyrite; muscovite, biotite, chlorite, dolomite, kyanite; calcite, vesuvianite, scapolite; plagioclase, quartz, cordierite, sphalerite, pyrrhotite; orthoclase, microcline, chalcopyrite, galena.

TABLE 7

	Estimated surface energy γ ergs/cm^2	Surface energy anisotropy (very approximate only) ϕ ergs/cm^2
Diamond	6000	1500
Spinel	1800	
Muscovite	1600	1000
Pyrite	1500	
Periclase	1500	1000
Rutile	1400	
Apatite	1100	300
Hematite	1000	
Garnet	1000	300
Orthopyroxene	750	250
Clinopyroxene	700	200
Hornblende	650	200
Fluorite	600	400
Quartz	500	200
Plagioclase	500	300
Forsterite	400	
Pyrrhotite	400	
Gypsum	350	
Sphalerite	350	
Aragonite	350	200
Calcite	300	200
Chalcopyrite	250	
Ilmenite	250	
Galena	200	100
Orthoclase	200	200
Halite	200	200
Graphite	100	50

A considerable degree of ionic mobility is required to form a face on a crystal in a solid matrix and then to preserve it as it advances through a solid matrix. This implies very slow growth. If energy gradients are sufficiently great to cause rapid growth then the slight energy difference between a rational and an irrational surface would be insignificant and the crystal would not be bounded by the lowest-energy surface possible. Metamorphic minerals can be placed in a "crystalloblastic" or "form" series depending on the tendency for the mineral to form idioblasts.

The unreliable measurements given in Tables 1, 2 and 3 suggest the following very approximate values for average surface energy (γ) and surface energy anisotropy (ϕ) in ergs/cm^2 (Table 7). A high absolute surface energy and a high anisotropy favour the formation of crystal faces.

Although there are exceptions, there is a distinct correlation between high packing index, low bonding index, high hardness, high surface energy, large size and development of crystal faces.

Force of Crystallization

The relations between a growing crystal and its solid environment are not clearly understood and Becker and Day (1905) advanced the view that a crystal can exert a pressure on its surroundings when increasing its size. Illustrations of this "force of crystallization" include frost heave, "boils" on dry lake floors due to growth of gypsum crystals in mud, splitting of slate laminae by pyrite crystals, lifting of paint off a brick wall by frost, and crumbling of limestone or sandstone by gypsum crystals. Experimental investigation has been made by Becker and Day (1905), Correns (1926), Shubnikov (1934) and Schuiling and Wensink (1962).

Some confusion exists as to the meaning of the terms "force" and "power" of crystallization (Turner, 1948, p. 154; Buckley, 1951, p. 468; Rast, 1965, p. 90). Force of crystallization is considered here in its original meaning (Becke, 1903) to be the force exerted by a growing crystal; the concept of Ramberg (1947), i.e. the excess of mechanical pressure which a crystal can bear without dissolution at a certain degree of supersaturation, is not followed. Becke correlated the tendency for a species to form idioblastic crystals with its power of crystallization and listed minerals in order of this power. Two separate concepts have thus been confused: the *force* of crystallization which the crystal exerts on its surroundings, and the *power* of crystallization (*form energy* of Eskola, 1939, p. 278) which controls its place in the "crystallographic series" and its ability to develop faces. It has been incorrectly suggested that crystals with a high power of crystallization not only form well-shaped crystals but also large crystals (Becke, 1903; Schuiling and Wensink, 1962). Both size and shape of a given crystal depend on its surface energy but also on the conditions (chemical and physical) of formation and the rate and/or duration of growth. A high surface energy seems to promote crystal faces and as it depresses nucleation, it also favours a large size.

The force of crystallization has been considered to be comparable in magnitude with the crushing strength of the crystal and to enable the crystal to grow in directions in which growth is opposed by external forces (Becker and Day, 1916). The concept has been extended to metamorphic rocks where it has been supposed that a growing porphyroblast can push the adjacent rock apart as it grows (the "concretionary growth" of Wegmann, 1935, and Ramberg, 1952). It has commonly been stated (Harker, 1939, p. 334; Sengupta, 1963, p. 95; Miles, 1945, p. 121; Agron, 1950; Misch and Hazzard, 1962), that where the schistosity of a rock bows around a porphyroblast, the porphyroblast has formed after the schistosity and has pushed it aside (by virtue of its force of crystallization). This explanation appears to be almost invariably incorrect and the relationship can generally be shown to be due to a later schistosity taking on a curved form around a pre-existing crystal. The distinction is an extremely important one in the chronological analysis of polymetamorphic rocks.

The effect of rigid objects on adjacent layering can be seen in the following:

(1) Differential compaction of bedding in clays around a pebble or fossil, or of layering in tuff around a crystal.
(2) Flow-layering in the groundmass of porphyry or rhyolite around phenocrysts.
(3) Curvature of cleavage or schistosity in the matrix of a conglomerate, or in a slate, around an enclosed pebble or rigid fossil, or in a sheared granite or porphyry around large porphyroclasts.

There are innumerable examples of rigid objects causing a later structure to curve around them and the curvature of cleavage around porphyroblasts in schists can almost invariably be simply correlated with other evidence showing that the porphyroblast is older than the foliation. On the other hand, it is extremely rare to find any evidence in simple metamorphic rocks, such as hornfelses, that a growing crystal can push aside the bedding, in fact, the preservation of an old fabric within helicitic crystals shows that growing crystals do not mechanically disturb other mineral phases.

Although the transformation of anhydrite to gypsum should be accompanied by an increase in volume, the "force of crystallization" is apparently not great enough to overcome the weight of any appreciable overburden (von Gaertner, 1932; Goldman, 1952, p. 10).

Rast (1965, p. 90) discussed the relation of force of crystallization to porphyroblasts, poikiloblasts and idioblasts and presented a convincing case in rejecting force of crystallization as a property enabling growing crystals to push the matrix aside and as a factor related to the size and shape of crystals.

Many of the experiments quoted in support of a force of crystallization are not analogous to the growth of crystals in metamorphic rocks. In the original experiment Becker and Day (1905) placed a weight on top of a large crystal of alum surrounded by a supersaturated solution and found that the weight was lifted due to the growth of new alum at the base of the large crystal. From the principles of Le Chatelier and Rieke one would expect the solubility of the salt to be higher at positions of higher pressure and that deposition would take place around the sides of the crystal where the pressure is lowest rather than at the base where the pressure is highest. The natural examples quoted earlier (frost shattering, gypsum boils, etc.) are all of processes which are driven one way by a gradient of such magnitude (supercooling, supersaturation) that they cannot be reversed by the slight pressure of the confining body.

The experiments of Becker and Day (1905) and others are to a certain extent misleading because a density gradient is set up in the solution and this forces deposition where supersaturation is highest and retards it where pressure is lowest. Capillary attraction and adsorption pull the solution into the narrow cavity between the crystal and its support and force crystallization there (Taber, 1916).

Schuiling and Wensink (1962) concluded that a crystal in a rock could displace adjacent layering on the basis of an experiment in which the growth of a crystal of copper sulphate from a saturated solution pushed aside layers of sand and glass beads. The analogy with metamorphism is poor because the work to be done is minute (lifting glass and quartz bodies with a density of $2\frac{1}{2}$ supported by a solution of density perhaps $1\frac{1}{2}$) and the driving power is very great.

The chemical experiments are not really relevant to metamorphic growth. A crystal grows under a large confining pressure and perhaps a directed pressure as well and the work necessary to mechanically displace its matrix is very large. At the same time, the energy gradients (analogous to the supersaturation of solutions) causing growth of the crystal are small. The crystal does not grow outwards by driving forward its boundaries as rigid impermeable walls forcing the matrix aside but by the addition of selected units which diffuse through the matrix to the interface at the same time as others diffuse away. The interface may enclose some impurities, others may be adsorbed on the interface and move out with it (although these will tend to poison the interface and retard its growth) and other unwanted constituents may be gently displaced away from the interface and take the place of material which is attached to the interface preferentially. Much of the material in the matrix not required by the growing crystal and not attracted to it will migrate outwards more quickly than the interface under chemical gradients of extraneous origin.

Dendritic Crystallization

Crystallization may be regarded as normal or *layer* when successive layers are attached to an advancing low-energy surface or *dendritic* when successive units are added to high-energy projections to give a tree-like structure (Saratovkin, 1959; Horvay and Cahn, 1961). Dendritic crystallization may begin with a slight departure from layer crystallization when a random protuberance is formed. If the energy gradient is large enough, accelerated growth takes place on the protuberance as it grows along local concentration paths. Slight irregularities on the sides of the protuberance may extend in the

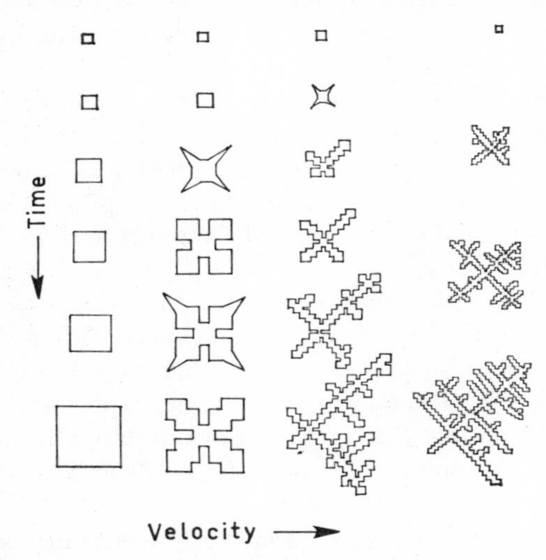

Fig. 44. Dendritic crystal growth. Growth in successive stages (from top of diagram downwards) for four rates of growth increasing from slow layer growth on the left to rapid dendritic growth on the right (Alexander, 1951, Fig. 3)

same way to give a geometrical network. Units of the network may meet and interfere with each other and continued crystallization will fill in the interstices to give a uniform crystal (Smith, 1963).

A dendritic crystal is one whose shape is markedly irregular with a skeletal or network form (Fig. 44), the symmetry of the crystal generally being made apparent by the angles between the rods and units which make up the whole. Despite its complex form the crystal structure is continuous and most bounding faces are rational. Dendritic crystals can be produced experimentally by rapid precipitation from gas or liquid under strong supersaturation gradients around a limited number of nucleii.

PLATE XI. (a) Two sets of (110) slip lines on experimentally deformed olivine (Raleigh, 1964). (b) Traces of (100) slip lines in kyanite. Kink bands with external rotation about [010] (Raleigh, 1964). (c) Deformation twins in plagioclase from charnockite (×25). (d) Conjugate deformation twins in calcite from marble (×15)

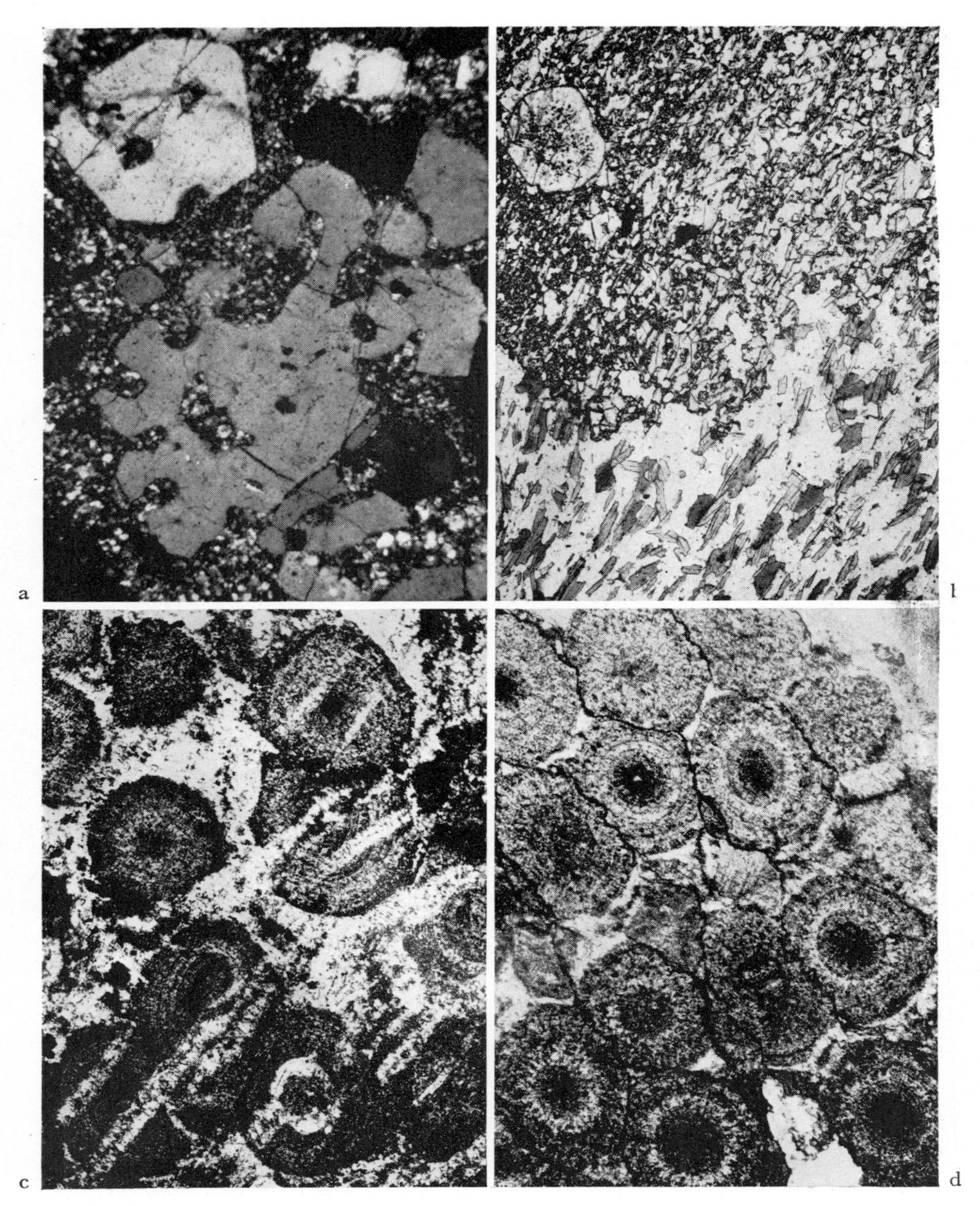

PLATE XII. (a) Embayed crystal form; quartz in porphyry (×23). (b) Aggregate-pseudomorph of staurolite after biotite flakes in schist, Little Falls, Minn. (×10). (c) Deformed ooliths; tensional ruptures marked by veins with comb structure, N. Devon (×18). (d) Deformed ooliths; shortening takes place by pressure solution producing stylolites, N. Devon (×18)

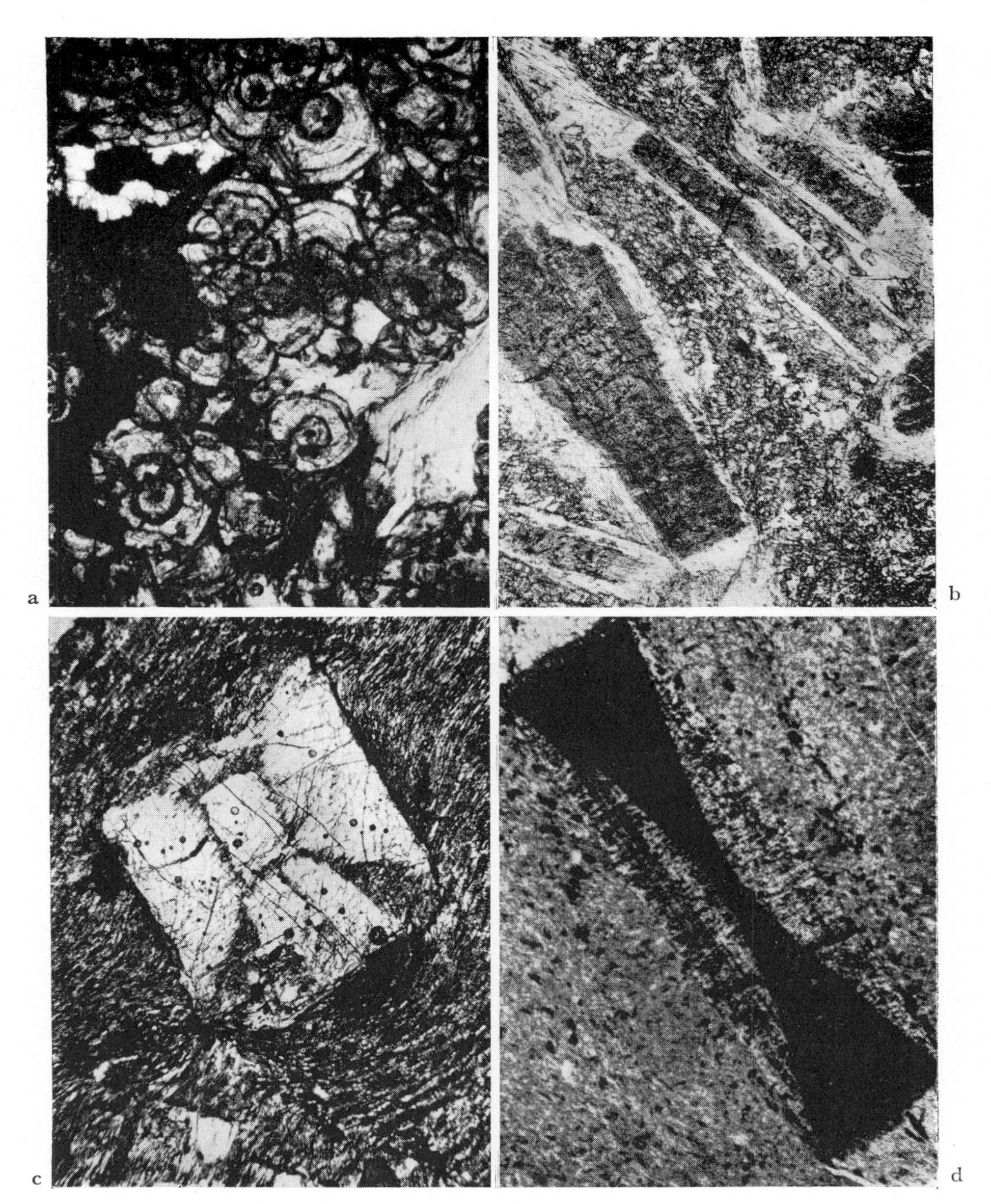

PLATE XIII. (a) Zonal structure in andradite; hornfels, Trial Harbour (×25). (b) Secondary rim of colourless amphibole around hornblende; epitaxial growth-zoning, epidosite, Goslerwand (× 15). (c) Chiastolite in hornfelsed phyllite; Santa Rosa Mtns., Nevada. Chiastolite has a post-tectonic relation to cleavage which it encloses helicitically. Slight deflection of cleavage is due to minor tectonism after formation of the chiastolite (×25). (d) Hour-glass structure in chloritoid, Ottrez (×10)

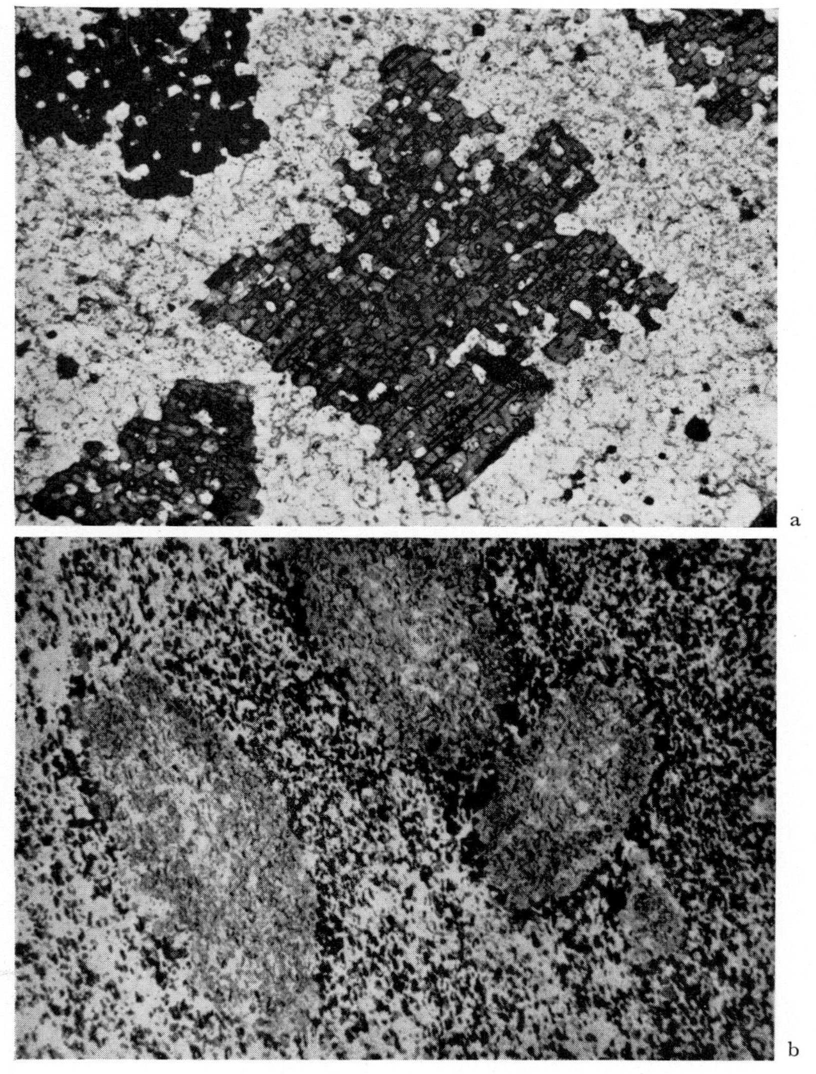

Plate XIV. (a) Poikiloblastic biotite in hornfels, Trial Harbour. The quartz inclusions are a little smaller than the quartz grains in the matrix (×55). (b) Cordierite porphyroblasts in hornfels, Trial Harbour. Biotite has been displaced and is concentrated at the rims (×37)

PLATE XV. (a) Fibrolite needles penetrate and replace a quartz crystal which is surrounded by a contorted and matted felt of fibrolite (×22). (b) Prisms and needles of sillimanite in granoblastic-polygonal aggregate of quartz in hornfels (×20). (c) Sillimanite in gneiss, Broken Hill. There is no boundary between adjacent crystals in which the (010) cleavage is parallel (centre), but a low-angle boundary appears where they are not quite parallel (top left) (×30). (d) Poikiloblastic orthoclase from gneiss, Broken Hill. Quartz inclusions are round. Biotite is circular in basal section but longitudinal crystals develop (001) with rounded ends. (Photographs (c) and (d) by courtesy of R. H. Vernon)

PLATE XVI. (a) Cruciform structure in chiastolite (×6). Arms are outlined by feather-like segments parallel to the prism faces. (b) Detail of inclusion pattern in corner of (a). The andalusite has been converted to white mica (×35)

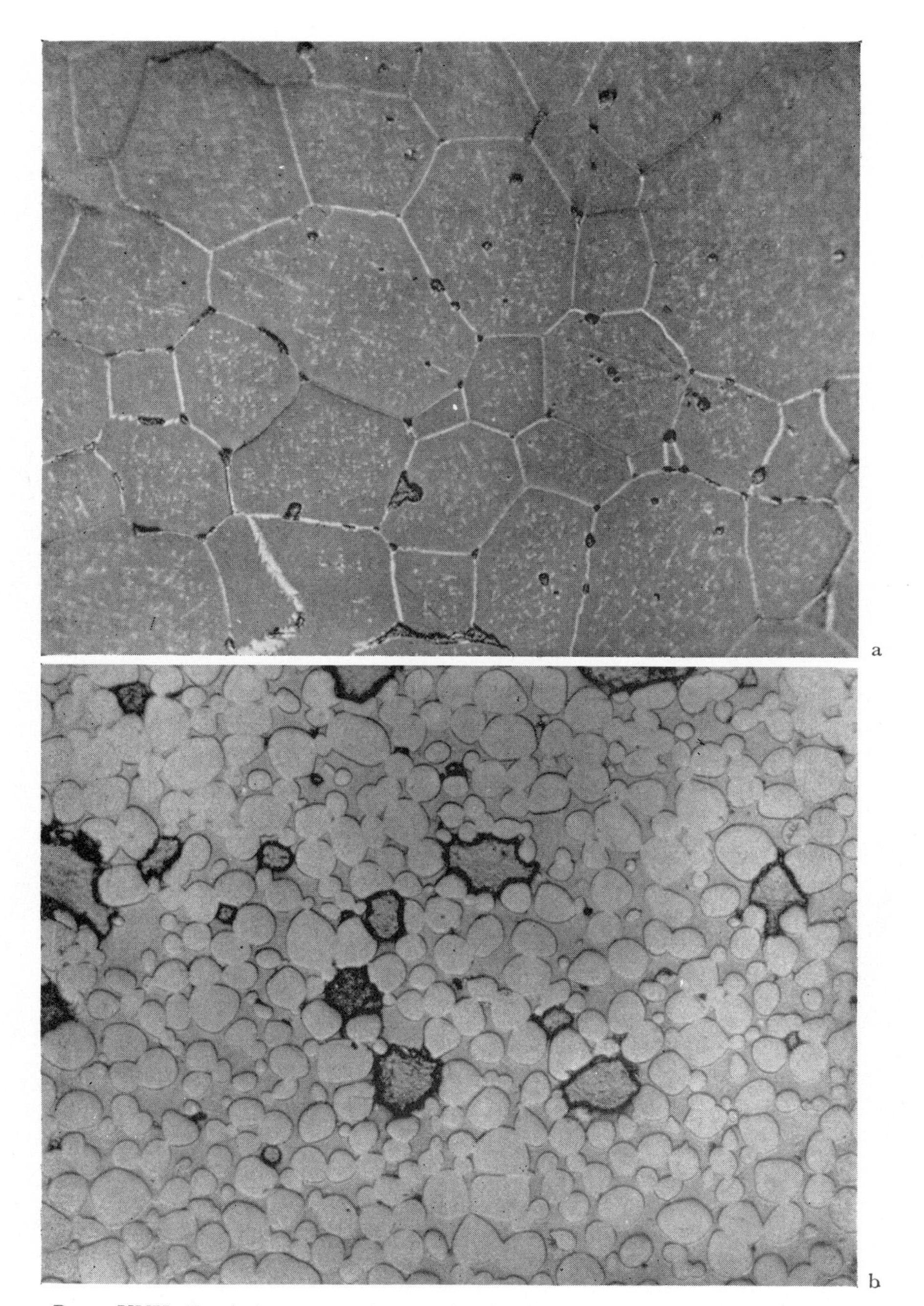

Plate XVII. Exsolution textures in ceramics. (a) Granoblastic-polygonal texture in periclase with straight boundaries and triple-points. Exsolved magnesioferrite forms a light-coloured phase as intragranular inclusions and an intergranular film (×400) (b) Lighter-coloured periclase forms round crystals in a matrix of slightly darker monticellite (×150). (Photographs by W. S. Treffner, shown as Figs. 3 and 4 in Treffner, 1964)

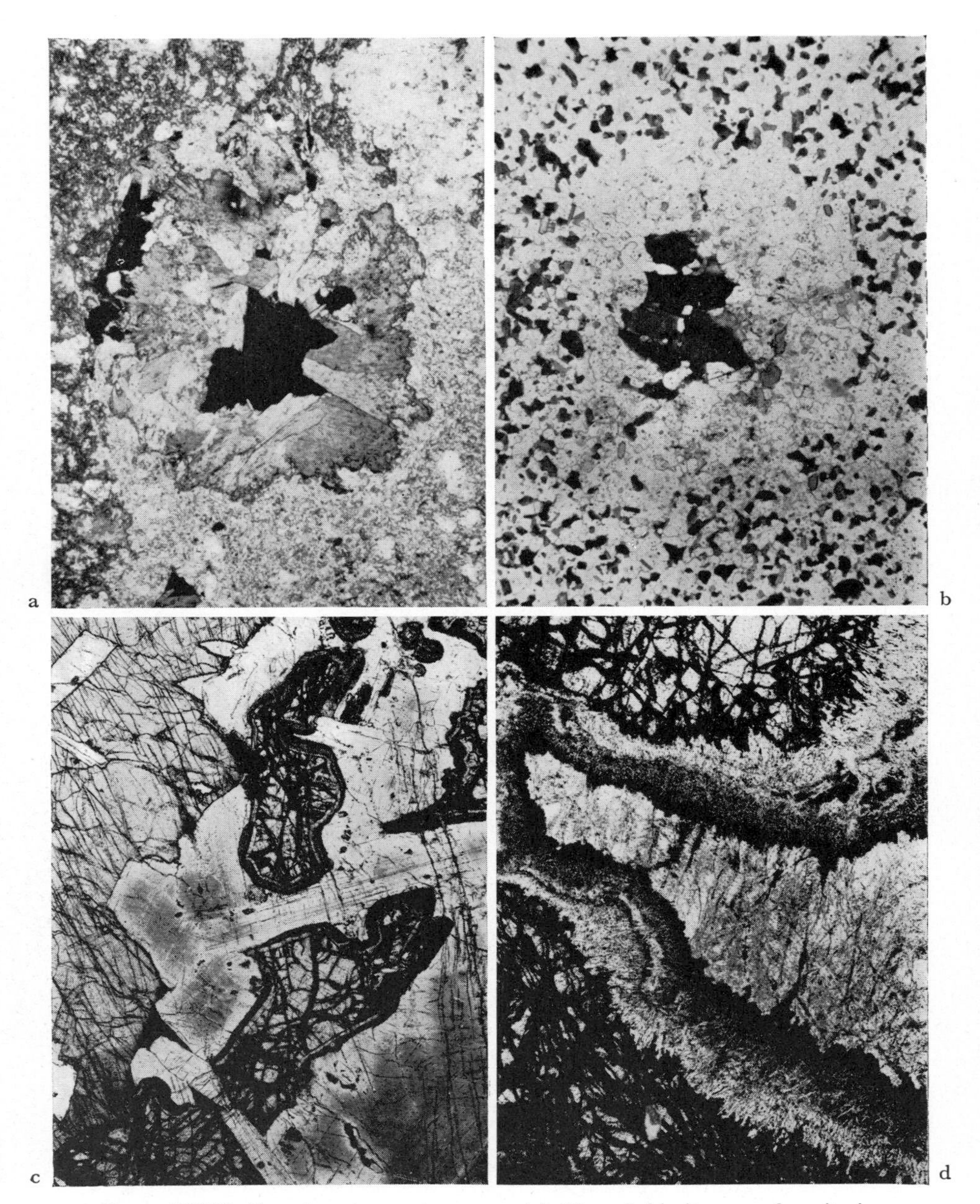

PLATE XVIII. Reaction rims and coronas. (a) Rim of chlorite around pyrite in schist, Mt. Lyell (×40). (b) Growth of biotite porphyroblast in fine-grained pelitic hornfels leaves a "halo" deficient in biotite. Tilbuster (×30). (c) Coronas around olivine, and "clouded feldspar" in gabbro, Oberkensbach, Hesse (×20). (d) Multiphase, complex corona between olivine (high relief) and plagioclase (centre) (×15)

One of the main characteristics of the dendrite is the persistence of a single nucleus orientation; nucleation difficulty may lead to a number of related free-growth forms which are less complex than true dendritic crystals (Fig. 45), e.g. "split" crystals (Grigorev, 1965) or "amoeboid" crystals.

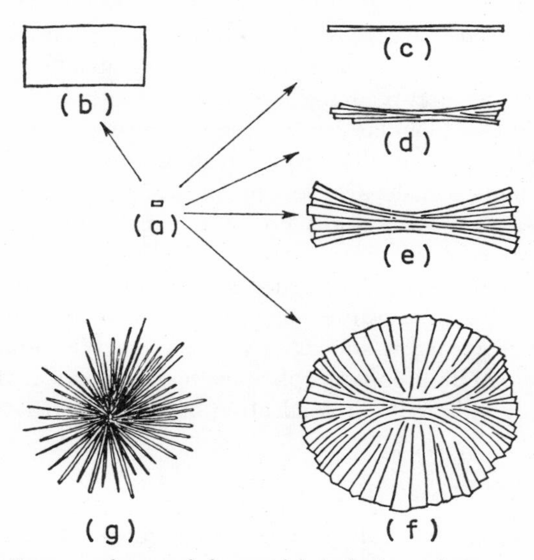

FIG. 45. Some free-growth crystal shapes. (a) A single nucleus may grow to (b) an idioblastic crystal, (c) needle (acicular crystal), (d) fascicular bundle, (e) bow tie, (f) spherulite. (g) A radiating acicular aggregate involves separate nucleii of different orientations

A *radiating acicular* aggregate differs in that it consists of differently oriented, physically separate, radiating needles which form from a large number of separate nucleii. The point from which they radiate is presumably a locus of seeded or catalysed (but not oriented) nucleation.

Geological examples of true dendritic crystals are not common; they occur in volcanic glasses as crystallites and skeletal crystals where they result from incipient crystallization of a rapidly cooled melt. They are rare in metamorphic rocks because such crystals have a very high surface area (and hence a high surface energy) per unit volume and are thus thermodynamically unstable. Chemical gradients sufficiently high to produce true dendritic crystals must be rare under metamorphic conditions.

Skeletal crystals of hematite in muscovite, described by Frondel and Ashby (1937), have all the characteristics of true dendrites and are accepted as such, but the suggestion by Rast (1965) that the inclusion pattern in chiastolite might be due to dendritic growth is discussed elsewhere.

"Dendrites" or dendritic growths with moss or fern-like form occur in joints and cracks where they result from the evaporation of minor infiltrations of iron and manganese solutions during weathering. These are a separate phenomenon and are not discussed here.

A poor dendritic form may be produced in metamorphic rocks where nucleation of a mineral is very difficult. A single nucleus may then slowly extend through the rock to give an irregular crystal with a single orientation. This does not, however, necessarily imply the strong supersaturation which is involved in the original meaning of dendritic crystallization. Rast (1965, p. 86) considered that acicular crystals require even higher supersaturations than those producing dendritic crystallization but even though acicular prisms have a high surface area per given volume, the crystal faces are perfect (low-energy) surfaces and such crystals may form very slowly under low supersaturations.

Similarly, skeletal crystals of garnet formed by deposition along intergranular boundaries in a rather fine-grained quartz-rich rock or large skeletal crystals of garnet or staurolite which replace a mesh of chlorite or biotite as an aggregate-pseudomorph may have a kind of dendritic form. These crystals imply difficulty in nucleation but not true dendritic growth.

ACICULAR AND RADIAL GROWTH

Acicular (needle-like) crystals, *fascicular* aggregates (like bundles of rods), *bow-tie* structures and *spherulitic* aggregates are all somewhat dendritic because of the predominance of growth over nucleation (Fig. 45) but are not due to true dendritic (branching) crystallization.

Acicular crystals develop from a single nucleus. A fascicular growth develops initially from a single nucleus followed by slight branching which may be due to catalyzed nucleation of units with slightly different orientations. The divergence of the needles increases to give a bow-tie or spherulitic structure. The sub-grain boundaries between the units are moderately coherent and there are no open spaces.

A spherulite is a round, spherically symmetrical, radiating crystal aggregate whose mutual internal boundaries tend to be structurally coherent. The term spherulitic should not be loosely applied to any radiating fibrous or acicular aggregate (of sillimanite, tourmaline, etc.) in which each crystal is a separate entity and has nucleated and grown individually.

The internal and external symmetry is definitive. Spherulites result from crystal growth governed by poor nucleation and are most common in media with high viscosity (glass, plastics, etc.). They are best known in devitrified glassy igneous rocks but also occur in some metamorphic rocks and are well known in industrial glasses. The most interesting recent work on the kinetics and mechanics of growth of spherulites has been done on plastics

(Keller, 1958; Mandelkern,1958; Garnick, 1965; Sauer, Morrow and Richardson, 1965).

The spherulitic structure is the result of the reorganization of disordered material with amorphous structure (glass, polymer, gel), or a finely-crystalline, randomly oriented aggregate, or even a uniform crystal, by the rotation of small units and their addition to a nucleus. Spherulites form by a combination of difficult nucleation and slow growth and the kinetics are governed largely by the nucleation.

The spherulite is an aggregate of crystals or fibres whose marked elongation is rational crystallographically, i.e. the radiating crystals in spherulites of quartz, chalcedony, calcite, tourmaline, mica, are each elongate in the same optic direction and give a "polarization cross" under the microscope.

Some spherulites (calcite, chalcedony) are imperfectly formed and the individual crystals are undulose with a kind of lineage structure. In others, the lattice is well formed and the boundaries between adjacent crystals have a high degree of coincidence; many are probably low-angle tilt boundaries marked by abundant dislocations.

Complete spherulitic growths are uncommon, e.g. the perfect spherulitic adinole of Dinas Head (Harker, 1939, Fig. 56c) but a *bow-tie*, *fascicular* or *radiating fibrous* habit develops in acicular or prismatic crystals such as micas, amphibole (Harker, 1939, Fig. 59b), tourmaline, sillimanite, kyanite, epidote and wollastonite under non-stress conditions, e.g. in hornfelses or post-tectonically in regional metamorphics. Nucleation and growth of similar textures in metal eutectics have been discussed by Hogan (1965).

Embayed Crystals

An embayed crystal is one in which the basic shape (anhedral to euhedral) is modified by embayments (Plate XIIa). These embayments may penetrate well into the centre of the crystal and have curved boundaries most of which are convex outwards from the crystal. Many embayments are lobate but a few have pointed ends. Embayed crystals commonly enclose subcircular or elongate areas which resemble the embayments in shape and internal structure. The enclosures commonly contain fine-grained, indefinite material which resembles the matrix around the crystal. The boundaries of the embayments are almost invariably curved and even though a few examples have some parts in a rational orientation to the host, *embayed crystals* are distinguished (Bain, 1925) from *skeletal crystals* which are bounded largely by crystal faces and which are due to dendritic crystallization.

Embayed crystals are common in acid volcanics but are rare in metamorphic rocks. The shape has been attributed to *dissolution* (corrosion, resorption), to amoeboid growth and to coalescence.

Embayed crystals in porphyritic igneous rocks such as rhyolite, obsidian, quartz or feldspar porphyry and also in isolated tuffs or "porphyroidal" sheared equivalents have been described and figured in standard texts such as Alling (1936, p. 151), Johannsen (1939, vol. 1, p. 21, Figs. 29, 30, 31 and 39) and Hatch, Wells and Wells (1961, pp. 77, 220, Figs. 92, 96). The shape is considered to be a modification of early euhedral, anhedral or skeletal phenocrysts by later attack by the magma. There seems no other reasonable explanation for such crystals in glassy lavas and indeed, Krank and Oja (1960) have reproduced comparable textures experimentally by partially melting granite.

Moorhouse (1959), Foster (1960) and Vance (1965) have summarized evidence for the dissolution hypothesis. Some embayed crystals of plagioclase (Kuno, 1950; Vance, 1965), pyroxene, olivine or quartz contain compositional zoning or external crystal faces which are transected by the embayments which therefore are due to resorption. In some examples it is clear that the embayed crystal is in a foreign environment and that it formed under much higher temperature conditions than the host rock, e.g. where the crystal is of a high temperature variety (mixed Na–K feldspar, sanidine) or is coarsely perthitic, or shows a relict outline after high-temperature quartz, or has inclusions of glass or devitrified glass.

Some rare crystals in both igneous and metamorphic rocks have an irregular or amoeboid shape which is a primary growth feature. The amoeboid growth process is not well understood but would appear to have a velocity between slow layer growth and rapid dendritic or skeletal crystallization. Zoning in such crystals is parallel to the embayment surfaces, some of which may be rational and develop crystal faces. Radial and concentric lines of inclusions or bubbles are typical.

This growth process is uncommon and appears to require a fluid medium. Crystals of this kind occur in volcanic rocks but also in some hornfelses which contain evidence of there having been an abundance of intergranular fluid.

Embayed crystals have been postulated as having formed by amoeboid growth or by coalescence in the solid state by Goodspeed (1937a, b), Misch (1949), Duschatko and Poldervaart (1955), Bradley (1957) and Elliston (1966) but this hypothesis is not supported by evidence of any weight.

COALESCENCE

The term is used here in the metallurgical sense to mean the process whereby adjacent crystals with somewhat different orientations become joined together to give a single crystal. In the early stages of the process the large secondary grain has a mosaic (growth) substructure and consists of lattice regions of slightly different orientation (sub-grains) so that it has undulose and patchy extinction but this disappears to give a crystal with uniform orientation. Coalescence occurs during recrystallization of annealed metals and is an homogenization process which is something like polygonization in reverse.

The principle was demonstrated by Walker (1948), Li (1962) and Hu (1962) who showed that metal sub-grains coalesce to give a large grain without the movement of a grain boundary. The boundaries between sub-grains fade away to give a strain-free region which can then grow by other processes.

The term has been used so loosely in geological literature that its value has been reduced, e.g. Mitchell and Corey (1954), Coombs (1954, p. 71) and Poldervaart and Ekelmann (1955). It is recommended that the term coalescence be restricted to the process by which separate differently oriented crystals of the same species join together to give a single crystal of uniform orientation. It is distinct from grain growth by nucleation plus growth or by normal grain boundary movement.

Geological evidence of coalescence is rare. It has been postulated that small quartz granules in mylonitized rock combine to give large grains by late recrystallization (Crampton, 1963, p. 367), and Rinne (1926) referred to *Sammelkristallisation*, i.e. the *integration* of small crystals in a fine-grained aggregate of gypsum (Goldman, 1952, p. 11; Carozzi, 1960, p. 145). The coalescence of sillimanite from finely felted fibrolite "mats" or "felts" and veins, to the "simple resolution of the felts into discrete rods, to the coarse growth of sillimanite prisms several millimetres in length, often as stellate aggregates enclosed in biotite and tourmaline" has been traced by Chinner (1961). Sillimanite generally appears to form fibrolite (Plates XV and XXVIII) first and larger prisms later although the process may be complex and the relationships between fibrolite and coarse sillimanite are not clear in many rocks (Tozer, 1955).

Coalescence in a monomineralic rock is postulated to produce large secondary crystals with mosaic substructure in a fine grained matrix, i.e. something like *mortar* texture. Some embayed crystals also show such features as undulose extinction, curvature, lamellae, twinning, mosaic structure and splitting of the crystals. These are attributed to later deformation by adherents to the dissolution hypothesis but are considered alternatively to be due to the growth and coalescence of small crystal units into larger ones (particularly Reynolds, 1936). The deformation substructure of quartz in deformed rocks has been discussed in some detail previously but it is not possible, at present, to give definitive differences between growth substructure (e.g. Plate Ia, b, c) and deformation substructure (e.g. Plate Id). Smith (1963, p. 472) suggested that "strain mosaic" texture can be distinguished from growth lineage mosaic texture by the nearly equant shape of the "strain" blocks but a blade shape (perpendicular to growth banding on crystal faces) of growth blocks. Details of mosaic structure in minerals are lacking and care must be observed in attempting to use Smith's criteria because the mosaic structure in some strained quartz tends to form elongate lenses (deformation bands) outlined by sub-grain boundaries approximately parallel to [0001]. The elongation of sub-grains in growth lineage on the other hand is related to external surfaces controlling nucleation or growth.

It is concluded that most embayed megacrysts (large crystals) in metamorphic rocks are relict and pre-metamorphic and that they were formed by igneous crystallization followed by partial dissolution rather than by solid state coalescence. The crenulate grain boundaries of crystals in granular aggregates such as hornfelses are not embayed boundaries in the sense used here and are due to grain boundary motion during solid state recrystallization.

Elongate Crystals

The shape of a crystal may not be due to the achievement of lowest free energy (a thermodynamic control) and the shapes of many elongate crystals are controlled by kinetics. An elongate form may be due to:

1. *Static growth*
   (a) Faster growth of certain lattice directions, e.g. elongation of amphibole, sillimanite, tourmaline.
   (b) Nucleation at a surface followed by growth perpendicular to the surface with mutual impingement (comb structure in veins).
   (c) Recrystallization of a previously deformed aggregate; the elongation is due to faster growth in directions controlled by mutual orientation or impingement, e.g. preferred orientations of annealed metals.
2. *Deformation*
   (a) Change from equant to elongate by movement on one or more glide planes.
   (b) Fracturing or slicing an equant grain into elongate fragments.
3. *Syntectonic crystallization*
   Crystallization of susceptible materials while undergoing movements.

The elongate shape of many minerals in tectonites, e.g. quartz in quartzite, may be due to two processes which are essentially identical.

(1) The shape of a grain surrounded by an intergranular fluid can be regarded in terms of Riecke's principle. The solubility of a crystal in contact with a supersaturated solution increases at points of increased pressure and decreases at points of reduced pressure. Thus material tends to dissolve on

the sides of a grain facing the compression and to deposit on the ends facing the tension. An equant crystal is thus made elongate by a kind of pressure solution. It does not change its lattice orientation.

(2) The shape of a crystal in a polycrystalline aggregate at elevated temperatures can be regarded in terms of Nabarro–Herring self-diffusion. Nabarro (1948) and Herring (1950) pointed out that self diffusion can allow a crystal to yield to an applied stress because diffusion takes place from grain boundaries where there is a normal compression towards those with a normal tension. The mechanism is thus the solid state equivalent of Riecke's principle.

Mutual Growth Forms

The simplest view of recrystallization is that of the formation of scattered nucleii which expand by growth through the surrounding grains until all of the old grains are replaced by new grains. Growth at identical rates from two separate nucleii will ultimately produce a plane interface halfway between them. Growth then becomes concerned largely with the impingement of one crystal upon another. If the crystals are not nucleated at the same time or if their rates of growth are not equal the interfaces between the crystals will be curved (*initial impingement interfaces*). In most aggregates only a small portion of the interface can form between any two grains before it is affected by adjacent grains.

A non-equilibrium *impingement texture* has curved boundaries, irregularly elongate crystals, triple points with unequal dihedral angles, quadruple points, and a range of grain sizes and shapes (Plate VII). Grain boundary movement allows the initially curved interfaces to straighten and the interface junctions to adjust giving minimum energy positions and the very important *granoblastic polygonal* texture. The term *granular* is applied to igneous and sedimentary rocks as well as metamorphic, and simply means that the rock is composed of units of about the same size. In keeping with the principle of using different terms for igneous and for metamorphic textures, the term "granoblastic" is preferred here. *Polygonal* refers to the general shape of the crystals which are interlocking ("crystalline" or "crystalloblastic" texture) with moderately straight boundaries tending to meet in triple-points; many of the crystals have six sides (Plates VIII and IX). The use of "mosaic texture" to mean granoblastic-polygonal was discouraged earlier (p. 34). The term "granuloblastic" (Binns, 1964, p. 297) is approximately equivalent.

The term *heteroblastic* refers to a rock composed of crystals with a range of sizes. Terminology is a little difficult because the size and the shape of a crystal are related to each other because of the necessity of space-filling.

The shapes of the crystals in a polycrystalline aggregate in textural equilibrium depends on whether there is one, or more than one crystal species present, on the anisotropy of the crystal lattices, on the presence of an intergranular fluid and on the relative interfacial energies of the different grain boundaries. At one extreme is the granoblastic polygonal texture of a monomineralic aggregate with low grain boundary energy and at the other is the complex texture of a multicomponent aggregate of species with markedly different properties.

An equilibrium texture in an assemblage of crystals in which differences in surface energy are small (quartz, calcite) will consist of polygonal grains of equal size and shape, with straight boundaries meeting in triple-points. If one species is present the triple points are equal. The interfacial angles will vary only slightly if there are two or more species with similar surface energies and small lattice anisotropy (Smith, 1948; McLean, 1957; Burke, 1959).

Two contrasting tendencies control the formation of a granoblastic-polygonal texture in monomineralic aggregates. With completely isotropic crystals the lowest grain boundary energy state is achieved by interlocking xenoblastic crystals with comparatively straight boundaries and 120° interfacial angles at the triple-points. In aggregates of crystals with strong lattice anisotropy, the grain boundary energy varies considerably within each crystal boundary and the lowest energy state may result in the formation of some rational faces. The grain boundary area is not minimal but the boundary energy is least because the energies of the rational faces are less than those of the irrational surfaces necessary to produce interfaces with 120° triple-points.

Some metamorphic minerals (e.g. quartz and calcite) are almost invariably xenoblastic and their shapes are largely controlled by a balance of grain boundary energies. Other minerals such as dolomite, amphiboles, micas, garnets, etc., have a strong tendency to be idioblastic or at least subidioblastic. The strong anisotropy of these crystal lattices has a greater effect over minimum surface energy requirements than do interfacial angles and total areas of interface; their granoblastic-polygonal textures contain few 120° triple-points, and many of the grain boundaries coincide with rational lattice directions. This is seen in Plates VIII–X where many grain boundaries are parallel to cleavages.

Crystal faces are not common in metamorphic minerals and, where present, are always the simplest dominant forms. There is little evidence that impurities affect the relative development of specific forms as they do in solution-grown crystals (Buckley, 1930; Davies and Rideal, 1961, p. 427) but perhaps the environment is so chemically complex that all faces are depressed. There is no great competition between faces belonging to different forms but in micas, pyroxenes and amphiboles, epidotes, staurolite, kyanite, sillimanite, etc., the shape is due to competition between basal pinacoids, pyramids or domes on one hand and non-basal pinacoids and prisms on the other. In most of these the crystal growth is dominated by the ease of adding extra chains or sheets.

Decussate texture is shown by interlocking, randomly oriented, somewhat elongate, prismatic or subidioblastic crystals, generally of one species (Fig. 53 and Plate XIX). It is shown best by amphibole, wollastonite or biotite aggregates in hornfels but a similar kind of texture develops during the post-tectonic crystallization of various amphibole-rich rocks. Harker (1939, p. 35) stated that the crystals lie in *all* directions (i.e. are not randomly oriented) and that crystals have grown so as to reduce internal stresses produced by growth. This concept is based on acceptance of the "forces of crystallization" exerted by growing crystals and this is discounted elsewhere.

Decussate texture can be regarded as an equilibrium boundary configuration somewhat similar to granoblastic-polygonal. Nucleii which are regularly spaced and randomly oriented grow until they impinge. Grain growth then takes place by grain boundary movement, destruction of smaller or imperfect (high energy) crystals and adjustment of grain boundaries. Triple-points are present but they tend not to be equiangular as most of the grain boundaries are rational, low-energy (commonly prismatic) planes and thus a final low-surface-energy texture is produced without equal dihedral angles.

A rare texture shown by some rocks, which are composed almost entirely of idioblastic crystals, has been called "pile of bricks" texture (Plate VIIId). Friedman (1965) coined the term *idiotopic* to relate "to the fabric of diagenetically altered carbonate rocks or to chemically precipitated sediments. It is a descriptive term for fabrics in which the majority of the constituent crystals are euhedral". There seems no reason why the term idiotopic should not be extended to cover the rare instances of this texture in metamorphic rocks. This texture involves considerable space-filling problems and may be restricted to low-temperature, low-pressure rocks such as dolomite, anhydrite rock and gypsum rock in which an intergranular solvent has been active.

Columnar impingement texture consists of parallel, elongate crystals (Plate VI), e.g. comb structure in quartz veins, cross-fibre chrysotile in veins, fibrous quartz in pressure fringes around crystals or between dismembered fossils and the columnar structure in cast metals (Christian, 1965, p. 582). Columnar impingement texture can form by recrystallization entirely in the solid state or by replacement and does not necessarily imply a solvent phase or fissure filling.

The texture is formed by nucleation at a surface followed by growth normal to that surface because of a temperature, concentration or pressure gradient. The nucleii appear to be randomly oriented but as the crystals grow outwards they impinge on each other laterally and mutually interfere (Grigorev, 1965, Fig. 168). Under these conditions it would be expected that crystals which happen to be so oriented that their direction of fastest growth is normal to the controlling surface will outgrow the others and dominate the texture so that a lattice preferred orientation accompanies the dimensional one. This is true for many cubic metals which are oriented so that $\langle 100 \rangle$ is parallel to the growth direction. Measurements by Pabst (1931) on quartz fibres in pressure fringes showed no lattice preferred orientation and apparently growth had not proceeded sufficiently far to allow the fastest growing columns to dominate.

Layer Growth and Zoning

A crystal growing by normal layer processes adds units to its outer surfaces and a uniform crystal results if the units are of constant composition and orientation. A defect structure such as lineage results if the orientations vary slightly. If the units change gradually in composition, the structure of the crystal may persist outwards but the composition will change. Such changes may be gradual or abrupt, continuous or cyclic, and will result in a compositional zoning of the crystal. Zoned crystals such as plagioclase, pyroxene, olivine, etc., are common in igneous rocks where they indicate incomplete reaction with a changing residual magma. Zoned crystals appear to be less common in metamorphic rocks, although cryptic zoning may be more common than is known at present.

Zoning in crystals in metamorphic rocks may be a relict condition in premetamorphic igneous crystals or may have been formed during metamorphism. Some compositional changes between successive zones are gradual and some are abrupt; the lattices of successive zones in the latter do not change gradually but may have contact regions of imperfect fit marked by epitaxial dislocations. Such a chemically inhomogeneous crystal is somewhat ordered with respect to the distribution of certain elements but there will be a concentration (or chemical activity) gradient across it. The stability would be increased (free energy reduced, entropy increased) if homogeneity were to be attained by diffusion. The lack of zoning in metamorphic crystals does not necessarily imply that homogenization tends to be achieved but rather that crystallization during metamorphism takes place under regular chemical and physical conditions. Zoned crystals indicate changes in the environment during crystallization and zoning is preserved because the driving energy tending to produce homogeneity is not great (structural and chemical differences are slight).

Zoning may be *multiple* where a number of zones are present or *simple* where a relatively homogeneous crystal is rimmed with a single layer of a different nature. Zoning in metamorphic crystals may be due to differences in composition, degree of alteration, abundance of inclusions, or possibly in structure (order–disorder).

Zoning is common in igneous plagioclase (Vance, 1962, 1965) but is comparatively rare in metamorphic plagioclase (Phillips, 1930; Misch, 1955; Barth, 1956; Greenwood and McTaggart, 1957; Jones, 1961; Rast, 1965; Cannon, 1966). Zoning may be *normal* (i.e. progressively more sodic towards the rim) *reverse* (progressively more calcic towards the rim) or *oscillatory* (alternately more and less sodic), Phemister, 1934. It was once thought that zoning was normal in igneous feldspar and reversed in metamorphic feldspar (Becke, 1897) but normal, reversed and oscillatory zoning have been found in metamorphic plagioclase (Goldschmidt, 1911; Wiseman, 1934; Misch, 1955; Rast, 1965). Rutland *et al.* (1960) considered that oscillatory zoning in the plagioclase of some Norwegian gneisses was a relict, igneous feature, but Cannon (1966) suggested that it might be of metamorphic origin.

Simple zoning with a clear outer rim is common in metamorphic rocks, some being due to late soda metasomatism. *Rapakivi* and other complex feldspar growths may be regarded as extreme varieties of zonal texture.

The shapes of the zones and the succession of compositions has been demonstrated to be very similar in separate crystals in metamorphic rocks (Greenwood and McTaggart, 1957) particularly in high-grade hornfelses and migmatites.

Idioblastic lime-garnets of calc-silicate hornfelses commonly contain regular zoning parallel to their crystal faces (Plate XIIIa) revealed by differences in birefringence and in refractive index (Frondel *et al.*, 1942; Ingerson and Barksdale, 1943; Holser, 1950). Some garnets have straight, clearly defined zone boundaries and a difference in orientation between successive zones giving a structure resembling twinning. The similarity of such zoning to that in igneous and some ore minerals (Edwards, 1960, Fig. 21) and the presence of abundant crystal faces suggests that these crystals form similarly to those in igneous rocks, i.e. in contact with a fluid phase which changes rapidly in composition.

Compositional growth zoning has been recorded in metamorphic garnet, staurolite and clinozoisite (Alling, 1936, p. 148), zoisite (Vogel and Bahezre, 1965), zircon (Poldervaart and Eckelmann, 1955), and also tourmaline, andalusite and corundum.

Cryptic zoning (microscopically invisible) has been demonstrated in almandine garnet from regional metamorphic terrains (Harte and Henley, 1966; Hollister, 1966) and promises to be common. Compositional changes (recognized by the electron probe) are gradational. An increase in the proportion of FeO and MgO and a decrease in CaO and MnO from centre to rim may well reflect a gradual increase in metamorphic grade.

A zonal structure is produced in garnet, andalusite (Plate XIIIc), cordierite (Harker, 1939, Fig. 11a) and feldspar (in metamorphic rocks) and dolomite (in dolomitized limestones) by differences in the pattern of inclusions. Some garnets have a core riddled with inclusions but a massive homogeneous rim (Rast and Sturt, 1957; Joplin, 1968, Fig. 242) probably due to initial rapid growth followed by slow growth. Some garnets (Spry, 1963b) have a core with rotational structure and a massive idioblastic rim and represent rapid syntectonic crystallization followed by slow post-tectonic crystallization.

An irregular zonal structure may be produced by differential or zonal (core, rim or intermediate) alteration. Examples include frameworks of garnet within feldspar (Poldervaart and Gilkey, 1954, Fig. 6), zonal structure in pyrite (Edwards, 1960, Fig. 106), partial alteration of garnet (Rast, 1965, Fig. 5), and zones in garnet due to inclusions or to alteration (Vogel and Bahezre, 1965).

Overgrowths and Epitaxy

Overgrowths of a secondary on a primary mineral during metamorphism generally result from some change in physical or chemical conditions so that equilibrium can not be established. In the simplest case, quartz may be added to the margin of clastic grains to give a silica cement converting a sandstone to quartzite. A slightly more complex example is the addition of a zone which is free from inclusions to a core riddled with inclusions as the rate of growth decreases. More complex again is the addition of a markedly different member of a solid solution series, e.g. colourless amphibole growing on hornblende (Plate XIIIb). These examples involve only slight, if any, changes of structure or composition and the zone or overgrowth is structurally continuous with the core; however, more complex examples are common.

Nucleation of new crystals is difficult and it is easier to enlarge an existing crystal than to produce a new one. Thus some crystals will act as "seeds" for other crystals; they become the *substrate* for a second phase *overgrowth*. The term *epitaxy* ("arrangement on") was introduced by Royer (1928) to denote the oriented growth of one crystal upon another. The phenomenon occurs because of structural similarities of substrate and overgrowth and Royer advanced rules stating that lattice planes in each must have similar structural patterns, interatomic distances and bonding. Structural similarities for experimentally observed growth of KBr on mica are shown in Fig. 46. The degree of "fit" between the structures is defined in percent as $100(b-a)/a$ where a and b are appropriate lattice vectors in each structure. It was originally considered that epitaxy would not occur if the misfit exceeded

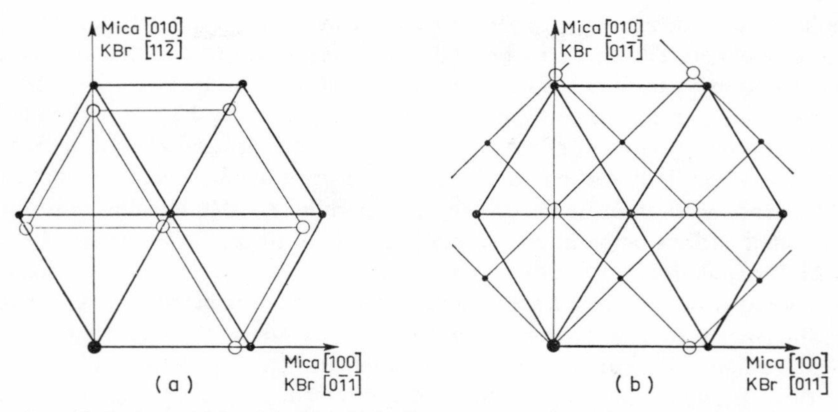

FIG. 46. Epitaxy. Epitaxial relations for KBr growing on mica. Close correspondence exists in the positions of K in (001) of mica (solid circles) and in KBr (open circles) in two orientations (Pashley, 1956, Figs. 3 and 5)

15% but extensive experiments show that large misfits may occur and that substrate and overgrowth may differ considerably in structure and bonding. Pashley (1956) concluded that none of the present hypotheses of epitaxial growth are satisfactory. For example, the growth of ice crystals on organic crystalline substrates is unusual in that the nucleating ability of the organic compound does not appear to be closely related to atomic correspondence in the two crystals, but to the presence and arrangement of hydroxyl-bonding groups and the presence of surface imperfections (Head, 1962; Fukuta and Mason, 1963). The mismatch between substrate and overgrowth may be relieved by periodic *epitaxial* dislocations, e.g. Plate IVa.

The laws governing epitaxy (producing *oriented overgrowths*) are basically the same as those governing certain exsolution textures (*oriented intergrowths*), structurally controlled replacement textures, and mineralogical transformations involving retention of structural continuity (*topotaxy*). Goldman (1952, p. 7) used the terms *syntaxial* for oriented overgrowths and *distaxial* for non-oriented overgrowths.

The importance of epitaxy as a control in crystallization has been widely studied (Pashley, 1956) and structural correspondences between substrate and overgrowth have been demonstrated in hematite on beta-alumina (Segnit, 1956), silicon on sapphire (Manasevit *et al.*, 1965), silicon on quartz (Joyce *et al.*, 1965) and many others.

The rock-forming silicates, oxides and carbonates show many examples of correspondence of structures which could lead to epitaxial overgrowth but the process is not always simple, e.g. the alumininium silicates have many aspects of structure in common but there is little evidence of an epitaxial relation between sillimanite, kyanite and andalusite. Kyanite and andalusite replace each other to give multi-crystal pseudomorphs, and Chinner (1961) and others have shown that when sillimanite is formed in a rock containing andalusite or kyanite, it is nucleated within biotite. Chinner (1961, p. 318) suggested that the sillimanite is nucleated epitaxially by biotite and stated that "the main features of the structure of sillimanite are chains of Al-O tetrahedra paralleling the *C*-axis, and linked together by chains of slightly distorted oxygen tetrahedra which also parallel *C*, the tetrahedrally co-ordinated cations being alternately Si and Al. In mica, the alternate sheets of Al-O octahedra and Si-O tetrahedra normal to the *C*-axis are composed of hexagonal arrays, the "sides" of each hexagon (each composed of a pair of octahedra or a pair of tetrahedra) being oriented parallel to (010) and to the two pressure-figure directions at 60° to this. Nucleation and growth of sillimanite on these trigonally arranged octahedra or tetrahedra could thus lead to a triangular arrangement with the same orientation in the host biotite." A similar triangular array of mullite needles is produced by the laboratory dehydration of pyrophyllite.

An epitaxial control of mineral nucleation may not be obvious but is probably important in producing mimetic textures during post-tectonic crystallization. Cubic crystals such as garnet (Powell, 1966) or pyrite show a preferred orientation of lattice and crystal faces in some rocks and this is probably due to nucleation on parallel flakes of mica.

INCLUSIONS AND INTERGROWTHS

The phases in an equilibrium assemblage might be expected to exist as separate crystals but two or more phases may coexist to give textures which range from the random arrangement of a *sieve texture* to the geometrical precision of an oriented intergrowth.

Intergrowths and inclusions are formed by four main processes:

(1) Growth of a porphyroblast to enclose residual foreign phases (poikiloblastic texture).
(2) Partial replacement of an early crystal by a late one leaving unaltered relics.
(3) Simultaneous crystallization of two or more phases (symplectites, eutectoids).
(4) Partial ordering of a disordered homogeneous phase (exsolution intergrowths).

Shapes of Minor Phases

The equilibrium shapes of minor mineral phases at grain boundaries or within grains depends on the interfacial energies, interfacial strains and the requirements for space-filling.

Idioblastic crystals, needles or radiating prisms may sprout into the major phase from the grain boundary when the minor phase has high specific surface energy and normally would form high-energy interfaces with the surrounding major mineral. Formation of crystal faces allows the achievement of a lower interfacial-energy.

Plates, flakes or prisms may lie along the grain boundaries of the major phase. This could be a kinetic relation (faster growth of the minor phase along the boundary due to easier diffusion), a mechanical relation (a flake along a surface of intergranular movement) or an equilibrium energy relation (achievement of a low-energy surface by forming a semi-coherent, rational, epitaxial type of interface).

The shape of intergranular particles has been correlated directly with grain boundary energy by Toney and Aaronsen (1962) who showed that large-angle (high-energy) boundaries cause the formation of xenoblastic crystals along grain boundaries. Coherent (low-energy) grain boundaries would have their energies greatly increased by this arrangement, thus the second phase tends to grow in special orientations, e.g. as plates, with maximum coherence with the two adjacent grains.

Irregular small crystals dispersed along the major phase boundaries may belong to a non-equilibrium texture or one in which the surface energy of the interface is low and not strongly dependent on orientation, thus the minor phase can take a variety of orientations.

The minor phase in a granular-polygonal aggregate shows a preference for location at boundaries and triple-points, either because the minor phase has pinned the grain boundaries and prevented them moving (and thus may not be an equilibrium feature) or because it achieves a lower total boundary energy. In addition, nucleation of new mineral phases is favoured at grain boundaries, triple-points and dislocations (Christian, 1965, p. 405). Growth of new calcite crystals in quartzite or of feldspar in feldspathized quartzite begins at quartz triple-points (Voll, 1960).

The shape of the minor phase in metals has been explained using a surface tension model by Smith (1948) and McLean (1957) and followed by Stanton (1964) for ores and Rast (1965) for minerals. When an inclusion of one species (B) is located at the junction of three crystals of another species (A) the original triple-point is replaced by three closely spaced triple points (Fig. 47). The shape of the inclusion depends on the interfacial angles at these triple-points. The interfacial tensions of the self-boundaries (γAA) and of the interphase boundaries (γAB) are related thus: $\gamma AA = 2\gamma AB \cos \frac{1}{2}\theta_2$. Thus when $\gamma AA = \gamma AB$, the inclusion will be convexly triangular, Fig. 47 (c) because $\theta_2 = 120°$. Where γAA is less than γAB the inclusion is globular with a limiting spherical case where γAA is infinitely less than γAB. Where γAA is greater than γAB, the inclusions become triangular (case (d) where

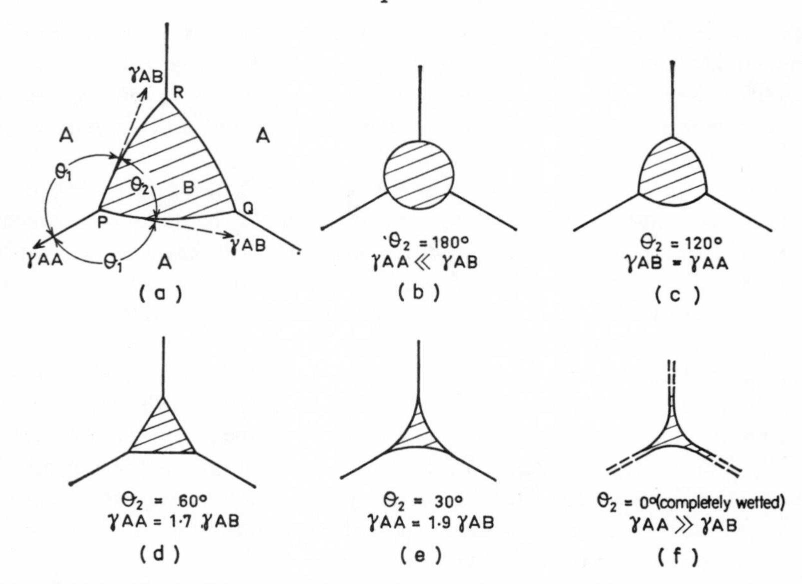

FIG. 47. Surface tension and inclusion shapes. (a) Inclusion of B at a triple-point between three grains of A with three triple-points P, Q and R. (b) to (f) The shape of the inclusion depends on the size of θ_2, which in turn depends on the relative magnitudes of γAA and γAB

$\theta = 60°$ and $\gamma AA = 1{\cdot}7\gamma AB$) and then cuspate (case (e) $\theta = 30°$ and $\gamma AA = 1{\cdot}9\gamma AB$) to the limiting case where $\theta = 0$ ($\gamma AA = 2\gamma AB$) and the inclusion completely wets the intergranular boundaries.

Interfacial tensions are not the same as specific solid–vapour or solid–vacuum tensions but might be very approximately related. It may possibly be true that the inclusion is globular where its surface tension is greater than that of its host and cuspate where it is less (Stanton, 1964, pp. 66, 72) but any such conclusion should be regarded with caution. The problems of the shapes of inclusion located entirely within crystals of the host are somewhat different from those of inclusions at boundaries and triple-points.

Explanations in terms of surface *tensions* are simple but a little unsatisfactory and grain boundary configurations of anisotropic silicates are better explained in terms of interfacial *energies*, as follows: if the structures of host and inclusion are very different, a high energy non-coherent interface will result; the energy is lowest if the area is lowest so that the inclusion will form a globule because a sphere has the lowest surface area for a given volume. The curved interface will be incoherent and will only be stable if it is insensitive to the orientations of the lattices. This argument implies that the inclusion will be globular even if its specific surface energy is *less* than that of the host.

If the atomic structures of host and inclusion are similar so that a low energy interface can be achieved by forming a planar interface in a favourable orientation with respect to both lattices then the total surface energy will not depend on the total surface area and a geometrical shape may be formed.

Under special conditions, i.e. where the minor phase is relatively abundant, is mobile, and the energy of the host inclusion boundary is extremely low, an interface of large area can be achieved, i.e. the "inclusion" becomes dispersed along the grain boundaries of the major phase and wets it.

In general a minor phase which is dispersed either along grain boundaries or within crystals of the major phase tends to take a globular shape; idiomorphic crystals of certain species are common, cuspate crystals rare and the dispersed continuous layer along the "wetted" boundary extremely rare. The specific surface energies or tensions of the various minerals are not known but estimates of relative tensions can be made, and it would appear that globular crystals occur whether the specific surface energy of the host is greater or less than the inclusion, e.g. globules of spinel in plagioclase, of quartz in cordierite, of feldspar in hornblende and of olivine and periclase in calcite. The main control here appears to be an attempt to achieve the lowest possible interface area.

Most inclusions which are bounded by surfaces which are rational to the host are not perfect negative crystals but have rounded corners. The sharp internal corners required by the negative crystal faces are probably regions of high lattice strain and hence would tend to be unstable. The ultimate lowest energy shape might be regarded as a compromise between a negative crystal and a sphere.

POIKILOBLASTS

If the host mineral has abundant tiny inclusions it may simply be cloudy but where the inclusions are large the crystal appears spongy and the texture is *poikiloblastic* (an alternative spelling is poeciloblastic but this is less common) a term analogous with *poikilitic* in igneous rocks. The terms *emulsion* and *sieve texture* are also applied (Plates XIVa, XVd). *Ice cake texture* consists of rounded or corroded fragments of an early mineral surrounded by a single crystal of a secondary mineral. This is a replacement texture. It is distinct from *cement texture* in which an early mineral has been mechanically broken and the dispersed fragments cemented by a second mineral.

The inclusions may be randomly disposed, as in spongy cordierite, or may form regular patterns. In the latter case, the pattern may be a relict texture and represent a rock structure such as a foliation which is older than the crystal (a *helicitic* texture, Plates XXVI, XXVII) or a structure produced during the growth of the crystal (*snowball* or *rotational* texture, Plate XXV). The inclusions may form a pattern which is related to the internal structure of the host, e.g. cross, hour-glass or zonal *growth structures* (Plate XIIIc, d) or oriented inclusions due to *exsolution*.

Collette (1959) and Howkins (1961) departed from established practice and used the term *helicitic* for both syntectonic and post-tectonic inclusion patterns. Even though it might not be possible to decide on the origin of all such patterns, the distinction between rotational (syntectonic) and helicitic (post-tectonic) structures is an essential one.

Inclusions in crystals are of two main types: fluid (liquid, or liquid plus gas) and solid. Most fluid inclusions in solids have a spherical shape because of minimum surface area and energy requirements. Under ideal conditions, however, the host–inclusion interface may achieve a still lower surface energy condition by developing crystal faces to give a negative crystal in which the faces belong to the host.

Solid phases are dominant as inclusions in metamorphic crystals, and liquid or liquid–gas inclusions are most common in crystals deposited from melts (igneous), solutions (veins), or by hydrothermal metasomatism. Smith (1963, pp. 63–65, 509) has described the textures formed by the incorporation of fluid during both dendritic and layer growth and noted the tendency for inclusions to lie either along lineage boundaries (more or less perpendicular to growth surfaces) or along growth planes (parallel to growth surfaces).

The three main processes which cause the formation of inclusions are adsorption plus absorption, exsolution, and alteration (replacement).

(1) ADSORPTION AND ABSORPTION

A crystal growing in a medium with the same composition (e.g. quartz in a quartzite) enlarges itself by using all units of adjacent material but a crystal growing in a medium of different composition can only use a certain proportion of the material available and must reject the rest. The growth of the crystal is thus allowed by the inward and outward migration of various components. The crystal may free itself of all foreign material and be fresh and clear but under certain circumstances inclusions of unwanted minerals may remain (Bunn, 1933).

Inclusions increase the total free energy and thus a poikiloblastic crystal is not in its most-stable or least-energy condition. The surface free energy will be increased both because the total surface area of a spongy crystal may be an order of magnitude higher than that of a solid crystal of the same volume and because the problems of fitting inclusions in the host lattice means that the host is somewhat strained in their vicinity.

It would appear that the formation of poikiloblastic crystals would be favoured by processes involving large energy differences and enough excess energy in the system to permit the formation of "high-energy" crystals. In general, spongy xenoblastic crystals suggest rapid growth, and homogeneous idioblastic crystals suggest slow growth.

However, the presence of inclusions does not necessarily mean an unstable or high-energy condition as inclusions may be adsorbed on the moving front of a crystal in a stable fashion. The foreign atoms, ions, molecular groups or mineral units may be attached to the surface by two processes, adsorption and absorption (De Vore, 1955a, b, 1963; Thomas, 1962).

Adsorption (physical adsorption): a *loose* attachment of foreign material to the surface by, for instance, van der Waals bonds. The material is not accommodated in the structural pattern and may be located at defects or in irregular films or aggregates.

Absorption (chemisorption): the process by which additional units are chemically bonded to the structure and form an integral part with it.

Whether or not a foreign unit is adsorbed, absorbed or rejected depends on its ease of attachment to the free surface, i.e. its similarity in structure or its bonding ability. Units of foreign materials which have similar structure, interionic distances, etc., may be deposited on the surface to give a homogeneous layer or *overgrowth* of new mineral (*epitaxy*).

Small units of foreign material may be adsorbed onto the crystal surface and then either be enclosed by the growing crystal or rejected by being held on the advancing crystal face. This process can be understood by consideration of a surface tension model in which adsorption is controlled by the relative surface tensions of the host and impurity, the contact angle between them, and the ability of the impurity to "wet" the surface. Figure 13 illustrates the behaviour of a drop of liquid at a solid surface. In (a) there is a considerable difference in surface tension between liquid and solid, the contact angle is low and the liquid does not wet the surface. In (b) there is a small difference in surface tension, the contact angle is high and the liquid wets the surface.

Consider the interface between a crystal and an impurity. If the surface tensions are similar (possibly because of some similarity in structure, bonding, etc.) the impurity will wet the surface, be tightly adsorbed and possibly enclosed as an inclusion as the crystal grows. If the surface tensions are markedly different, the impurity will not wet the surface. Units with the same composition as the crystal will be attached to the surface, i.e. will wet it, in preference to those of the impurity and so they can be attached *between* the impurity and the crystal. Thus the impurity may be held loosely at the surface as the crystal grows, i.e. the crystal "rids itself of inclusions" or "pushes the inclusion aside".

A *surface-energy* model also explains inclusions. If the impurity can form a low-energy interface with the growing crystal, then under conditions of rapid growth, i.e. high free-energy gradient, the activation energy of attachment is easily overcome. The crystal builds up around the inclusion and there is little driving force for its removal because it is bounded by a low-energy immobile interface.

Consider the model (Fig. 48) of the growing crystal of composition *ABC* in a matrix of a mixture of materials *AB*, *BC*, *AD* and *E*. The crystal grows

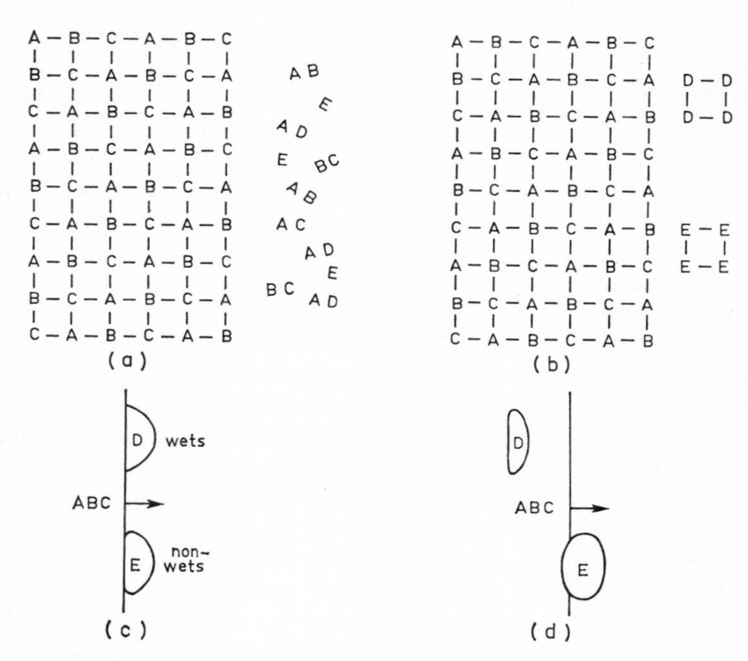

FIG. 48. Adsorption and crystal growth. (a) The crystal *ABC* in an environment of *AB*, *E*, *AD*, etc., grows by adding *A*, *B* and *C*. (b) Impurities of *D* and *E* might be adsorbed onto the surface, (c) *D* "wets" the surface, but *E* does not. (d) Advance of the face encloses *D* poikiloblastically, but *E* is rejected

most easily by attaching groups of *AB* or of *BC*; it grows less easily by breaking up the *AB*, *BC*, or *AD* groups into individual *A* and *B* ions. If an *AD* is broken up, the *D* ion may be rejected or may be retained in the lattice in solid solution. The material *E* is considered to be completely foreign. In the hypothetical example shown, *D* is adsorbed and retained as an inclusion but *E* is rejected and moves outwards.

Ramberg (1952) accepted the possibility that a growing crystal could push the surrounding minerals aside, provided that the interfacial energies were sufficiently high; thus he postulated that garnet pushed mica aside because the interfacial energy of a garnet–mica interface is high, but quartz is enclosed because the garnet–quartz interfacial energy is low.

Strong adsorption of impurities on crystal faces is not only caused by structural or chemical similarities but may arise because of special surface characteristics of the impurity. Carbon (as graphite or as amorphous carbon in metamorphic rocks) has a strong surface activity due to its large surface area and to free van der Waals bonds, and carbon (as well as iron oxides) shows a great tendency to be strongly adsorbed on crystals of andalusite, albite, chloritoid and garnet and then to be enclosed within the crystals. Adsorption of impurities on all grain boundaries may hinder grain growth and cause the grain size of a rock to be smaller than normal. If impurities are adsorbed preferentially on certain crystal faces they may slow the advance of these with respect to others and thus modify the habit of the crystal.

Wetting is affected by surface agents and by bonding. Silica and glass are readily wetted by anionic agents but readily become non-wetted by solutions with low concentrations of cationic agents because the cations are adsorbed onto fixed negative charges of silicate groups on the interface (Davies and Rideal, 1961, p. 438).

Inclusions cause a certain amount of strain in the surrounding host depending on the energy of the interface, i.e. on the degree of lattice mismatch. Rosenfeld and Chase (1961) described garnet which is anisotropic around inclusions and suggested that this strain was due to different amounts of contraction by host and inclusion on cooling, and that the strain was related to the pressure and temperature of crystallization.

Cordierite (and to a lesser extent andalusite, staurolite and garnet) characteristically forms large crystals with irregularly distributed inclusions (quartz, mica, zircon, iron oxides, etc.). Cordierite's inability to rid itself of inclusions of quartz is probably due to the low-energy interfaces between quartz and cordierite (suggested by their similar low density, packing index, structure and hardness). Mica tends to form higher energy boundaries and is commonly rejected by cordierite porphyroblasts to be concentrated at the margins (Plate XIVb).

Perhaps the most common reason for such poikiloblastic textures is a combination of such circumstances as

(1) Rapid growth of the host.
(2) Ability of the host lattice to absorb the strains imposed by the impurity.
(3) High chemical abundance of the impurity throughout the system.
(4) Some surface affinity between host and inclusions.
(5) Difficulty of nucleation of the host.

Helicitic structure is a special case of the envelopment of certain inclusions by a growing crystal (particularly albite, garnet and staurolite). It refers to a straight or curved inclusion pattern which is a structural element of the rock older than the crystal and which has been preserved during the growth of the crystal (Fig. 61, Plate XXVIa, b).

The internal structure represents an S surface and is referred to as S_i (S-internal); the S-surface outside the crystal is the S_e (S-external). The schistosity or bedding S_e may pass continuously through the crystal and out the other side (i.e. S_i is concordant with S_e) showing that the crystal was formed after S_e. The external foliation abuts the crystal as though the crystal outline had been cut out with a biscuit cutter. Such crystals are typical of the post-tectonic crystallization stage in many schists. S_i and S_e are discordant in many polymetamorphic schists where S_i represents an older foliation and S_e a newer one. Growth of the host crystal is so gradual that the inclusions are enclosed without displacement and still show their previous orientation in detail (Sturt and Harris, 1961; Simpson, 1964).

It has been suggested that the *size* of the inclusions in some helicitic crystals (particularly of quartz in garnet or amphibole) may not change appreciably as they are enveloped in the growing porphyroblast (Harris and Rast, 1960a; Rast, 1965). Where quartz inclusions marking S_i are much smaller than those in the matrix marking S_e, it may indicate that the grain size before the formation of the porphyroblast was smaller than that after. This explanation is reasonable where the grainsize of the enclosed quartz is estimated in groups of quartz crystals. Where there are isolated single-crystal inclusions, the variation in size may simply be due to differential dissolution of the inclusions.

Hour-glass and *Maltese cross* structures are regular geometrical patterns of inclusions which have been enveloped during growth; the patterns are related to the host crystal structure (Plates XIIId, XVI). Frondel (1934) and Buckley (1934, 1951, p. 417) demonstrated that such patterns could be produced in crystals growing in solution and attributed them to differential adsorption of impurities on various crystal faces. For example: in the face-centred-cubic NaCl structure, {100} faces contain alternate positive sodium and negative chlorine ions and have no net charge. On the other hand, {111} faces contain close-packed ions which are either all cations or all anions, depending on where the structure is cut. Thus {100} and {111} have different adsorption behaviour towards charged impurity ions.

Consider a hypothetical crystal consisting of two forms {100} and {111}; impurities are adsorbed on {100} but not on {111} (Fig. 49). As the crystal grows, inclusions will be retained on successive positions of {100} but not {111} and a Maltese cross pattern arises. Adsorption of this kind may, or may not, lead to a decrease in the growth rate of the face.

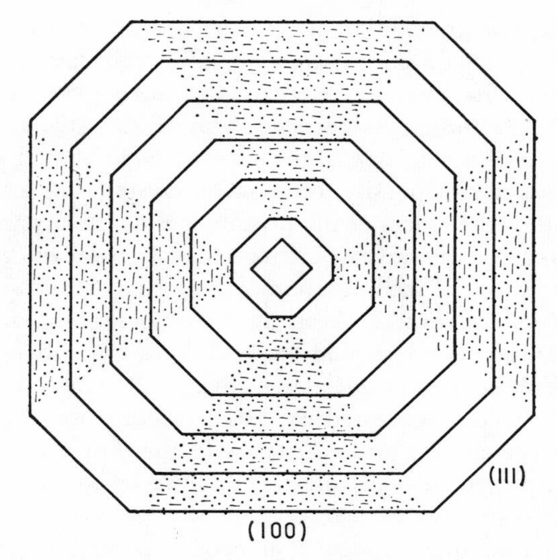

FIG. 49. Maltese cross inclusion pattern. Successive stages in the growth of a crystal accompanied by adsorption of impurities on {100} but not {111} produces a Maltese cross (after Buckley, 1951, Fig. 125)

The Maltese cross or hour-glass pattern in chloritoid and chiastolite appears to be due to preferential adsorption (Frondel, 1934). Andalusite forms orthorhombic prismatic crystals with the angle (110) $\wedge$ $(1\bar{1}0)$ of 89°, elongate parallel to *C*. Basal sections of the variety chiastolite show various inclusion patterns (Harker, 1939, p. 43), including:

(1) A central area whose outlines are parallel to the prism faces and which is either clouded with inclusions or forms a clear quadrilateral.

(2) A cross extending from the centre to the prism edges (i.e. the lines parallel to *C* at the intersection of the {110} faces.

(3) Concentrations at the outer prism edges (corners).

The inclusion-filled regions may be structureless but two patterns are found. Most common is a feathered texture due to the arms of the cross being composed of small planes parallel to {110} (Plate XVI and Fig. 50). Less common is a helicitic structure in which both core and arms are composed of dark wisps with similar appearance and orientation to the matrix foliation (Plate XIIIc).

Three hypotheses have been advanced to explain the feathered chiastolite cross.

1. Harker (1939, p. 43) gave an explanation which has apparently been accepted without question for the past thirty years as Deer, Howie and Zussman (1962, vol. 4, p. 133) quote it.

Harker considered that "the dark area in the centre represents the nucleus of the crystal, unable at that stage to free itself from inclusions. Subsequent growth has been effected by a thrusting outward, which was most effective in the directions perpendicular to the prism faces. Much of the foreign matter brushed aside in this growth accumulated at the edges of the prism, and was enveloped by the growing crystal. The arms of the dark cross represent thus the traces of the prism edges as the crystal grew.

"The streaming out of trains of inclusions away from the centre presents almost a visible picture of the operation of the forces of crystallization."

The analogy of the crystal "brushing" the inclusions aside like a ship moving through water is not a good one and there is no obvious mechanism for propelling inclusions sideways along advancing crystal faces.

2. Rast (1965, p. 86) suggested that the structure was due to dendritic growth at the corners of the andalusite, the feathery projections being relics of the dendritic projections. An illustration by Alexander (1951, Fig. 3) of dendritic growth resembles the feathery inclusion patterns and lends support. Rast considered that the homogeneous central zone was formed by slow layer growth, then an increase in supersaturation caused dendritic growth at the corners while layer growth contained in the middle of the faces. The supersaturation was reduced so that growth ceased at the corners and enlargement of the centres of faces produced a cross-shaped crystal.

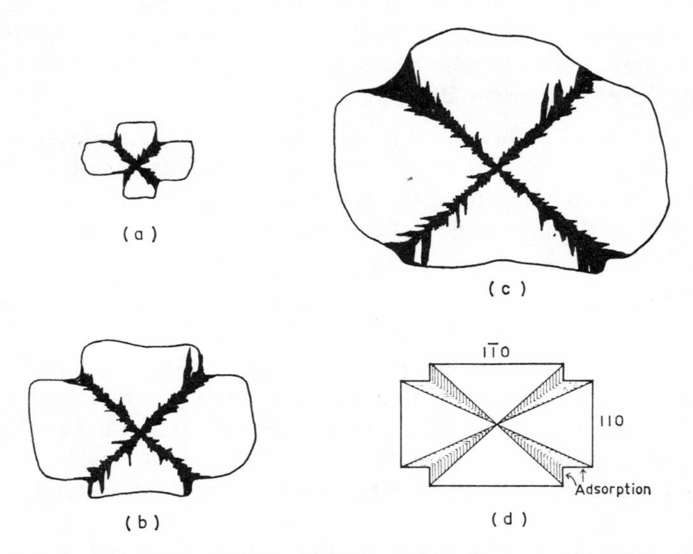

Fig. 50. Chiastolite. A crystal (Plate XVI) is considered to grow through stages such as (a), (b) and (c). The inclusion pattern and shape is due to preferred adsorption on (110) and ($1\bar{1}0$) and depression of growth at the prism corners (d)

3. It is proposed here that the structure is due to preferred adsorption of carbonaceous particles against the prism faces and particularly at the prism edges. Carbon adsorbs strongly and appears to "wet" many materials easily. The greater concentration of carbon at the crystal corners is associated with a re-entrant region which suggests that the corners did not advance as rapidly as would be required to complete a simple prism, presumably due to "poisoning" by the carbon. Stages in the growth of the crystal can be reconstructed as in Fig. 50, if simple layer growth is assumed. The reconstruction suggests that poisoning was more prominent earlier in the life of the crystal.

The abnormal adsorption of impurities in chiastolite may be in some way related to its unusual atomic structure. The other aluminium silicates (kyanite, sillimanite) contain chains of alumina octahedra (6-fold coordination) but in andalusite the electrostatic valence rule is not strictly fulfilled and some alumina is in 5-fold coordination.

Intergrowths due to Replacement

The complete replacement of one mineral by another may be recognized only if a pseudomorph is formed. Partial replacement will lead to some kind of intergrowth between host and metasome but similar textures are produced by replacement, exsolution and simultaneous crystallization because the factors governing the three processes are similar.

The main study of replacement textures has been in ores rather than in silicate minerals; textural interpretations have been summarized by Bastin (1950) and Edwards (1960) but there is considerable disagreement in interpretation and Stanton (1964) has shown that some so-called replacement textures are probably due to simultaneous crystallization of minerals with different surface energies.

Typical textures include the following:

(1) *Rim replacement.* The alteration of the outer part of the crystal. The attack is controlled by the distance from the grain boundary (Edwards, 1960, Figs. 99, 100), e.g. peripheral uralitization of pyroxene, altered rims of feldspar (Voll, 1960, pp. 522–4) and peripheral feldspathization of quartz grains in quartzite. Rim replacement should be distinguished from a growth rim in which crystallization occurs *upon* or *around* a crystal.

(2) *Core replacement.* Alteration attacks the centre of the crystal but leaves the outer part unaltered (Edwards, 1960, Figs. 103, 104), giving *atoll structure.* Rast (1965, p. 95) attributed atoll structure in garnet to partial replacement controlled by an original compositional zoning.

(3) *Zonal replacement.* A layer of secondary mineral is formed partway between the margin and the core of a crystal (Edwards, 1960, p. 118 and Fig. 106). It is rare in metamorphic rocks but a possible example is given

in Poldervaart and Gilkey (1954) where garnet has partially replaced plagioclase.

(4) *Guilded penetration replacement* (Bastin, 1950, p. 38). The metasome makes use of pathways of enhanced migration such as cracks or cleavages. *Vein replacement* produces *non-dilational* veins which should be distinguished from *dilational* veins which have matching walls. Replacement commonly takes place along cleavages (Edwards, 1960, Fig. 115), e.g. chloritization of biotite, sericitization of feldspar or the serpentinization of olivine to give *mesh texture serpentine.* Cleavage-controlled replacement is a special example of *lattice-controlled replacement* in which relative ease of diffusion in certain lattice directions influences the migration of foreign ions or solutions and the deposition of secondary mineral. Under these circumstances the lattice may also control the orientation of the interface between the primary and secondary mineral and the orientation of the lattice of the secondary mineral to give an *oriented intergrowth.* Some replacement perthites of this kind cannot be distinguished from exsolution perthites.

Distinguishing Characteristics between Types of Oriented Intergrowths

Oriented intergrowths in which host and inclusion have some structural relationship are favoured by the presence in host and inclusion of the same or similar atoms forming similar patterns with similar atomic spacings (lattice vectors). The correspondence in atom type, pattern and vector may be very close but considerable departures are possible, particularly at high temperatures. A certain degree of misfit across an interphase boundary (either of an intergrowth or overgrowth) can be compensated for by a series of *epitaxial dislocations* (Plate IVa).

Oriented intergrowths may be formed by simultaneous crystallization, exsolution, or partial replacement. The distinction between exsolution and replacement textures presents a very important problem. A decision as to the origin of certain conformable ore bodies and of various rocks of igneous appearance may hinge largely on whether a sulphide intergrowth or a perthite was formed by exsolution (hence the crystal was originally a homogeneous solid solution with its implication of high temperature) or whether it formed by replacement (with the possibility of a low to moderate temperature). Stanton (1960, p. 32) maintained that sulphide solid solutions, such as sphalerite–chalcopyrite, segregate so quickly on exsolution (giving separate crystals rather than oriented intergrowths) that exsolution intergrowths should not be preserved in ore deposits by cooling from high temperatures. He considered (ibid., p. 34) that these textures "could represent low-temperature segregation during crystallization of the amorphous or cryptocrystalline sulphide mixtures".

The formation of an oriented intergrowth, by either exsolution or replacement, is governed by the same requirements of minimum interfacial energy and strain. In general, one might expect replacement to take place under much higher energy (supersaturation or chemical activity) gradients than exsolution, and hence to be more irregular in appearance. Emulsion, patch, plume or ice cake textures may well form by replacement, but it is very unlikely that perfect, delicate, parallel-sided blades or parallel strings (in one, even two or three directions) could be formed by replacement.

If replacement has not proceeded far, the host will be more abundant than the metasome and will retain a uniform structure within which the metasome is distributed. At an advanced stage, the metasome predominates and the primary mineral remains as isolated relict patches, all of which have the same orientation, e.g. replacement patch-perthite.

Dendritic replacement texture may be due to growth of a metasome along a cleavage and then out into the body of the host, but conversely may be a relict of the original crystal after almost complete replacement (the metasome acts as the "host", Bastin, 1950, p. 44).

The shape of the primary/secondary mineral interface has been regarded as significant in determining which mineral is host and which is metasome. It has been said that the metasome is idioblastic towards the replaced host or that the embayments on a scalloped margin are convex towards the metasome. Edwards (1960, p. 129) pointed out that these criteria are unreliable and Stanton (1964) showed that they are controlled more by mutual grain boundary energy than order of crystallization.

The simplest mineralogical examples of oriented intergrowths occur between magnetite, ilmenite and hematite (Gruner, 1929; Bray, 1939). Layers of the minerals intergrow so that (111) of magnetite is parallel to (0001) of ilmenite or hematite. The basic explanation (allowing for a little oversimplification) is that the oxygen tends to form close-packed structures with the cations in interstitial "holes" and the triangular pattern of close-packing for (111) in cubic close-packed and for (0001) in hexagonal close-packed structures are the same (Fig. 51). The "fit" between units of these minerals is not perfect but a general relationship may be summarized as follows:

Spinel (111) $/\!/$ hematite (0001) $/\!/$ ilmenite (0001) $/\!/$ rutile (010)

The interface thus can be a low-energy coherent surface of oxygen atoms shared by the magnetite lattice on one side and hematite or ilmenite on the other. These compound crystals are formed by both exsolution and replacement (oxidation, etc.).

Magnetite ($FeO.Fe_2O_3$) and spinel ($MgO.Al_2O_3$) form intergrowths with plates of the inclusion parallel to (100) of the host. Both minerals belong

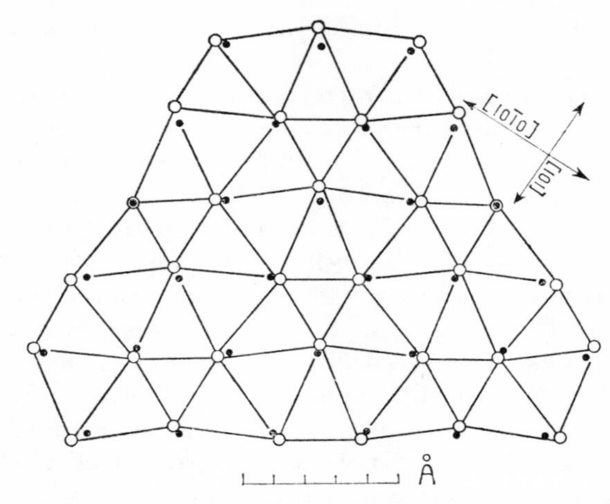

Fig. 51. Oriented intergrowths between ilmenite and hematite are allowed by structural corresondence. Pattern of oxygen atoms in (0001) ilmenite or hematite (open circles) and those in (111) of magnetite (closed circles) (after Gruner, 1929, Fig. 1)

to the spinel group and are isomorphous. They show solid solution with interchange of Fe^{2+} and Mg^{2+} ions and Fe^{3+} and Al^{3+} ions. The oxygen lattice is common to each and a mixed crystal can be disordered or show some ordering into a superlattice. The corresponding interionic distances are a little different for the two end members, e.g. *a* in magnetite and in spinel are 8·396Å and 8·103Å, respectively. Cooling leads to an increase in order and the solid solution exsolves into two almost identical structures as interleaved layers by diffusion of cations.

Triangular oriented arrays of sillimanite in biotite (Tozer, 1955; Chinner, 1961) are possibly due to replacement but are best regarded as due to epitaxial or catalysed nucleation allowed by structural similarities.

Oriented zoisite prisms (Hall, 1937) or magnetite inclusions (Richards, 1950) in mica can not be explained on the grounds of exsolution, alteration or simultaneous crystallization and are probably a special case of poikiloblastic growth, possibly by the "skating" process of Frondel (1940a). The magnetite forms a triangular pattern with blades constituting a single skeletal crystal with (111) parallel to (0001) of the mica, and one face diagonal parallel to *b* of the mica.

Oriented intergrowths of various sulphides are permitted by close structural similarities (sulphur atom patterns), e.g. in (111) of sphalerite, chalcopyrite, bornite and chalcocite; in (100) of sphalerite and (100) or (001) of chalcopyrite; in (0001) of pyrrhotite and (111) of pentlandite or (111) chalcopyrite; in chalcopyrite (111) and cubanite (111); and in silver (111) and dyscrasite (0001) etc.

PERTHITES

Perthites, antiperthites and peristerites are special cases of oriented intergrowths. *Perthite* is a feldspar intergrowth in which inclusions of plagioclase are enclosed in potash feldspar (Goldich and Kinser, 1939; Rosenqvist, 1950; Gates, 1953; Sen, 1959). *Antiperthite* is the reverse case where inclusions of potash feldspar are enclosed in plagioclase. *Peristerite* refers to plagioclase in the compositional range of about An_3 to An_{17} which consists of exsolved microscopic or submicroscopic mixtures of An_{0-2} and $An_{26\pm2}$ (Laves, 1951; Brown, 1962). Much of the following discussion of perthites also refers to peristerites which are formed by exsolution from homogeneous high-temperature plagioclase.

Definitions of perthite and antiperthite are based on the role of host and inclusion; alternative definitions which depend on the relative proportions of the two feldspars are less satisfactory. It is easier to decide which is host and which is inclusion in most specimens than to determine relative abundances. Although uncommon, species have been described in which the minor constituent acted as host (de Saenz, 1965, p. 109). Feldspars containing equal proportions of feldspar so that it is not possible to decide which is the host have been called *mesoperthite* (Michot, 1951) and *isoperthite* (Rudenko, 1954).

Tuttle (1952) advanced a general classification according to size:

Sub-X-ray perthite inclusions less than	15Å (perpendicular (201))
X-ray perthite inclusions less than	1μ
Cryptoperthite inclusions less than	1–5μ
Microperthite inclusions less than	5–100μ
Perthite inclusions less than	100–1000μ

The general classification on size and shape follows that of Alling (1938) for igneous plutonic feldspars (Fig. 52), e.g. stringlet (hair or guttate), string, rod, bead, band, film, patch and plume (flame) perthites.

The intergrowths are commonly strongly oriented and are best seen on (010) cleavage faces where the inclusions lie almost parallel to the basal plane. The included phase has been described as lying close to $(60\bar{1})$ or $(70\bar{1})$ or to (100), (010), (001), (110), (320) or even $(13{\cdot}0\bar{2})$. Watt (1965, p. 17) stated that vein perthites have a variable size and orientation, film perthite is perpendicular to (010) and at $-75°$ to (001); and that strings are parallel to (010) and at $-73°$ to (001).

Perthites and antiperthites are most common in plutonic igneous rocks (Anderson, 1928; Alling, 1936). They are restricted to the higher grade metamorphic rocks, being most common in gneisses, granulites and charnockites

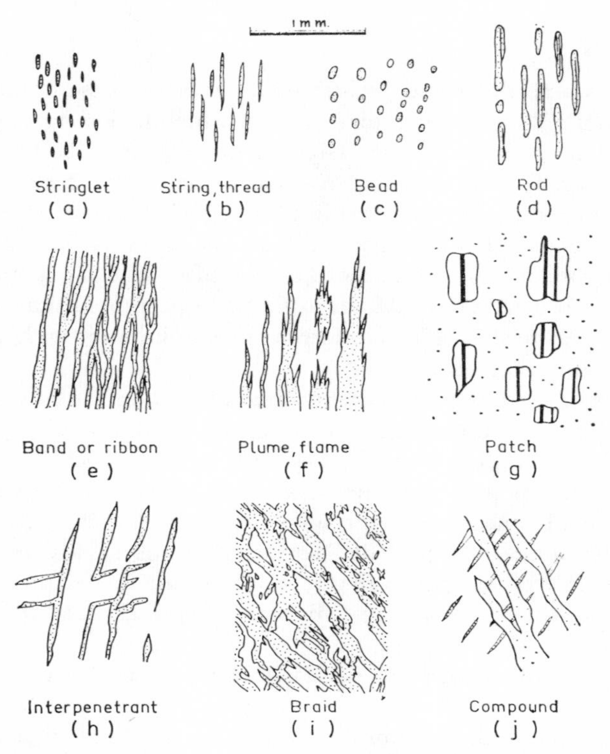

Fig. 52. Perthite. Classification mainly on shape (modified after Alling's 1938 classification of igneous perthites)

(Eskola, 1952; Rosenqvist, 1952; Subramaniam, 1959; Heier, 1961; Watt, 165; de Saenz, 1965). Crowell and Walker (1962) figured hair mesoperthites in metamorphosed syenites of the Amphibolite Facies, but perthites are rare in rocks of the lower Amphibolite Facies and appear to be unknown in the Greenschist Facies. They occur in high grade hornfelses (Drever, 1936) but are not common. Thermal metamorphism of coarse feldspathic rocks may produce perthites by exsolution. Most antiperthites are in the oligoclase-andesine range (less common in albite and almost unknown in labradorite) and most are found in granulites or charnockites. Evidence suggesting that exsolution is promoted by deformation was reviewed by Chayes (1952).

Perthites and antiperthites are considered to result from three processes:

(1) Exsolution from an originally homogeneous mixed feldspar.
(2) Partial replacement of one feldspar by another.
(3) Simultaneous crystallization of two feldspars (Robertson, 1959).

No reliable criteria have been found for distinguishing between these three genetic types but it is generally believed that most are due to exsolution. A constant proportion of host to inclusion in separate grains strongly favours an exsolution origin as replacement would be expected to produce varied proportions (Hubbard, 1965).

Perthites attributed to exsolution have all the characteristics of oriented intergrowths due to low temperature exsolution from an originally homogeneous, high temperature phase (Laves, 1951, 1952a) (The term perthite will be taken to include antiperthite in the following.) The original homogeneous phase is a mixed potash–soda–lime feldspar. During exsolution, either the potash feldspar or the plagioclase phase dominates as the host, and the minor phase is progressively liberated as inclusions. The size and shape of the included phase depends on the ionic mobility (i.e. temperature) and on the surface energy of the interface. The exsolved phase may form globular inclusions where the lattices of host and inclusion are independent and a minimum interfacial energy requirement leads to the interface becoming as small as possible. Feldspars are composed of a linked framework of SiO_4 and AlO_4 tetrahedra with either large (K) or small (Na, Ca) cations in the interstices and the structures of the two feldspars are so similar, however, that high coincidence and hence low interface energy boundaries can be produced by the formation of oriented inclusions.

The original homogeneous feldspar may have a completely disordered distribution of potash, soda or feldspar (although this is unlikely even in volcanic rocks) but some initial order is probably present. A fall of temperature produces internal strains because of differences in cation size and this strain provides the driving energy for diffusion. Movement, chiefly of potash one way and of soda the other, produces regions in the lattice which are more like the potash feldspar molecule or the plagioclase molecule. The strain *within* these regions is decreased but the strain *between* them increases until the regions separate as distinct homogeneous phases. The structures of the two will be similarly oriented and there will be a high degree of atomic correspondence across the interfaces. The process is at least partly reversible and perthites can be homogenized by heating (although the activation energy is very high, e.g. Jagitsch and Mats-Goran, 1954). The temperature of homogenization is a lower limit to the temperature of crystallization of the original feldspar.

A high temperature is necessary to produce the original mixed feldspar and a slow rate of cooling is necessary to allow exsolution; exsolution perthites are thus common in plutonic igneous rocks and high grade metamorphic rocks and would not be expected in low grade types. It is necessary, of course, to distinguish between perthites in which feldspar formation and exsolution both took place during metamorphism and those in which metamorphism produced exsolution of a feldspar of pre-metamorphic origin.

The size and shape of the inclusions is a function of ionic mobility and appears to depend mainly on the kinetics of exsolution. Rapid cooling prevents exsolution and preserves the strained metastable mixed feldspar in volcanic rocks. Somewhat slower cooling allows slight exsolution with thin lamellae and a strong degree of structural continuity. Slower cooling allows complete exsolution and the development of thicker, less regular inclusions. The highest degree of ordering would be the complete rejection of the minor phase from the host lattice and its concentration as grains on intergranular boundaries. Smith (1965, p. 1986) described complete rejection of spinel from magnetite in rocks from the Sanidinite Facies. Hubbard (1965, p. 2040) considered that the mantle of alkali feldspar surrounding antiperthite in a charnockite from Nigeria was due to complete exsolution.

The replacement mechanism (Orville, 1962) involves the entry of ions (chiefly the cations K^+, Na^+ or Ca^{2+}) into an essentially homogeneous crystal to replace one feldspar by another. A strong chemical activity gradient at high temperatures will allow a reconstructive transformation which is energetically inefficient and which may produce a recognizable replacement texture. The primary feldspar may be cut by irregular intersecting veins of secondary feldspar which begin at the crystal boundaries and decrease in size towards the centre. Characteristic patch-perthites with similarly oriented relict blocks of original feldspar are typical. Replacement under a lower energy gradient however, produces oriented intergrowths which, superficially at least, appear to be identical with exsolution textures. Replacement may take place preferentially along cleavages to give wispy inclusions which may be indistinguishable from exsolution lamellae lying along rational lattice planes (which may also be cleavages).

Comparison of exsolution and replacement intergrowths in various magnetite, ilmenite, hematite, rutile, sphene, leucoxene and goethite mixtures (Rao and Rao, 1965) indicates that an exsolution origin is favoured by a crystallographic control of the form of the intergrowths, an absence of enlargement where blades or lamellae cross, and a lack of relationship of included minerals to grain boundaries or to mineral contacts. However a relationship between the distribution of inclusions and the outer margin of the crystal must not be taken as proving a replacement origin for the inclusions. It is common in metal solid solutions to find that exsolution has taken place in the centre of the crystal but that the rim is free from inclusions. This can be shown to be due to a lack of nucleii (in this case vacancies) near the rim and in some instances the marginal material diffuses to the grain boundary and precipitates there instead of exsolving in the outer part of the crystal.

The plagioclase of some metamorphosed igneous rocks is cloudy and turbid (*clouded plagioclase*; Bailey, 1916, p. 204). MacGregor (1931a) and Pichamuthu (1959) attributed clouding to exsolution from plagioclase during metamorphism and the phenomenon is generally accepted (Bailey, 1950, p. 303) as indicating that a rock has been metamorphosed.

Poldevaart and Gilkey (1954, p. 88) concluded that clouded feldspars are common in thermal metamorphic aureoles but that not all the plagioclase there is clouded nor is clouding confined to this environment; it follows that clouding alone cannot be used as evidence that metamorphism has occurred. Slight clouding may be due to exsolution but strong clouding, which has only been observed in intermediate plagioclase compositions, is due to the introduction of iron from outside of the crystal.

Clouding is most common in plagioclase, but the same effects may be encountered in apatite, spinel, olivine, pyroxene, amphibole, mica, quartz, potash-feldspar, garnet, calcite and serpentine. It is due to the presence of very numerous tiny dark particles whose identity cannot be ascertained with certainty even under the microscope (Plate XVIIIc). The particles have been described as dust-like, short rods, thin hair-like growths, needles, or sub-microscopic, and are generally said to consist of opaque iron oxides but other minerals such as spinel, garnet, biotite, hornblende and rutile may also be present in clouded minerals.

The requisites for plagioclase clouding appear to be an elevated temperature for a prolonged period, the presence of water and a supply of iron. These conditions may be met when igneous rocks are affected by thermal or low-stress regional metamorphism but also by deuteric activity.

Burns (1966) correlated the degree of clouding of the plagioclase in the Scourie dykes with the H_2O content of the feldspar and attributed it to "post-magmatic hydrothermal activity". The dark feldspar in some charnockites and metagabbros differs from normal clouded feldspar in that no inclusions can be seen under the microscope; electron microscope examination shows the presence of defects and the feldspar contains 2250 ppm of water (Bridgewater, 1966).

CHAPTER 7

Textures of Thermal Metamorphism

THERMAL metamorphism near igneous contacts involves crystallization at moderate to high temperatures and low to moderate confining pressures with a general absence of directed pressures. Movements accompanying the intrusion of granitic magma may cause deformation during thermal metamorphism but this is not common.

The term *hornfels* was originally applied only to the fine-grained and dark-coloured thermal metamorphic rocks but is now used as a general class name for all thermal metamorphic rocks. Its use as a verb (abhorred by Harker, 1939, p. 40, as an "affront" to the English language) is widespread.

The common minerals of hornfels are quartz, feldspar, calçite, forsterite, cordierite, dolomite, mica, epidote, amphibole, pyroxene, andalusite, garnet and sillimanite. The final, most stable texture in an aggregate of various mixtures of these minerals depends largely on their relative surface energies and abundances but is not particularly influenced by the original rock type. The lack of deformation, however, is favourable for the preservation of relict (palimpsest) textures and these dominate non-equilibrium textures, particularly at low to medium grades. The rocks are typically *crystalline*, e.g. consist of interlocking crystals which have largely formed in their present positions.

Microscopic textures include the following (Fig. 53 and Plate XIX):

Granoblastic (granular): containing equidimensional generally xenoblastic crystals of approximately equal size.

Polygonal: containing crystals with polygonal shapes (many of them five- or six-sided) and fairly straight boundaries meeting at triple-points. Many hornfelses, marbles and quartzites have a *granoblastic-polygonal* texture. The term "mosaic texture" has been applied in this way but "mosaic" has been used in such diverse ways (Johannsen, 1939, p. 224; Niggli, 1954, pp. 242, 244; Voll, 1960, p. 518) that it is better to restrict it to the more widely accepted sense, i.e. the defect structure in single crystals illustrated in Fig. 9.

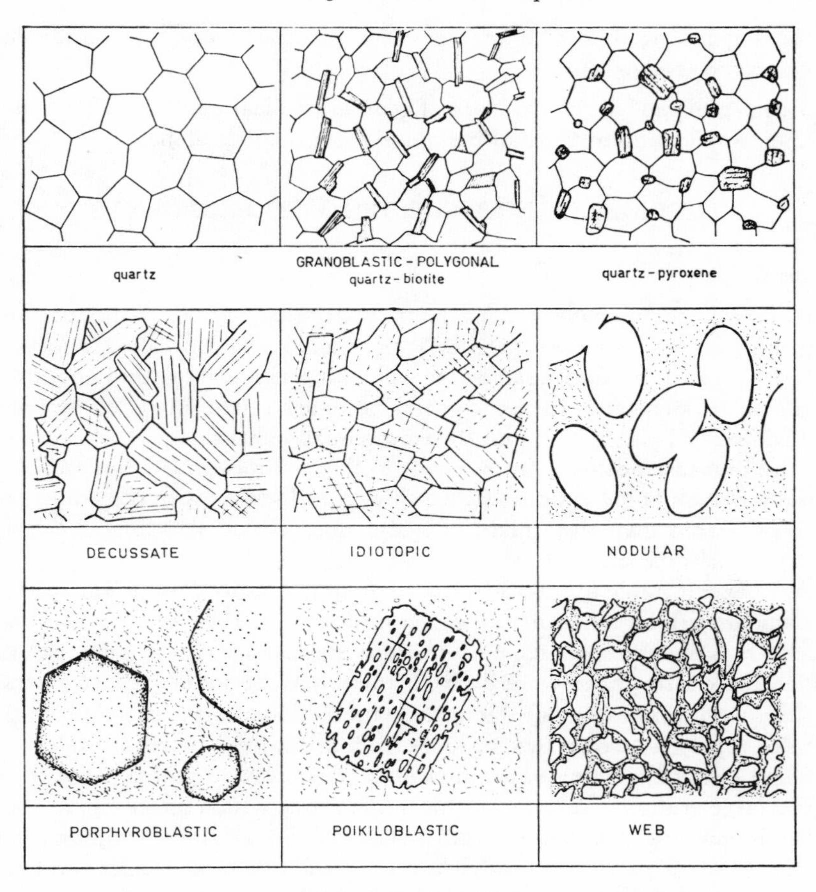

FIG. 53. Common metamorphic textures, mainly thermal

Decussate: a granoblastic texture in which the crystals tend to be subidioblastic, prismatic and randomly oriented.

Porphyroblastic: containing porphyroblasts. *Maculose* or *knotted* rocks contain scattered porphyroblasts, generally all of the same species and of similar size, in an otherwise fine-grained hornfels matrix.

Blastoporphyritic, blasto-intergranular, blasto-ophitic, blasto-amygdaloidal: metamorphic textures dominated by relict pre-metamorphic igneous textures, still recognizable as having been porphyritic, intergranular, ophitic and amygdaloidal respectively.

Blastopsephitic, blastopsammitic, blastopelitic: metamorphic textures dominated by pre-metamorphic relict sedimentary textures, still being recognizable as having belonged to conglomerates (or breccias), sandstones and mudstones respectively.

Terms prefixed by "blasto-" are of general application and are not limited in their application to thermal metamorphism textures alone.

THERMALLY METAMORPHOSED SEDIMENTS

Quartzites

Thermal metamorphism converts a *quartz-sandstone* (or orthoquartzite) into a *quartzite* (or metaquartzite) which could also be called a quartz-hornfels. A rock is called a sandstone when it breaks along grain boundaries; it is called a quartzite when the grains become so tightly bonded by recrystallization or by a cement that the rock breaks across them.

Pure quartz-sandstones are resistant to thermal metamorphism and show little alteration at low grades. Minor clay impurities between the grains may recrystallize to give tiny flakes of white mica which lie along the intergranular boundaries.

As quartz is stable at most metamorphic temperatures, there is no internal free-energy driving force to cause recrystallization. However, if the quartz has been previously strained, either within the rock itself or in the rock from which the clastic grains were derived, there may be sufficient lattice strain to cause polygonization or recrystallization. Thus large undulose grains may be replaced by one, two, or more fresh unstrained quartz crystals.

The driving force due to grain boundary energy is small because of the initial large grain size, however, quartz-sandstone with a grain size of about 1 mm is commonly recrystallized to saccharoidal quartzite containing grains 3 mm across. This requires rather a high grade of metamorphism and possibly the presence of an intergranular fluid.

The end product is a granoblastic quartzite with a polygonal texture consisting of interlocking quartz crystals with rather straight grain boundaries meeting in triple-points with equal interfacial angles. This is the equilibrium texture for a polycrystalline aggregate with rather low lattice anisotropy and there seems little reason to doubt that it was produced by a process somewhat analogous to primary recrystallization in annealing. The lack of attainment of the equilibrium texture is indicated by the departure from straight boundaries, equal grain size and equal interfacial angles. Such quartzites may contain a few quartz crystals which are about five times larger than the majority and these presumably formed by secondary or abnormal recrystallization.

Little is known of the details of the nucleation and growth process. Crystals may grow from:

(1) True nucleii, i.e. very tiny centres, located probably at grain boundaries or positions of high lattice strain.
(2) Strain-free units in a polygonized crystal.
(3) Original clastic grains.

Cherts are metamorphosed readily to fine to coarse quartzites under the driving force of relatively high grain boundary energy (Plate VIIa, b). Slight heating or even faulting is apparently sufficient to provide the required activation energy for recrystallization.

Most of the growth phase of the recrystallization (but not of the initial nucleation) can be seen in thin section (Plate VIIa, b). The chert develops a *grumous* (patchy or clotted) texture due to recrystallization beginning at isolated centres or along veins or fractures so that the original crystals of the matrix are replaced by secondary crystals which have a radius from 10 to 100 times larger and which form groups. The crystals within the groups impinge on each other and their growth almost ceases, but at the same time each group grows outwards into the matrix until the groups themselves impinge on each other. The advance of the front of the group is visualized as follows: a newly crystallized grain (or even a favoured original grain) may grow by moving its boundary outwards in all directions and the boundary cease to move when it is in contact with three or four primary grains whose lattice orientations are unfavourable for further movement. A new crystal then forms at the interface with the unrecrystallized chert either from a nucleus or from a favourably oriented original crystal, and its boundary expands outwards through adjacent crystals to give a new secondary crystal. Boundary migration continues until the boundary is again impeded and the process is repeated.

This initial phase produces a fine to medium-grained aggregate from the initially very fine-grained chert when all the groups have met each other; crystallization may continue by minor grain boundary adjustment to give general coarsening and a granoblastic polygonal texture.

The grain size of the recrystallized chert will increase with time and temperature but not directly proportionally. The ultimate grain size will be independent of diffusion in the monomineralic aggregate and will be largely unaffected by difficulties in nucleation. It will be controlled by:

(1) The spacing of the initial nucleii. Quartz nucleates easily but in cherts nucleation may be accelerated by impurities which act catalytically.

(2) The spacing of tiny impurity particles which reduce grain boundary mobility. There is little retardation in the initial stages when the length of the moving grain boundary is of the same order as the distance apart of the impurities. The mobility of a grain boundary, however, is reduced when it contains closely spaced particles after crystal growth.

(3) Grain boundary area. The large grain boundary area initially present in the fine-grained rock gives a large driving force for growth but this decreases during recrystallization as the total grain boundary area is reduced.

(4) Mutual interference of grains. Grain growth into the matrix will initially be rapid but will slow down within clusters where crystals interfere with each other. The clusters themselves will expand rapidly outwards into the matrix, but growth will become slower as the clusters meet each other. This phase will end when the initially fine-grained aggregate has been replaced by a moderately uniform group of new crystals. Further growth to give larger crystals will be very slow due to mutual interference.

(5) The total thermal energy budget. Thermal energy of metamorphic origin provides the initial activation energy, the energy required to build up new lattices and to move grain boundaries during recrystallization. The rate of recrystallization will depend mainly on the maximum temperature held for an appreciable length of time (but see p. 125). The effect of differences in rate of heating is not known but it may be significant if the rate of recrystallization is either faster or slower than the rate of heat transfer in.

Minor impurity phases will normally be caught on moving grain boundaries; they may be swept along with the boundary, or may pin the boundary itself. The shape of the minor phase depends on its surface energy relative to quartz. It will remain as discrete small bodies if it is rather immobile and if there is an appreciable difference in surface energy so that it does not wet the quartz. Thus tiny mica flakes or epidote grains will be dispersed as xenoblastic or subidioblastic grains. Garnet, sillimanite and tourmaline may become idioblastic and the last two even penetrate the quartz as needles (Plate XVa, b).

Calcite appears to have the ability to wet quartz, i.e. the adhesion of a calcite–quartz interface is greater than the cohesion of a quartz–quartz grain boundary. Calcite therefore spreads along the quartz grain boundaries to cement the grains in a sandstone. The low grade metamorphism of calcareous sandstones, or the introduction of calcite into a sandstone at low temperatures, commonly causes peripheral replacement of quartz or feldspar by calcite and gives sutured boundaries. Carbonate forms large crystals up to a centimetre across enclosing quartz crystals poikiloblastically.

The metamorphism of quartz–feldspar aggregates (feldspathic sandstone or arkose) is generally similar to that of quartz-sandstone because of the similarity in surface energies of quartz and feldspar. High-grade metamorphism produces a massive, tough, hornfels of igneous appearance with a granoblastic-polygonal texture in thin section. Some variation in behaviour may be caused by the alteration of potash feldspar to a white mica pseudomorph or of plagioclase to "saussurite", calcite or white mica. There is some tendency for feldspar to be subidioblastic when associated with quartz.

Metagreywacke

The recrystallization behaviour of greywacke, subgreywacke and other poorly sorted sediments is considerably more complicated than that of quartz-sandstone because of:

(1) The abundance of chemically reactive clastic minerals, rock fragments, matrix, etc.
(2) The original texture, with its wide range of grain sizes.

Recrystallization is driven by surface energy and free energy of reaction in the fine-grained pelitic matrix; driving forces in the large clastic fragments are varied and smaller.

The earliest stage of metamorphism causes recrystallization of the matrix but leaves the larger mineral or rock fragments unaltered. The rock is dark, tough and massive with a basic igneous appearance in hand specimen. In thin section the matrix has been recrystallized to a fine-grained aggregate of quartz with a typical hornfelsic texture of randomly oriented mica flakes or amphibole fibres. The formation of mica and amphibole from clay, micas, chlorite, calcite, etc., is reconstructive and involves chemical reaction, nucleation and growth. Recrystallization of quartz is variable; some is original fine-grained clastic quartz which has been recrystallized but some has been liberated by chemical reaction.

At medium grades the large clastic grains still retain original outlines but become recrystallized internally. Quartz grains are polygonized, chert is coarsened, igneous fragments recrystallized, feldspar altered, and fragments of clastic pyroxene or amphibole pseudomorphed by amphibole, mica or chlorite. The texture of the matrix is hornfelsic and medium grained.

Increased ionic mobility at high temperatures in high-grade metamorphism allows a general increase in disorder. Chemical gradients tend to be flattened out, reaction takes place between separated species, and grain boundaries respond to small surface energy drives. The rocks in general *tend* to become homogeneous and equigranular with an almost igneous aspect. Evidence of lack of equilibrium may be seen in reaction rims around some clastic fragments which differ too much in size or composition from the matrix to be completely absorbed. The texture in detail is somewhat irregular because the various minerals differ in surface energy, abundance and mobility and this leads to differences in the ability to nucleate, to grow, to achieve crystal faces, or to adjust adjacent boundaries. Quartz, feldspar and pyroxene tend to form a granoblastic-polygonal aggregate; mica and amphibole form randomly oriented laths (Plate XIXa, b), garnet ranges from spongy xenoblastic crystals where growth has been rapid and impeded by adjacent crystals to solid idioblastic crystals where growth has been slower. Porphyroblasts, particularly of cordierite and andalusite are formed in some greywackes but as they are more common in meta-pelites they will be discussed there.

Conglomerate-hornfels and Hornfelsed Breccia

Conglomerates and breccias behave similarly to grewackes during thermal metamorphism. The matrix, if any, recrystallizes first to hornfels containing relict clastic fragments. At higher temperatures the clastic fragments recrystallize as well and may react with the matrix. Zonal structures are produced around some large fragments, particularly calcareous boulders, fossils or concretions.

Pelitic Hornfels

The characteristic feature of the argillites (mudstones, claystones, siltstones, pelites, shales) is a fineness of grain and a richness in clay minerals. Elevation of temperature causes profound textural changes driven by the high surface energy of the fine-grained particles and the high free energy of unstable minerals.

Argillaceous rocks are recrystallized to tough, fine-grained hornfels, the simplest consisting of quartz and mica in approximately equal proportions. Quartz forms a granoblastic-polygonal aggregate with randomly oriented mica flakes. The formation of a two-phase aggregate between minerals with differing surface energies (e.g. quartz mica, plagioclase–pyroxene, feldspar–hornblende, scapolite–diopside) such that one tends to be xenoblastic and the other idioblastic or subidioblastic, involves a compromise between several processes. The equilibrium texture will be controlled by the most abundant phase, (i.e. the most abundant grain boundaries) not necessarily by the phase with greatest surface energy. The texture of a quartz–mica hornfels can be regarded as consisting of either polygonal quartz with mica flakes lying along randomly oriented quartz–quartz grain boundaries or conversely as consisting of randomly oriented mica flakes which have pinned the quartz grain boundaries (Fig. 53). Triple-points between three crystals of the same species with low lattice anisotropy (quartz, feldspar, scapolite) are simple and have approximately equal interfacial angles. Triple-points between species with greater surface energy anisotropy are complex, in that interfacial angles are not necessarily equal, and the interfaces are not random but are commonly rational lattice directions. Triple-points between three grains of different species may tend to have interfacial angles which depend on the relative surface energies, but the grain boundary between two crystals of quartz and feldspar will meet an interface of mica, pyroxene, amphibole, or garnet at angles near 90°.

Some mineral species tend to be bound partly or wholly by crystal faces and an idioblastic form depends largely on:

(1) High ionic mobility due to elevated temperature.
(2) An extended period of slow crystallization.
(3) Perhaps the presence of an intergranular fluid which allows a surface rearrangement to the slightly lower energy crystal face.
(4) The surrounding crystal matrix being in effect "soft", i.e. with low surface energy and isotropy and thus with mobile boundaries whose configuration is easily dominated by the idioblastic crystal which is a species with high surface energy and anisotropy.

Some mineral species tend to form porphyroblasts to give a *porphyroblastic*, *maculose* or knotted texture. Cordierite and andalusite generally form a few porphyroblasts rather than many small crystals, and the accompanying tendency for both to form spongy poikiloblastic crystals is probably due to rapid growth. Small cordierite "spots" tend to be readily altered to indefinite mineral aggregates (Bosman, 1964) which have been regarded as embryonic crystals in an incipient stage of formation. *Nodular texture* is produced by closely spaced, even intersecting, oval porphyroblasts particularly of cordierite or scapolite.

The relation between grain size and ease of nucleation is demonstrated by the development of porphyroblasts. A mineral phase with high activation energy of nucleation (i.e. one which is difficult to nucleate) will begin crystallizing at only a few separate points throughout the rock, thus the hornfels will develop spots at an early stage. These may grow into knots and finally into large porphyroblasts.

The formation of a porphyroblast in a multicomponent rock implies that a certain region is converted to a composition which is different from that of the rock as a whole. This in turn means that certain constituents will be expelled from the region and others will enter it. These processes may give rise to a number of distinctive "zonal" textures:

(1) The growth of a porphyroblast may lead to impoverishment of certain constituents in its vicinity, e.g. a light coloured halo may surround a porphyroblast of a ferromagnesian mineral (Plate XVIIIb). Growth of the

crystal will slow down because the required materials must migrate from greater distances.

(2) The matrix close to cordierite porphyroblasts in some hornfelses is much richer in biotite than that further away (Plate XIVb). A growing crystal must either expel unwanted materials as it grows or retain them as inclusions. The material normally diffuses outwards as ions but in certain circumstances the porphyroblast appears to move crystals adhering to its external surface.

(3) A reaction rim may form around a crystal which either supplies nutrient to the species crystallizing around it or acts as a nucleation catalyst (Plate XVIIIa).

Just as the final equilibrium mineral assemblage in a metamorphic rock depends on the bulk chemical composition and on the physical conditions of metamorphism and is independent of the origin of the rock, so an equilibrium texture depends only on the properties of the crystals present and on the physical conditions. Complete equilibrium textures are, however, not common and most hornfelses contain sufficient palimpsest features to indicate their origin except at highest grades, e.g. a hornfels derived from a mudstone differs considerably in texture from one derived from a slate. Bedding is preserved in most hornfelses as compositional layering and there appears to be little tendency to homogenize the rock (Plate XXIVa). Delicate sedimentary structures may be preserved. Recrystallization of a homogeneous pelite gives randomly oriented crystals of metamorphic minerals but the presence of planar inhomogeneities controls the ease of ionic migration and of growth. A well-developed shaly parting is enough to cause mimetic crystallization of mica along the bedding to produce a weak preferred orientation. High-grade recrystallization of a slate, phyllite or schist may produce a massive hornfels with large, randomly oriented crystals of mica but the old foliation is hard to destroy completely. The foliation in a slate or phyllite acts like bedding in a shale and controls the formation of mimetic mica flakes which may grow preferentially along both bedding and foliation. Porphyroblasts of spongy cordierite tend to form ovoids with their long axes parallel to bedding.

Large parallel mica flakes constituting the foliation in a schist which is thermally metamorphosed may be stable mineral phases under the new conditions and thus have no great tendency to recrystallize so that a considerable part of the foliated texture may still remain to give a variety of *contact schist*. Thermal metamorphism of regionally or dynamically metamorphosed rocks is a type of post-tectonic recrystallization. Thermal metamorphic minerals replace regional minerals generally as randomly oriented or granular aggregates; thermal porphyroblasts of andalusite or cordierite grow across the foliation and do not push it aside.

Preferred orientations of minerals in hornfelses are due to:

(1) Mimetic crystallization, e.g. growth of mica along bedding, old foliation or veins.

(2) Parallel growth (impingement) texture due to local nucleation at an interface (bedding, fossil surface, vein, etc.).

(3) Secondary "annealing" texture, particularly in thermally metamorphosed quartzites which have previously undergone dynamic or regional metamorphism and which have a lattice or dimensional preferred-orientation.

(4) Slight deformation accompanying igneous intrusion.

(5) Deformation later than the igneous intrusion and its accompanying metamorphism.

Marble and Calc-silicate Hornfels

The thermal metamorphism of a pure limestone to a marble is a simple case of the recrystallization of a fine-grained, single-phase, stable polycrystalline aggregate of a species with rather low surface-energy anisotropy. Very fine-grained limestones recrystallize patchily like cherts and achieve an equilibrium granoblastic polygonal texture with equiangle triple-points. There is little tendency for the development of idioblastic crystals although many grain boundaries tend to be close to rational lattice directions (Plate VIII).

Fossils or fossil fragments in limestones behave similarly to other inhomogeneities such as pebbles in greywacke or amygdales in basalt. A calcitic fossil in a pure limestone is a negligible compositional discontinuity and is readily obliterated by slight recrystallization. A calcitic fossil in a contrasting matrix may persist to high grades if it is large and if the compositional difference is great (Joplin, 1968, Fig. 6b). Reaction between the calcite of the fossil and alumina, iron, magnesia and silica of the matrix may produce a type of layered reaction rim. The fossil may be replaced by metamorphic minerals to give a multi-crystal pseudomorph with the second generation crystals either randomly oriented or with a parallel or fibrous growth due to nucleation at the crystal–matrix interface (Plate XXIVb).

Thermal metamorphism of a dolomite produces a dolomite-marble with a granoblastic polygonal texture. Where calcite and dolomite recrystallize together, the dolomite is idioblastic towards the calcite because of its greater surface energy anisotropy; it forms rhombs in a polygonal matrix. Certain circumstances, particularly the dolomitization of limestone, produce an aggregate of idioblastic dolomite crystals with an *idiotopic* texture. Metamorphism of dolomite causes dedolomitization in which calcium carbonate forms a calcite marble and the magnesian component forms a variety of secondary minerals. High temperatures produce periclase (MgO) which forms globular crystals or octahedral idioblasts in the calcite matrix. Periclase is hydrated to brucite, $Mg(OH)_2$, which forms multicrystal pseudomorphs as in *pencatite* and *predazzite* (brucite-marble) if the P_{H_2O} is raised.

Recrystallization of a quartzose dolomite at low temperatures produces a marble containing irregular, embayed quartz crystals with scalloped outlines. At higher grades the magnesia and silica react to give forsterite crystals in a calcite matrix (*forsterite-marble*). Where a new phase such as olivine, periclase or quartz crystallizes in a calcite matrix, it may form round, subidioblastic or idioblastic crystals. Round crystals of olivine may be sufficiently abundant to overlap and give a nodular texture (Plate Xa). Hydration of olivine to serpentine produces single- and multi-crystal pseudomorphs. *Ophicalcite* (serpentine-marble) formed by serpentinization of a forsterite marble is texturally different from one produced by carbonation of serpentine. The relatively low surface energy of serpentine–carbonate interfaces allows serpentine to form intergranular wisps sheafs and spherulitic aggregates.

Tremolite appears to have difficulty in nucleating and forms felted masses which coalesce into large crystals or relatively few large (up to 2 cm) randomly oriented prisms in a calcite matrix. Slow growth and strong lattice anisotropy commonly allows the formation of idioblastic crystals. Diopside nucleates easily and tends to form abundant small granules within the calcite aggregate. The granules are round to subidioblastic and show a tendency to occur at calcite triple-points, or on calcite grain boundaries (particularly at bends where they have possibly "pinned" the moving boundary). Slow growth allows the formation of large subidioblastic or idioblastic forms.

Wollastonite has a considerable lattice anisotropy and tends to form prismatic crystals which, however, are not as elongate or idioblastic as tremolite which it resembles. Wollastonite is one of the few silicates which forms growth twins and the common multiple twinning in the randomly oriented crystals of non-deformed hornfelses probably form similarly to annealing twins. Random nucleation and growth may form an interlocking *decussate* texture but this tendency is opposed by the difficulty in nucleation. This difficulty is expressed in a variety of ways:

(1) Wollastonite tends to form large crystals in radiating sheafs or spherulites.

(2) It replaces fossils as a multicrystal pseudomorph with the fibres growing inwards from the outer wall of the fossil on which they have nucleated to give a *columnar impingement* texture.

(3) Metasomatic introduction of silica into a pure calcite rock produces a wollastonite–calcite interface which moves inwards, wollastonite being

nucleated continuously at this interface. The *columnar impingement* texture with low angle grain boundaries is a fairly low-energy stable texture which tends to be preserved.

Garnet displays a variety of shapes in calcareous hornfelses. Normally it has an idioblastic form against calcite but has non-rational boundaries against diopside and wollastonite. It tends to be difficult to nucleate and forms large crystals. It will form perfect single-crystal pseudomorphs of fossils which implies difficulty in nucleation; however, in many rocks it forms abundant tiny grains and nucleation may have been catalysed by some impurity. The common garnet in calc-silicate hornfelses is a grossular-andradite solid solution. Some crystals show a colour zoning with andradite-rich cores and grossular-rich rims, and many examples are weakly birefringent with zoning and twinning.

Vesuvianite and the epidotes (epidote, zoisite and clinozoisite) tend to be subidioblastic, and to have crystal faces against calcite but not against garnet, pyroxene, wollastonite, etc. Plagioclase and orthoclase are common as xenoblastic crystals. Sphene is characteristically idioblastic and forms diamond or wedge-shaped crystals. Prehnite is difficult to nucleate and has a variable lattice structure; it forms characteristic sheafs, "bow-ties" and spherulitic aggregates.

The initial effect of second phase impurities in an aggregate is to limit the grain size by restricting grain boundary migration. A small amount of an impurity can depress grain growth markedly if it is very fine-grained and evenly distributed, e.g. carbonaceous matter in fine-grained limestone. However, if the size of the second phase increases and the number of crystals decreases then a stage will be reached when the impurities cease to pin the boundaries. Calcite grows into large crystals, probably because its high symmetry allows easy addition of new units and hence easy grain growth, and once the calcite boundary ceases being anchored by the second phase, it may grow into very large poikiloblastic crystals (cf. Fontainbleu sandstone and gypsum sand crystals).

Calcite and dolomite recrystallize easily and react to give coarse-grained metamorphic minerals at comparatively low temperatures. Some calcareous sediments have recrystallized to coarse *calc-silicate hornfels* where adjacent sandstones are only slightly baked or indurated and not recrystallized.

Some impure dolomites (and limestones which have been metasomatized or "silicated") have bulk compositions comparable with those of basic igneous rocks, and either thermal or regional metamorphism may convert them into amphibolites which are indistinguishable mineralogically from those derived from a basalt. Relict bedding, amygdales or patches of igneous texture may give a clue to the origin but if a homogeneous equilibrium texture has developed, it may be difficult or impossible to make the distinction.

The name *skarn* has been given to hornfelses rich in iron, magnesia and lime and containing minerals such as pyroxene (rich in hedenbergite), garnet (rich in andradite), calcite, magnetite, pyrrhotite and various other sulphides and calc-silicate minerals. The rocks are generally formed by the iron-magnesia-silica metasomatism of limestone at granite contacts and form the host rock of many important ore bodies. The grainsize is large and variable and the textures are so irregular and indicative of non-equilibrium that they are not capable of a simple generalized interpretation.

THERMALLY METAMORPHOSED IGNEOUS ROCKS

Most of the mineralogical changes accompanying the thermal metamorphism of igneous rocks are retrograde in the sense that metamorphic recrystallization takes place at temperatures below those of original magmatic crystallization. The main effects of thermal metamorphism are the destruction of the original igneous texture and the replacement of the igneous assemblage by a metamorphic one. This generally means the replacement of anhydrous by hydrous silicates but some metamorphic assemblages (pyroxene–plagioclase, hornblende–plagioclase, quartz–orthoclase–biotite) are the same as the original igneous assemblages although the chemical composition of the individual mineral species may differ.

It has been considered (Harker, 1939, p. 12) that mineralogical changes during metamorphism are progressive but although this may be true for regional metamorphism, there is considerable evidence (Almond, 1964) in some thermal aureoles that this does not apply and that a high-grade assemblage can be produced directly from original igneous minerals without the intermediate stages of low and medium grades, e.g. metamorphic olivine may be derived from igneous olivine not necessarily from serpentine or some other low-grade phase. This means that the rate of heating was greater than the rate of mineral reaction.

Basic Hornfelses

The general principles of textural change are best illustrated by a discussion of the metamorphism of basic igneous rocks because these show a wide range of changes which have been extensively studied.

The texture of the original basic rock is dominated by plagioclase as a criss-cross or interlocking network with pyroxene fitting between to give an intergranular, intersertal or ophitic texture. The texture tends to be even-grained in gabbros and dolerites but becomes markedly porphyritic in basalts where the phenocrysts are olivine, pyroxene or plagioclase. Additional possible constituents are olivine in gabbros, mesostasis in dolerites and amygdales containing calcite or zeolites in basalts.

The rocks furthest from the intrusion may show the first stage of metamorphism involving changes to the ferromagnesians but not the plagioclase thus leaving the igneous texture intact. The only macroscopic effect is a slight change in colour and lustre. The pyroxene may simply show slight clouding and exsolution lamellae or may be altered to hornblende pseudomorphs while olivine is pseudomorphed by serpentine or chlorite. At this stage the plagioclase may become "chalky" and clouded with clay-like inclusions but is not recrystallized. It becomes *clouded* at a slightly higher grade.

Rocks closer to the intrusion, but still classified as low grade hornfelses (albite–actinolite hornfels), may become completely recrystallized to give an assemblage dominated by an actinolitic amphibole and albite. The plagioclase commonly retains its shape and consequently the igneous texture is still preserved (Plate XXV) to give blasto-intergranular, blasto-intersertal, blasto-ophitic, blasto-porphyritic, etc., textures; the flow structure of basalts may be preserved and pseudomorphs are common (Joplin, 1967, Fig. 11).

Medium grade metamorphism (Hornblende–Hornfels Facies) is marked mineralogically by the formation of a lime-bearing plagioclase and perhaps a diopsidic pyroxene in addition to hornblende. Most of the texture is destroyed because of recrystallization of the plagioclase although traces may still be shown by amygdales, relict phenocrysts or lines of ferromagnesian granules.

The original grain size is important. Fine-grained igneous rocks are more easily recrystallized than coarse ones, presumably because the higher surface area gives a greater grain boundary energy driving force and also permits easier intergranular diffusion or passage of intergranular fluids. A dolerite or gabbro may be texturally unaltered with only slight mineralogical changes whereas adjacent basalt is completely recrystallized with a new assemblage and texture.

The medium-grade equilibrium basic hornfels is typically a dark-coloured, fine- to medium-grained, hornblende– or pyroxene–hornfels with a granoblastic-polygonal texture. However, equilibrium is by no means common in such rocks. The texture may be patchy with alternate areas of medium and fine grain or with aggregates of pyroxene–andesine and hornblende–oligoclase side by side.

High-grade (Pyroxene–Hornfels Facies) basic hornfelses are medium- to dark-coloured (depending on the grain size) rocks with a granoblastic-polygonal texture. Olivine, orthopyroxene, clinopyroxene and labradorite give an assemblage similar to that of the original igneous rock and the tendency for pyroxene and plagioclase to form slightly elongate prismatic crystals also gives the rock a somewhat igneous aspect. The difficulty in distinguishing between a fine-grained pyroxene–plagioclase hornfels and an igneous rock is exemplified by the *beerbachite* which was first thought to be a gabbro-aplite by Chelius (1892) and then shown by Klemm, and by Macgregor (1931b) to be a basic hornfels.

Many accessory minerals appear to be resistant to metamorphism. Apatite and zircon are particularly refractory and may retain their identity even in high-grade hornfelses. Tourmaline and zircon may develop oriented (epitaxial) overgrowths.

Relict igneous minerals persist unaltered and without reaction rims into some high-grade hornfelses and it may require detailed chemical and optical work to decide whether a given crystal of olivine or orthopyroxene is of igneous or metamorphic origin. It is unwise to assume that a mineral is a member of an equilibrium assemblage simply because of the lack of any obvious disequilibrium texture.

Striking non-equilibrium textures such as reaction rims or coronas (p. 104) of three somewhat distinct varieties are recognized in igneous hornfelses:

(1) Simple, monomineralic reaction rims around a mineral which has either acted as a nucleus or as a source of material or both, e.g. sphene around ilmenite or hornblende around magnetite.

(2) Simple or compound, multi-layer polymineralic coronas. These are the classic coronas which form around a ferromagnesian mineral (particularly olivine) and particularly at olivine–plagioclase boundaries (Plate XVIIIc, d). Even though minerals such as orthopyroxene, garnet and spinel indicate rather high temperature metamorphism, textural disequilibrium is shown, not only by the corona, but by the general retention of the igneous texture.

(3) Simple to complex, mono- or polymineralic reaction rims around amygdales (or phenocrysts) (e.g. Plate XXVa). A strong compositional gradient exists from a calcite-filled amygdale to the basic matrix. Migration of lime outwards and of silica, iron, magnesia, etc., inwards may preserve the amygdale as a layered oval body containing calc-silicate minerals such as grossular, diopside, scapolite, plagioclase, prehnite, hornblende, and wollastonite (Macgregor, 1931a; Harker, 1939, pp. 105–10; Spry, 1953; Cann, 1965).

Metasomatism may produce unusual textures. Introduction of lime and carbon dioxide, of soda and chlorine, or of boron, may produce large amounts of calcite, scapolite or tourmaline (or axinite) respectively. Both scapolite and calcite have a strong tendency towards poikiloblastic growth

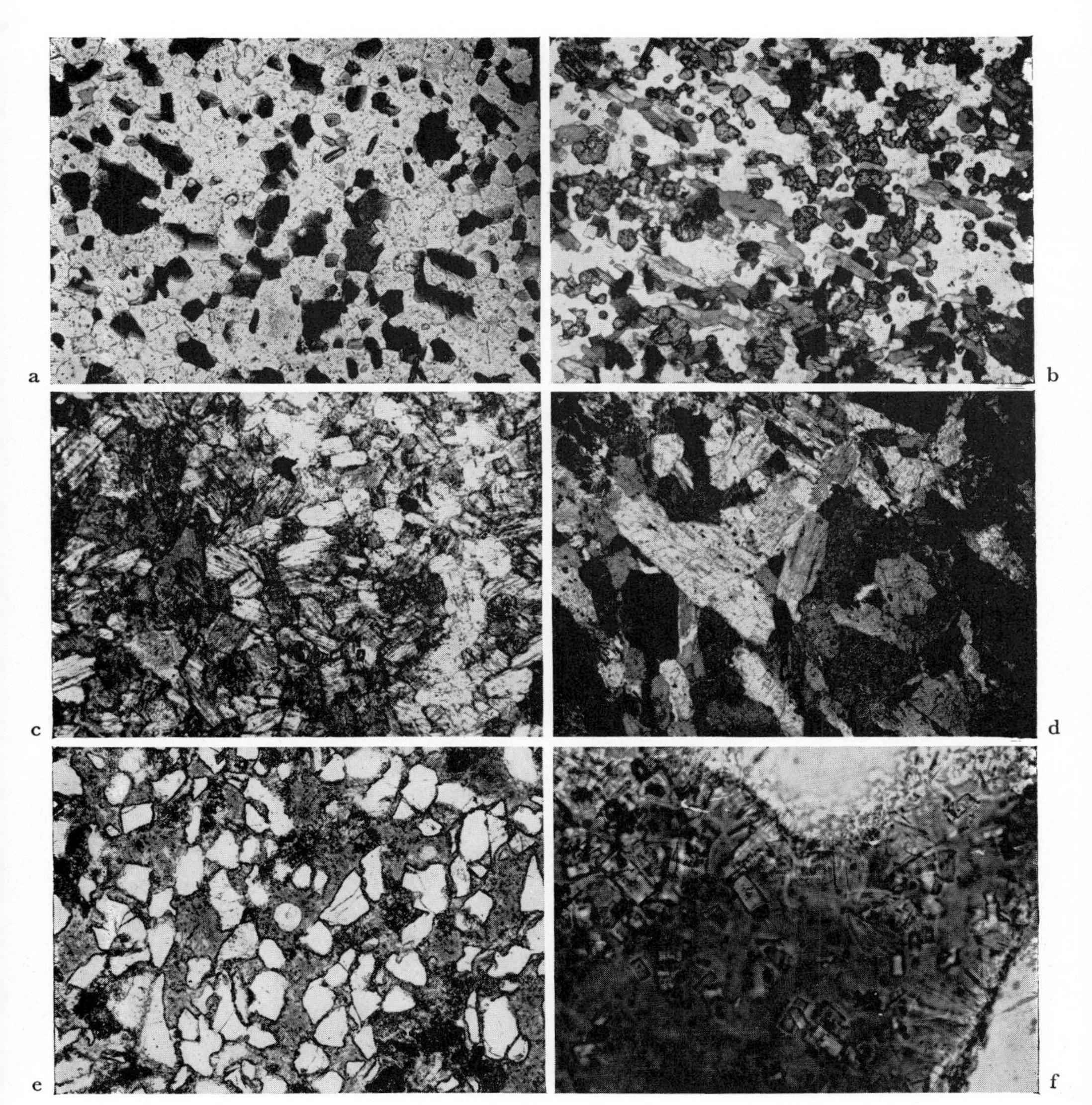

PLATE XIX. Textures of hornfelses. (a) Granoblastic polygonal aggregate of quartz and biotite, Tilbuster (×120). (b) Granoblastic aggregate of quartz, biotite and pyroxene, Trial Harbour (×50). (c) Decussate aggregate of biotite, Trial Harbour (×30). (d) Decussate aggregate of post-tectonic hornblende in schist from Lingoutte, Vosges (×11). (e) Web texture in buchite, Apsley. Colourless corroded relicts of quartz surrounded by glass (×11). (f) Large quartz relicts fringed with tridymite and surrounded by glass containing rectangular cordierite and opaque spinel; buchite, Apsley (×200)

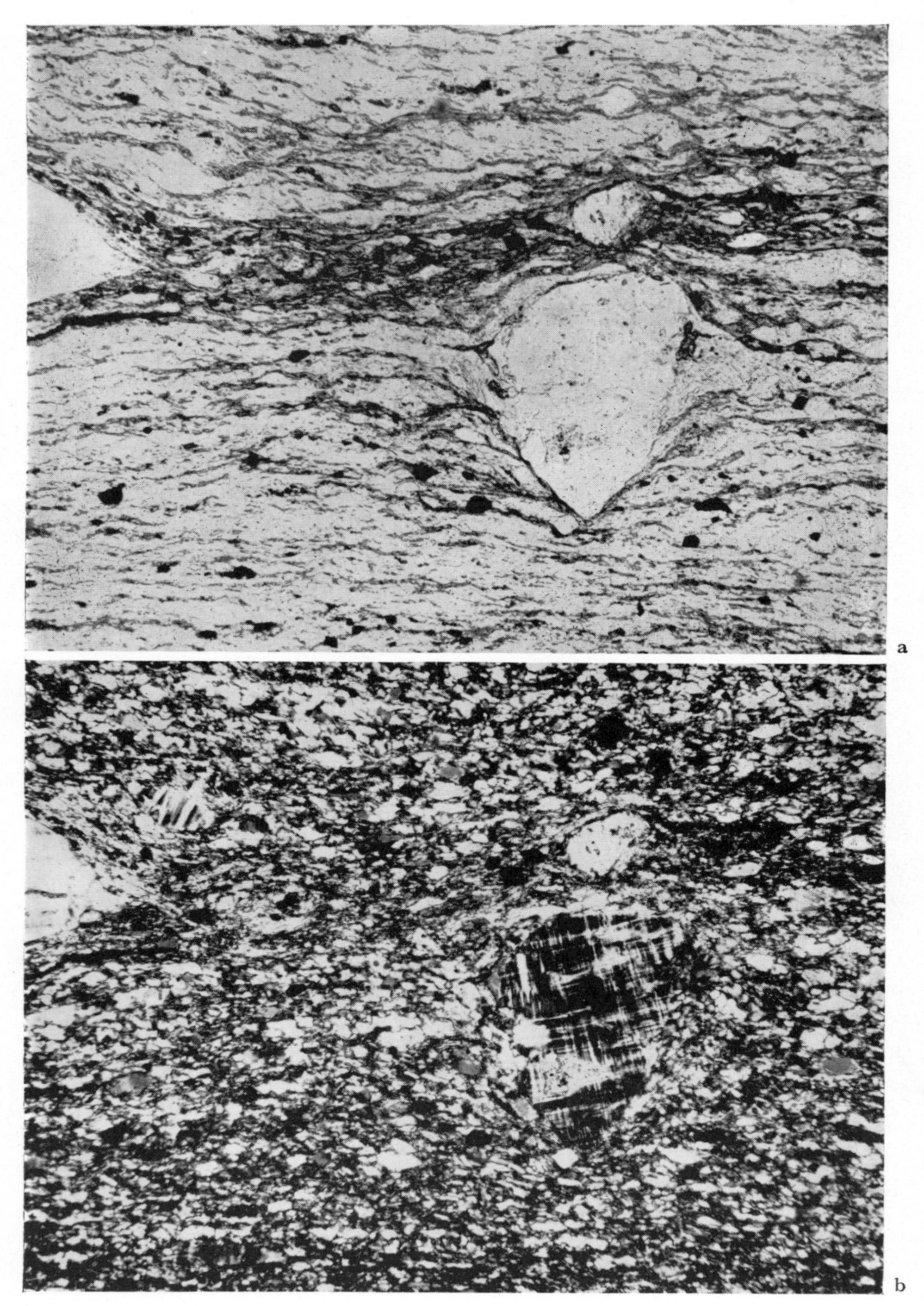

PLATE XX. Mylonite, Moine thrust (×45). (a) Plane-polarized light. (b) Crossed polarizers. Porphyroclastic microcline, streaky foliation, some recrystallization of matrix. (Photographs by J. Christie; shown in Plate VII, Christie, 1960)

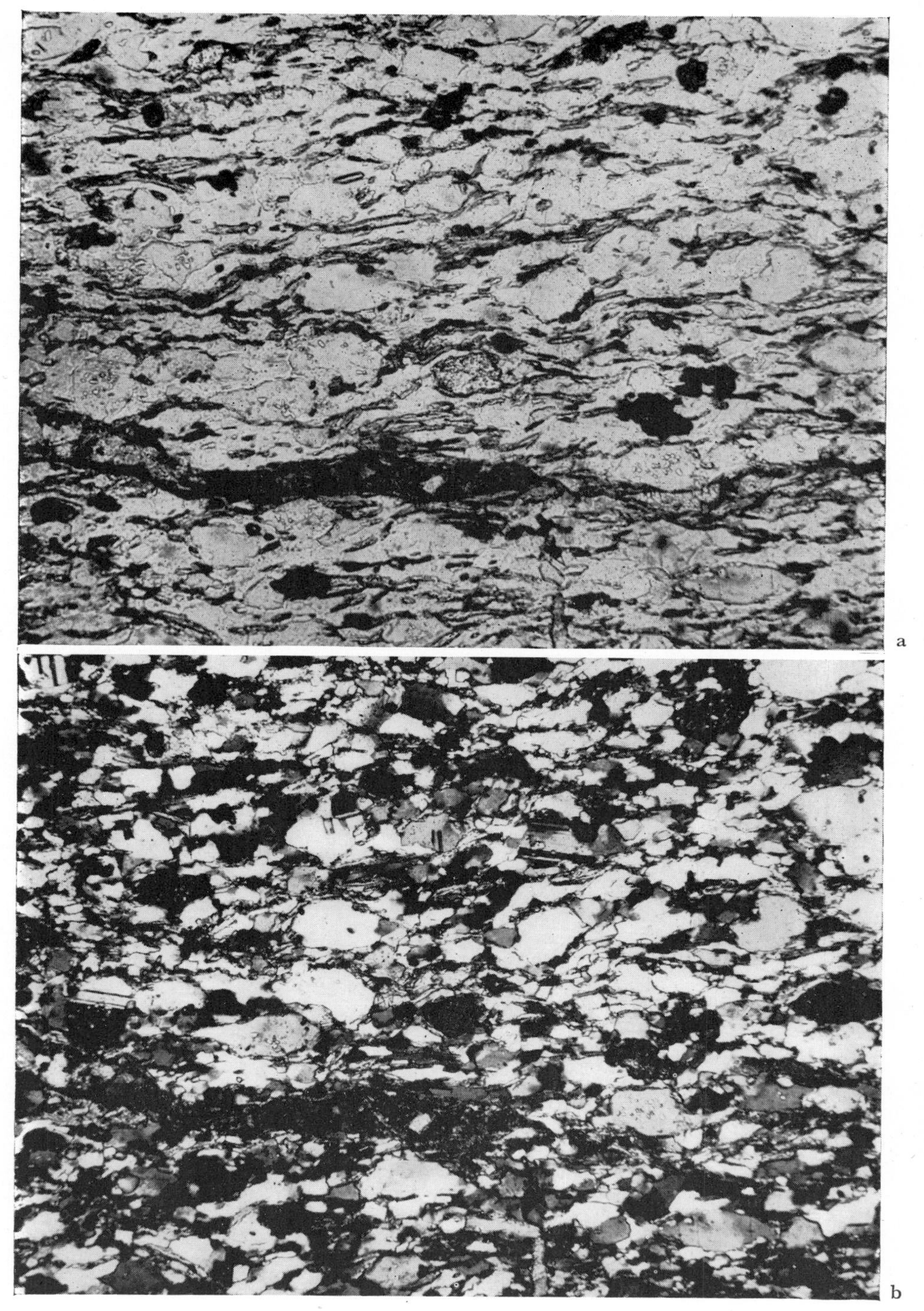

PLATE XXI. Blastomylonite, Moine thrust (×45). (a) Plane-polarized light. (b) Crossed polarizers. Relict foliation and considerable recrystallization. (Photographs by J. Christie; shown in Plate VII, Christie, 1960)

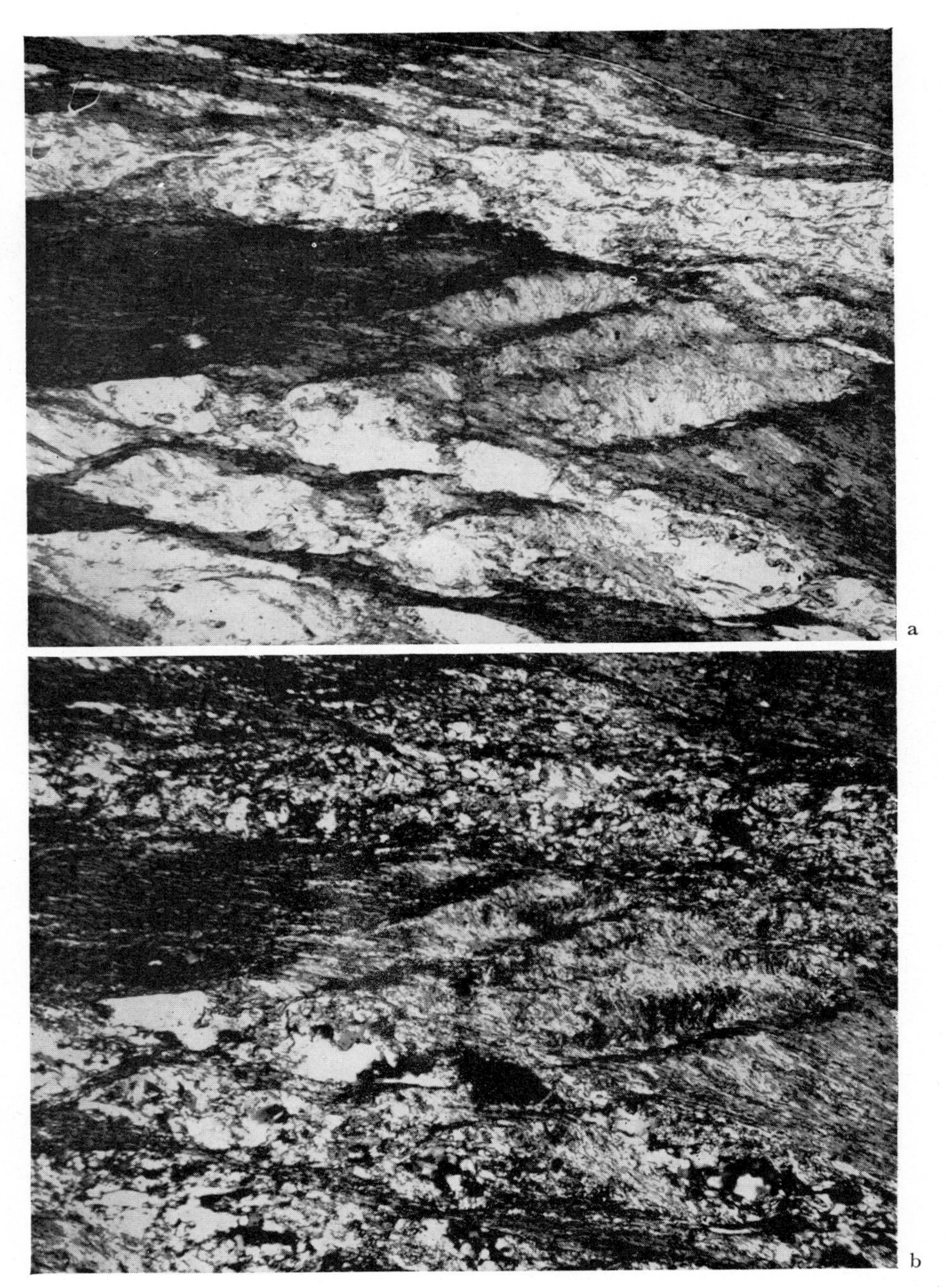

PLATE XXII. Phyllonite (×45). (a) Plane-polarized light. (b) Crossed polarizers. Characteristic lenticular structure, foliations of different ages, granulation of quartz and feldspar. (Photographs by J. Christie, shown in Plate VIII, Christie, 1960)

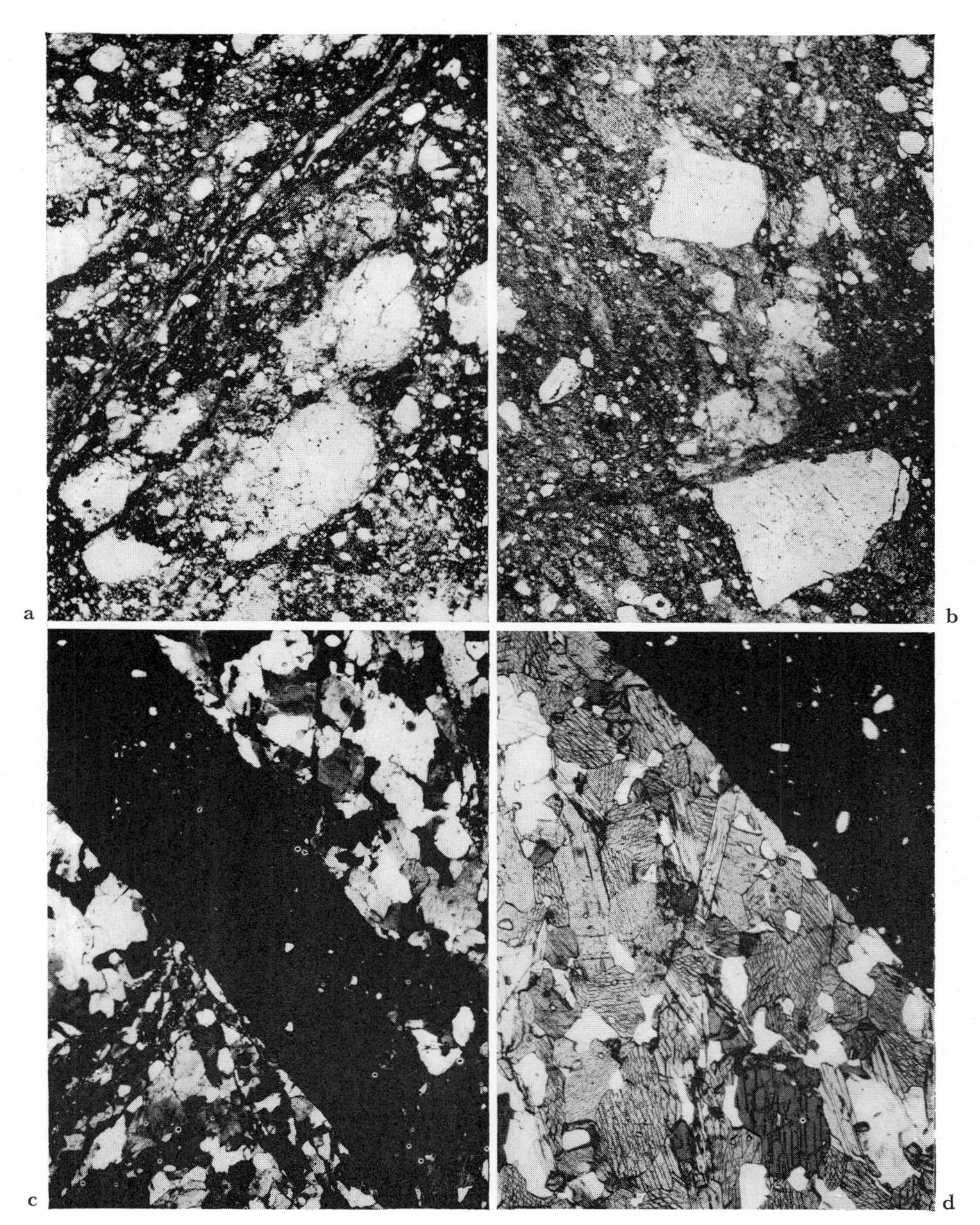

PLATE XXIII. (a) Protomylonite, Trekkoskraal Point, Cape Province (×12). (b) Cataclasite (×12). (c) Pseudotachylyte (black, isotropic) cutting sheared gneiss, Loch Tollie (×12). (d) Pseudotachylyte (black) cutting amphibolite, Loch Tollie (×12)

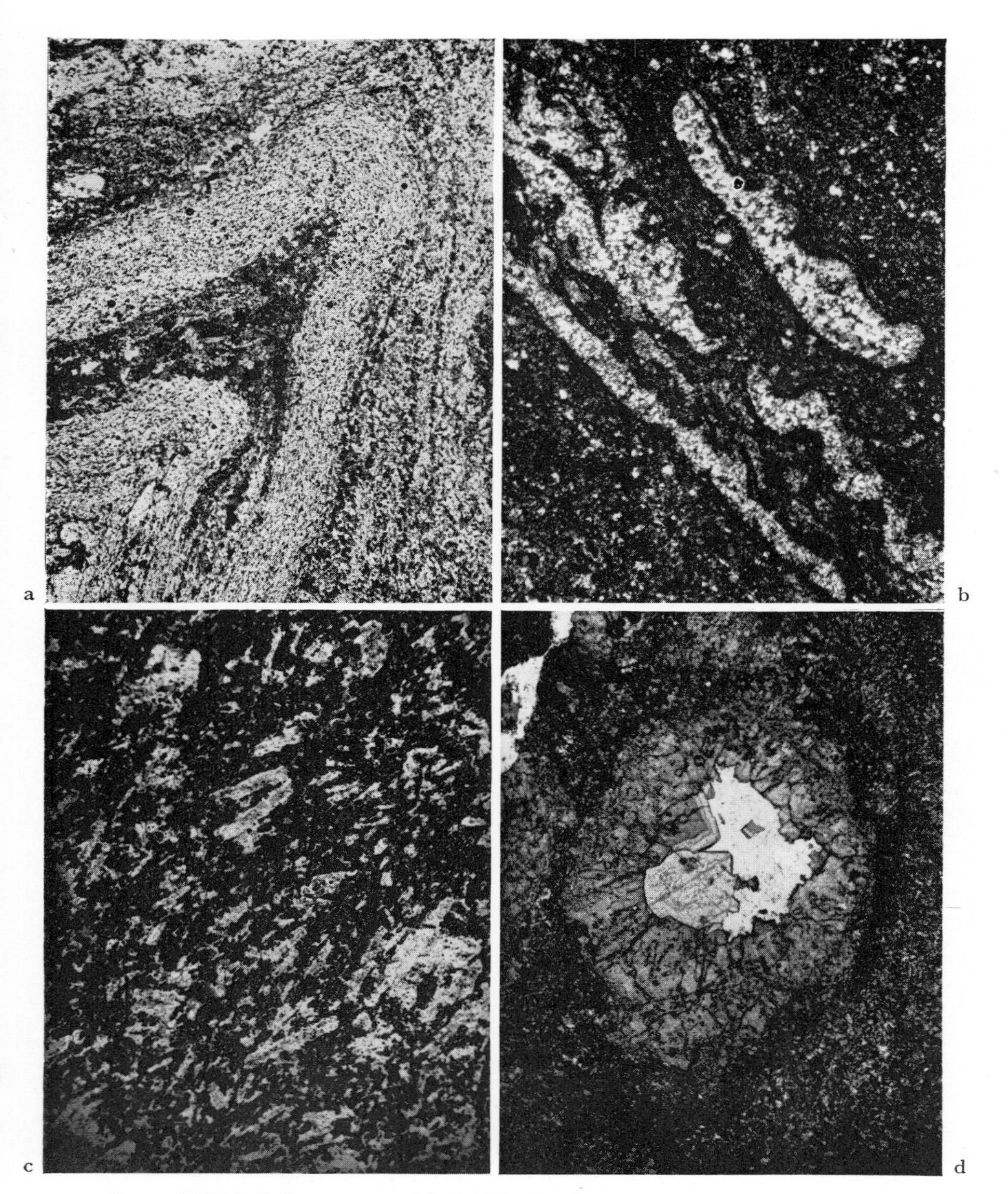

PLATE XXIV. Relict textures. (a) Fold in banded schist which has been later thermally metamorphosed. Some biotite is randomly oriented, some has grown along pre-existing foliation. Granville Harbour (×3). (b) Pseudomorphs of fossils in calc-silicate hornfels are composed of wollastonite. Proctors Road, Hobart (×20). (c) Relict igneous texture in thermally metamorphosed basalt, Trial Harbour (×35). (d) Amygdale in thermally metamorphosed basalt, Tilbuster. Outer rim is grossular, centre is scapolite and calcite (×15)

under these circumstances and replace many crystals of fine-grained plagioclase in meta-basalts to give a coarse, granoblastic, polygonal mass with sieve texture, enclosing tiny granules of pyroxene or hornblende which still outline a relict igneous texture.

Strong lime-metasomatism of basaltic rock produces medium-grade calc-silicate hornfels consisting of large bladed crystals of calcite and plagioclase which enclose pyroxene and garnet poikiloblastically and higher grade varieties consisting of coarse, irregular or granoblastic aggregates of wollastonite, garnet, pyroxene and plagioclase (Spry, 1953).

Ultrabasic Hornfelses

Anhydrous ultrabasic rocks such as peridotite and pyroxenite are generally resistant to thermal metamorphism and whether or not they are altered seems to depend largely on the availability of H_2O (or CO_2) and its ability to penetrate the rock. Previous alteration of the ultrabasic rock to a massive or schistose serpentinite however increases its susceptibility to metamorphism.

A granular ultrabasic rock consisting of olivine and/or pyroxene has a polygonal texture prior to metamorphism. At low grades the anhydrous silicates are altered to hydrous magnesian silicates, particularly spiky antigorite but also talc and magnesian chlorite. The flaky minerals commonly form pseudomorphs with a moderate degree of lattice retention to produce *mesh texture* serpentine.

Low-grade metamorphism of massive serpentinite may cause only minor recrystallization as many of the minerals are stable under the new metamorphic conditions. The main alteration is the replacement of a lamellar chrysotile by a coarse antigorite. The metamorphism of schistose serpentinite is generally similar except that the foliated texture tends to be destroyed by the random growth of chlorite, antigorite or amphibole.

Lack of equilibrium is shown in many such rocks by complex mineralogy and texture, oriented overgrowths of metamorphic minerals on igneous substrates, etc. Rearrangement or metasomatism cause considerable migration of constituents and produces veins or "blotches" of silica (quartz and chalcedony) magnesite, brucite, magnetite, etc. This inhomogeneity persists through medium grades and a variety of textures and mineralogies occur depending on the bulk composition and the ease of ionic movement. Variations include:

(1) A regular polygonal texture of cordierite and randomly oriented ragged anthopyllite prisms in granular cordierite–anthophyllite hornfels. Nucleation difficulty in the amphibole may cause it to form sheafs or spherulitic aggregates.

(2) A granoblastic to decussate texture in a monomineralic aggregate of slightly elongate amphibole or diopside.

(3) A porphyroblastic texture with large randomly oriented idioblastic amphiboles set in a fine-grained decussate aggregate of chlorite flakes.

It is not easy to distinguish between medium- and high-grade ultrabasic hornfelses but some additional varieties occur close to igneous contacts where they apparently belong to the Pyroxene Hornfels Facies. Monomineralic hornfelses consisting of granoblastic-polygonal olivine, orthopyroxene, or clinopyroxene have an igneous appearance and can only be distinguished as metamorphic by detailed chemical mineralogy.

Some hornfelses have complex textures due to a combination of original features and the effects of a range of surface energies. Spinels (magnetite, hercynite and spinel) show a variety of shapes: they form large idioblasts, round granules at the grain boundaries of other minerals, or small closely packed regularly shaped aggregates of round grains (suggesting that the spinel is probably the remains of a large crystal of another species which has been completely replaced). The round, globular form is presumably due to the high surface energy of spinel with respect to the surrounding phase. The lattice-anisotropy is apparently not sufficient to allow spinel to assume an idioblastic form except under conditions of high ionic mobility.

Acid to Intermediate Hornfelses

Original textures of acid to intermediate igneous rocks range from coarse-grained hypidiomorphic granular (plutonic rocks) or porphyritic (hypabyssal and volcanic rocks) to glassy (volcanics).

The plutonic rocks are particularly resistant to thermal metamorphism except at very high grades. Most early granites, for instance, are texturally completely unaltered by later granite, and metamorphism may be restricted to changes on the ionic scale, e.g. loss of radiogenic argon. Mineralogical changes consist of clouding, sericitization or saussuritization of feldspar, chloritization of biotite or hornblende, and perhaps the uralitization of pyroxene. Strained quartz may be polygonized and feldspar solid solutions may produce perthites or anti-perthites by exsolution, or may recrystallize to granoblastic or symplectic aggregates of two feldspars.

The metamorphism of granitic rocks by basic intrusions, however, involves high temperatures. Granite (or granitic gneiss) which is cut by dykes, or acid xenoliths in basic rocks, may be profoundly altered and partial melting may even be produced. The textures are complex and do not appear to represent equilibrium. Igneous crystals are polygonized and partially or wholly recrystallized. Exsolution textures and pseudomorphs are common. Textures tend to be irregular with reaction and migration along and across grain boundaries producing reaction rims, various symplectites (myrmekite, graphic and granophyric intergrowths), and replacement perthites.

Fine-grained igneous rocks are more reactive and the small grains in the ground-mass recrystallize easily. Glass is extremely unstable and recrystallizes to a granoblastic quartz–feldspar or feldspar aggregate. Phenocrysts are replaced by secondary minerals to produce pseudomorphs in a blastoporphyritic matrix.

THERMAL METAMORPHISM AND MELTING

Thermal metamorphism under the high-temperature conditions (900° to 1100°C or so) belonging to the upper part of the Sanidinite Facies may partially melt hornfelses. Such conditions are rare and are normally restricted to xenoliths or to the walls of volcanic necks along which basic magma has passed. Melting is recognized where the liquid produced remains as a glass to give a glassy hornfels or *buchite*. Glass has been found in buchites derived from sandstones (Wyllie, 1961; Ackermann and Walker, 1960; Butler, 1961; Spry and Solomon, 1964), granitic rocks (Knopf, 1938; Larsen and Switzer, 1939) and a variety of xenoliths (Lacroix, 1893; Richarz, 1924).

Melting of calcareous sediments (and some acid igneous rocks) may however produce a liquid which crystallizes readily thus destroying the evidence of melting. Melting is probably more common (particularly in rheomorphic veins) than has been commonly suspected.

Conversion of rather impure quartzose sandstone to buchite can be followed in some detail (Spry and Solomon, 1964). A sandstone consists mainly of quartz grains with possible minor K-feldspar, plagioclase, muscovite, clay mineral aggregates (pellets or clay fragments), a little clay as an intergranular film and accessory clastic magnetite, ilmenite, rutile, zircon, tourmaline, apatite and garnet.

Normal high-grade metamorphism of this rock would recrystallize it to a quartzite with a little feldspar and perhaps intergranular sillimanite but under very high-grade conditions the process is different. The minerals melt one by one, or in pairs, in order of increasing melting points.

The clay minerals melt first. If a low melting point pair (such as kaolinite plus a chloritic clay) are present in the pellets, these will melt (at 850°C?) to give almost spherical globules of liquid which ultimately solidify to brown glass. The intergranular kaolinite or illite film melts at a slightly higher temperature (900°C?) to a liquid which wets the quartz grains and occupies the grain boundaries. The surface tension of this liquid pulls the grains together and increases the closeness of packing. The partial melting of hornfels is analogous in many ways to the *sintering* of ceramics and experimental studies of mineral reactions, melting, diffusion, shrinkage, etc., in sintering are relevant to the Sanidinite Facies.

The proportion of liquid increases to perhaps 15% at about 950°C when micas melt and to about 20% at around 1000°C when first orthoclase then plagioclase disappears. The liquid now occupies all the pore space in the rock and increases in quantity by peripheral melting of quartz. The rock increases in density from about 2·2 to 2·6 and thus decreases in volume by possibly 5 to 10%. This decrease may cause columnar jointing in the buchite (Spry and Solomon, 1964).

The proportion of liquid, and hence of glass, increases with temperature, i.e. as the igneous contact is approached. It would presumably increase with time to an equilibrium quantity.

Buchites derived from siliceous coarse-grained rocks (sandstone, granite, etc.) consist of angular, corroded quartz crystals surrounded by a continuous matrix of glass giving a *web texture* (Fig. 53 and Plate XIX).

The textural units in buchites are glass, crystals and isotropic globules.

1. Glass may be:
   (a) Clear and free from inclusions.
   (b) Clouded with crystals such as well-formed spinel, cordierite, corundum, hypersthene and mullite.
   (c) Weakly birefringent due to devitrification, with a spherulitic, fibrous or graphic texture.
2. The crystals may be:
   (a) Original unmelted crystals. These range from moderately refractory (hence partially melted), e.g. quartz, garnet and tourmaline, to very refractory (and hence persistent as clastic grains to high temperatures), e.g. zircon, graphite and rutile.
   (b) Unmelted metamorphic crystals which were produced in the solid state by reaction. These are probably not common because the rate of heating in these rocks is probably higher than the rate of reaction and thus melting begins before normal metamorphic minerals are formed.
   (c) Crystals produced by incongruent melting of original clastic grains. Feldspars and micas melt incongruently to give glass plus spinel or grids of mullite needles.
   (d) Crystals precipitated from the melt. The silicate-liquid melt may solidify entirely to glass, but rather slow cooling may favour the precipitation of solid crystalline phases. These are igneous minerals, are commonly euhedral and include cordierite and tridymite.
3. Buchites contain isotropic globules with a variety of colour, size, composition and origin. Varieties include:

(a) Coloured glass globules enclosed within the general matrix of glass. These appear to represent individually melted grains. The viscosities of the mineral melt and the general matrix melt are so high that mixing is prevented.
(b) Tiny opaque or deeply coloured glass globules which appear to represent a truly immiscible melt phase.
(c) Spherical, ovoid, irregular or embryo-shaped globules containing colourless, yellow, red, or green isotropic opal, radiating fibrous chalcedony, or a colloform, crustified mixture of both. These appear to be silica-filled amygdales formed by vesiculation of the liquid melt when the pressure of dissolved gases exceeded the confining pressure.

A vague, slightly contorted planar structure in some glasses appears to represent some kind of flow structure, but bedding and other sedimentary structures are commonly very well preserved and show that no appreciable deformation has taken place. Extreme contortion of some normal hornfelses at igneous contacts suggests that the rocks behaved "plastically"; this does not necessarily imply partial melting but may simply be due to lowered viscosity or cohesion at high temperatures and high P_{H_2O}.

Melting of clay-rich shales appears to take place at somewhat lower temperatures than quartz-rich sandstones because of the lower melting range of the clay, hence vitrified shales tend to have a higher proportion of glass than adjacent sandstones. Bedding may be perfectly preserved as a fine banding and light to dark grey *porcellanites* are formed. The web texture is not well developed because of the lack of quartz and the glass tends to be dark because of pigments such as graphite, iron oxides and spinels. Bedding in fused shales is marked by layers of glass of different colour, lines of tiny opaque grains or by layers containing different sizes or different amounts of relict quartz.

Vitrified breccia may preserve the original clastic texture particularly where the matrix melts at lower temperatures than the fragments to give angular hornfels fragments in a glassy matrix.

Thermal metamorphism of acid rocks by dolerite may produce a porphyry-like rock containing strongly corroded quartz relics in abundant fine-grained graphic or granophyric matrix (Wilson, 1952). No glass is present but it is possible that the matrix is a crystallized quartz–feldspar melt. It is generally implied that rheomorphic veins (veins of sediment or sedimentary hornfels cutting an igneous body) were formed by a process of mobilization which did not involve melting. A feldspathic sandstone or arkose could, however, contain a considerable proportion of melt phase at 1000°C which is by no means an impossible figure at dolerite contacts. Such a liquid might devitrify into a cryptocrystalline mass or crystallize into a quartz–feldspar graphic or granophyric intergrowth.

Wyllie and Haas (1966) have discussed the melting and crystallization of melts derived from calcareous sediments. Experiments, particularly with ceramics, show that calcitic melts crystallize very easily and do not form glasses so that unequivocal textural evidence indicating melting is not to be expected in marbles. Lime-rich clays, however, might be expected to melt at accessible temperatures and to preserve glasses but silicate glasses rich in lime, water and carbon dioxide, however, should devitrify readily and are unlikely to be preserved, e.g. very coarse-grained, pegmatitic and spherulitic marble or calc-silicate hornfels (Schuilling, 1961; Temple and Heinrich, 1964).

CHAPTER 8

Preferred Orientations

Most deformed metamorphic rocks possess some kind of preferred orientation, i.e. a fabric in which the units prefer certain orientations to others. Two kinds are discussed here:

(1) *Dimensional preferred orientation.* Inequidimensional units such as elongate, flaky or prismatic crystals have a tendency towards parallelism of one or more of the morphological axes.

(2) *Lattice preferred orientation.* The crystal structures ("lattices") and hence optic directions and various crystallographic features (faces, cleavage, etc.) have related orientations.

Preferred orientations in many deformed rocks are expressed as mesoscopic planar or linear structures which allow the rock to be split more easily in some directions than others. A slate or similar low-grade rock splits along its *cleavage*; a similar kind of structure in higher grade rocks (phyllites, schists, gneisses, amphibolites) is called a *schistosity*. The term *gneissosity* has been applied to gneisses but this unmelodic word is little used at present. *Foliation* is used here to include *cleavage* and *schistosity*; this is in conformity with present general practice although it is contrary to that of Harker (1939, p. 203) who used foliation to mean compositional layering (referred to here as *layering* and also called *banding* or *striping*). The differences of opinion concerning the foliation-layering terminology were discussed in detail by Wilson (1961), and the practice of Turner and Weiss (1963, p. 28) is followed in regarding foliation as essentially consisting of "metamorphically produced penetrative surfaces of discontinuity in deformed rocks, including structures known as schistosity, cleavage, and so on". These foliation surfaces are reflected microscopically in a variety of ways (Turner, 1948, p. 275) such as:

(1) A system of closely spaced parallel fractures, e.g. fracture-, strain slip-, or crenulation-cleavage (Plate XXIXb, c).

(2) A system of closely spaced, nearly parallel but intersecting and anastomosing fractures, e.g. cleavage in some semi-schists.

(3) A preferred orientation of inequant bodies:

(a) Parallel mica flakes in a slate, phyllite or schist giving a fissility (the *lepidoblastic schistosity* of Harker, 1939, p. 201).

(b) Parallel, elongate or prismatic amphibole in an amphibolite or greenschist giving a fibrous kind of lineation (the *nematoblastic schistosity* of Harker, 1939, p. 201).

(c) An alignment of elongate and/or flat bodies such as flattened mud flakes, pellets, ooliths or pebbles (a *phacoidal* cleavage).

(4) A direction of splitting due to the parallelism of the mineral cleavage in constituent grains, e.g. of mica in a schist, calcite in marble or amphibole in an amphibolite.

Lineation is a common structure in deformed rocks. Cloos (1946, p. 1) defined lineation as "a descriptive and non-genetic term for any kind of linear structure within a rock. It includes striae or slickensides, fold axes, flow lines, stretching, elongated pebbles or ooids, wrinkles, streaks, intersections of planes, linear parallelism of minerals or components, or any other kind of linear structure of megascopic, microscopic or regional dimensions." Turner and Weiss (1963, p. 101) restricted the meaning somewhat to "linear structures, penetrative in hand specimens or in small exposures", and classified lineations into four types:

(1) Parallel, elongate aggregates of individual minerals forming streaks, trains, or cylindrical, prismatic or spindle-shaped bodies.

(2) Linear preferred orientation of boundaries of prismatic or tabular crystals.

(3) Axes of microcrenulations on a set of *S*-surfaces.

(4) Intersections of *S*-surfaces.

A *dimensional preferred orientation* of inequant crystals in a foliated rock is easily recognized and its importance is readily appreciated (Plates XXIXa, XXXb, f). However, many deformed metals, ceramics and rocks (such as quartzite and marble) consist of approximately equidimensional grains whose atomic structures ("lattices"), crystallographic axes and optical directions show various degrees of preferred orientation (*lattice preferred orientation*).

The preferred orientation is one aspect of the strain of a metamorphic rock, the main one being the distortion of markers such as bedding or fossils. It would appear that most such fabrics must be attributed, either directly or indirectly, to permanent deformation, although it is possible (Flinn, 1965, p. 62) that a preferred orientation might arise by a preferred nucleation or preferred growth mechanism during elastic (reversible) strain.

The relation to deformation is shown by:

(1) The general lack of preferred orientations in undeformed rocks. Primary or non-deformational preferred orientations may be produced during the formation of a rock, e.g. aligned phenocrysts in a porphyritic igneous rock, flakes of mica or clay parallel to the bedding of a sediment, and slight parallelism of *C*-axes of quartz in a sandstone due to alignment of elongate grains on bedding planes. Such preferred orientations may be distinguished from those produced tectonically by their degree of development, and their relation to known tectonic, igneous or sedimentary structures. In some complex cases however it is not possible to decide whether the earliest of a number of foliations in a rock is simply bedding or is a tectonic foliation along the bedding, or whether an old lineation is a sedimentation structure or an early tectonic feature (Spry, 1963c).

(2) Preferred orientations, foliations and lineations are, by contrast, common in rocks which show independent evidence of deformation, e.g. folds, deformed pebbles, ooliths or fossils. The similarity in orientation of a fold axis and an axial plane foliation, an axial mineral lineation and the symmetry planes of petrofabric diagrams implies a common cause.

(3) Preferred orientations can be produced experimentally by deformation such as the compression, tension, rolling and torsion of single crystals and polycrystalline aggregates of metals. Experiments on the deformation of single crystals and polycrystalline aggregates of calcite, dolomite and quartz under various conditions show that dimensional and lattice preferred orientations somewhat similar to those found in natural rocks can be produced by deformation.

The correlation between the preferred orientation and the total strain in a rock is qualitative rather than quantitative, and attempts to correlate the degree of preferred orientation with the intensity of deformation of metamorphic rocks have not met with success.

The textures of regional metamorphic rocks tend to be complex and inhomogeneous, i.e. composed of regions or *domains* which are themselves essentially homogeneous but which differ in some way from adjacent domains, e.g. lenses, layers, etc. The size, shape and orientations of different mineral species, or of different generations of the same species, may differ considerably. Thus a schist may consist of equant crystals of quartz with a lattice preferred orientation, plus parallel flakes of mica with both a lattice and a dimensional preferred orientation, and garnet with a random orientation. The relations between the dimensional and the lattice preferred orientations are complicated because although it is possible for a granular mineral such as quartz or calcite to have a lattice preferred orientation without any accompanying elongation and consequent dimensional preferred orientation, inequant minerals such as mica and amphibole have a relation between their atomic structure and their crystal shape such that if the outlines of a number of crystals are parallel, at least some elements of their crystal lattices must also be parallel.

Dimensional preferred orientations tend to be simple (except in polymetamorphic rocks where a number of foliations, etc., are present) and the long, mean and short dimensional axes of crystals tend to be parallel to each other; the long axes of the crystals are commonly but not invariably parallel to the long axes of elongate pebbles and to fold axes, and the long and mean axes lie in the axial plane foliation.

The dimensional preferred orientation may be poor because the minerals tend to be equant (e.g. quartz) rather than elongate (e.g. amphibole), because the amount of deformation is small, or because post-tectonic crystallization has partially or wholly destroyed the older syntectonic fabric.

It may be difficult to distinguish between the last two possibilities although the crystals are poorly directed and all look much the same in slightly deformed rocks but many of the crystals are athwart a relict foliation which has been *overprinted* by new growth in rocks with post-tectonic crystallization.

It should be remembered that the common method of cutting thin sections (normal to the lineation) minimizes the dimensional preferred orientation because the sections show the short and mean axes of the minerals. The texture may be quite different parallel to the lineation where crystals show their long and short axes, or parallel to the foliation where some domains show very little strain.

The patterns of preferred orientations are determined by the production of *petrofabric* diagrams (Turner and Weiss, 1963, pp. 194–255) which are plots of some significant lattice feature (e.g. the *C*-axis [0001] of quartz or calcite, the pole to the basal cleavage of mica [001], the *C*-axis of amphibole as defined by the intersection of the prismatic cleavages; *e* twin lamellae $(01\bar{1}2)$, and $(01\bar{1}2)$ cleavage in calcite).

Half a century of *petrofabric analysis* has given a great deal of information on the patterns of preferred orientation but the mechanisms by which they have been produced and their tectonic significance are still not clear (Flinn, 1965). This aspect of deformed rocks is complex and no attempt will be made here to cover the great body of facts, observations, experiments and hypotheses of the field of *structural petrology*; the reader is referred to the comprehensive review of Turner and Weiss (1963). The dimensional and lattice preferred orientations in metamorphic rocks are, however, an important aspect of texture and it is not possible to interpret the texture of a rock without considering the development of the preferred orientation in a general way.

The symmetry of the diagrams is one of the most striking and significant features. Classification of the various kinds of symmetry has been discussed at length by Paterson and Weiss (1961) but their detailed categories are difficult to apply to real diagrams which fall into three main classes:

(1) *Orthorhombic:* the diagram may be divided by three mutually perpendicular planes of symmetry.

(2) *Monoclinic:* the diagram can be divided by one plane of symmetry.

(3) *Triclinic:* the concentrations are sufficiently well defined to be recognized as real maxima and/or girdles (i.e. the pattern is not random) but the diagram cannot be divided symmetrically.

The very high *axial symmetry* type does not seem to be particularly significant except in the case of the diagram for [001] mica in a rock with a single planar foliation.

The decision as to the significant symmetry of many diagrams is not easy to make because the patterns tend to be somewhat irregular and poorly defined. Diagrams may be "almost orthorhombic", or "triclinic but with tendencies towards orthorhombic". One person may regard a diagram as basically orthorhombic but with monoclinic tendencies, whereas another may consider it as monoclinic with orthorhombic tendencies. The decision usually means ignoring the presence of minor maxima and equating angles between maxima or girdles which are 10°–20° different.

The symmetry of a diagram is important because of a considerable body of evidence which implies that it indicates (or at least is related to) the symmetry of the tectonic movements producing it, e.g. an orthorhombic fabric indicates non-rotational (flattening) movements, a monoclinic fabric indicates rotational or shearing (rolling) movements, and a triclinic fabric indicates either complex strain or overprinting of several movements in different directions.

MECHANISMS FOR THE PRODUCTION OF PREFERRED ORIENTATIONS

The total fabric of preferred orientation in metamorphic rocks containing several different minerals consists of many elements. It seems unlikely that the fabric of rocks such as schists containing both dimensional and lattice preferred orientations in diverse minerals such as quartz, mica and hornblende, could have been formed by a single process.

Preferred orientations of various kinds can be produced at three different stages in the metamorphism of a rock:

(1) Pre-metamorphic crystals are *mechanically* rotated in some way to assume related orientations without their identity being lost or their internal

structure and boundaries significantly changed. This mechanism operates in the cold-working of metals and the experimental deformation of marble at low temperatures. Geologically, crystals formed before deformation (pre-tectonic crystals) are oriented during the syntectonic stage.

(2) Crystals are formed initially in a preferred position during crystallization of a randomly oriented aggregate which is being strained, e.g. during

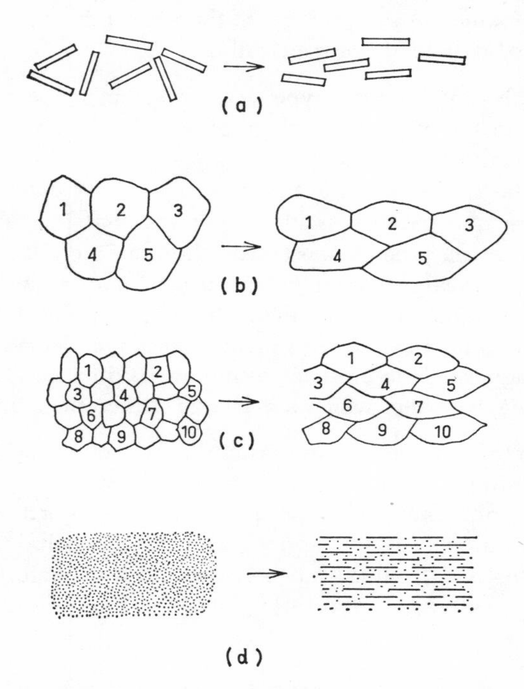

Fig. 54. Preferred orientation mechanisms. (a) Mechanical rotation of elongate crystals by flow of the matrix. (b) Plastic deformation by intragranular glide. (c) Preferred growth. Selective growth of certain orientations (grains numbered 1 to 10) at the expense of others. (d) Preferred nucleation, i.e. the formation of nucleii in certain restricted orientations during crystallization

the hot-working of metals and the experimental deformation of marble at high temperatures. Geologically, crystallization takes place during deformation, e.g. syntectonic crystallization in regional metamorphism.

(3) Crystals are formed in preferred positions because of some pre-existing orientation. Experimental examples include the annealing of cold-worked metals and recrystallization of natural marble containing an old fabric. Geologically, mimetic textures are formed during post-tectonic crystallization, during thermal metamorphism which is superimposed on earlier

regional or dynamic metamorphism, or during the later phases of multiphase regional metamorphism.

Dimensional preferred orientations can be produced by transforming equant grains into elongate shapes by the removal of material from one part of the lattice and its deposition in another. This may follow Rieke's principle (Russell, 1935; Turner and Verhoogen, 1960, p. 476) if an intergranular solvent is present, or take place by diffusion in the dry state following the Nabarro–Herring principle (p. 159).

Mechanical rotation can take place by the rotation of comparatively rigid, non-equidimensional bodies in a less rigid matrix, or by the deformation of constrained crystals with glide properties.

Simple Mechanical Rotation

Squeezing clay or compacting a sediment tends to rotate clay or mica flakes to a position normal to the compression. Such a preferred orientation was produced experimentally by Sorby (1853, 1856). Clastic mica flakes in sediments are rotated bodily into parallelism with a cleavage in pelites or even in sand-grade clastic rocks. Such a rotation requires a considerable degree of freedom because a high degree of preferred orientation appears to be produced by very small strains. It is possible that it takes place when the sediment is still wet and soft; such a cleavage might begin to develop during "dewatering" under small stresses. This mechanism probably has little general application to metamorphic foliations although it probably contributes to fissility in a slate where tiny clay flakes may have rotated mechanically. A possible criticism is that deformation greater than about 54% is necessary to produce a preferred orientation comparable with that of a typical slate.

Experiments of Means and Paterson (1966) support the concept of the formation of slaty cleavage by mechanical rotation. The experiments can be interpreted as indicating that tiny (10 to 20 microns) randomly oriented mica flakes are formed by crystallization during stress and that they subsequently tend to rotate to become perpendicular to the compression. The flakes adhere to each other wherever (001) faces come in contact (synneusis) to give large flat aggregates (30 to 80 microns). These resolve themselves into single sericite flakes by coalescence to give slaty cleavage. Further growth tends to be mimetic and to emphasize the dimensional (and lattice) preferred orientation.

Elongate mineral fragments ("sliced quartz") or mineral aggregates are aligned parallel to the foliation in quartzites with mortar texture, in flaser gneisses and in mylonites, but most of these elongate relics were formed *in situ* and do not appear to have been rotated mechanically into the foliation.

The mechanical rotation mechanism operates on inequant units and is basically one which produces a dimensional preferred orientation. It produces a lattice preferred orientation in minerals such as mica but not in equant crystals.

Ramberg and Ekstrom (1964) described a very unusual dimensional preferred orientation of pyrite porphyroblasts in slate in which the cube faces were parallel to each other and symmetrically disposed about the cleavage. It is known that rigid cubes mounted in a rubber block are rotated by compression of the block so as to make the cube faces symmetrical to the compression (Collette, 1959) and it is possible that the pyrite was oriented by such a process, but it seems more likely that the preferred orientation is due to "seeding" of pyrite nucelii on parallel mica flakes.

Polymetamorphic rocks such as phyllonites characteristically contain an old, tightly folded and translated foliation cut by a later planar one marked by parallel mica flakes (Plate XXIXc). The micas of the old foliation have been mechanically rotated so that they lie along the new foliation and are mixed with new micas from which they may, or may not, be distinguishable by size, shape, composition or optical properties.

ROTATION AND GLIDE

Deformation of a single crystal with a glide plane takes place by slip when the component of the stress parallel to the slip direction and plane (the resolved shear stress) is sufficiently great. Unconstrained slip causes a change in shape of the crystal but no change in orientation of the lattice. A dimensional preferred orientation may be produced in a polycrystalline aggregate by each crystal shortening in the direction of the compression by some mechanism of "plastic flow" such as slip on oblique planes.

If a single crystal is deformed while it is constrained by the anvils of the squeezing device, or if a crystal in a polycrystalline aggregate is constrained by adjacent crystals, and can no longer change its shape freely, then internal slip is accompanied by rotation of the crystal as a whole (Turner, 1952).

Slip along the glide planes can be regarded as a kind of rolling process (simple shear) which may cause rotation of internal markers such as old lamellae. At the same time, the crystal is forced to rotate as a whole in the opposite sense and thus its lattice (and optical) orientation is changed. *Internal rotation* (with reference to internal markers) thus takes place in the opposite sense to *external rotation* (with reference to external coordinates). A single glide plane rotates towards a position perpendicular to the compression or parallel to the tension.

When a granular polycrystalline metal (and probably calcite, dolomite, pyroxene and olivine) is deformed, the crystals undergo gliding (with dislocation movement) on those slip planes with sufficiently high resolved shear stress. Deformation of the whole aggregate proceeds homogeneously under certain conditions and the continuity of the whole is preserved as each grain is held by its neighbours and all grains are deformed an equal amount. Gliding of each constrained crystal causes internal rotation of the lattice. The final result is that the grains become inequant with a simple preferred orientation of the long, mean and short morphological axes and a rather more complex preferred orientation of the atomic structures.

A polycrystalline body cannot deform plastically by slip without forming voids at grain boundaries and triple points unless there are at least five slip systems in each grain. Void formation may be prevented by the operation of additional processes such as grain boundary migration. There is some evidence that the number of slip systems may be markedly reduced, perhaps to two, if dislocation climb and cross-slip are active. However, some important rock-forming minerals (e.g. quartz) have such limited slip ability that it is very doubtful if they deform homogeneously; grain boundary migration and intergranular slip must take place. Grain boundary slip is complex depending on the coherence of the various boundaries and the behaviour of the sub-surface layer and although the grain boundaries constitute a slip array containing an almost infinite number of slip systems, homogeneous deformation by slip requires five systems *within* the grains. Grain boundary slip in a polycrystalline aggregate of crystals with limited slip behaviour must lead to the formation of voids and resultant fracture (Day and Stokes, 1966) except at high temperatures where diffusion is important. Intergranular slip takes place along the boundaries and also by gliding on slip planes just below the boundaries. The contribution of grain boundary sliding is a function of the stress, strain, temperature and composition of the metal (Brunner and Grant, 1960; Day and Stokes, 1966). Sliding during creep is generally held to occur in significant amounts only if the temperature exceeds $\frac{1}{2}T_m$ (where T_m is the melting point in degrees Kelvin) but Gifkins and Langdon (1965) have detected very low temperature sliding in metals and Flinn (1965) has summarized the considerable evidence for grain boundary slip in deformed rocks.

A lattice preferred orientation arises from gliding because the glide planes rotate to some special position with respect to the stress. Crystals with a single glide plane, (e.g. hexagonal materials such as corundum or ice, or layer crystals such as micas, clays, molybdenite or graphite) rotate so as to cause the slip planes and/or slip directions to be parallel to each other and the slip plane to be perpendicular to the principal stress axis in non-rotational strain, or parallel to the movement surface in simple shear. Where the optic axis or *C*-crystallographic axis is normal to the glide plane a lattice preferred orientation automatically follows.

Crystals with high symmetry may contain a number of slip surfaces which constitute a glide system, e.g. the three $\{01\bar{1}2\}$ planes in calcite or the three $\{100\}$ planes in galena. Slip along the plane with the highest resolved shear stress will cause external rotation of the crystal until the shear stress rises sufficiently to cause slip on one or more other planes in the slip system. The crystal changes its external shape and rotates to an equilibrium position where gliding takes place simultaneously on several planes. The attainment of this equilibrium position by a number of grains produces a lattice preferred orientation. The external change of shape of the grains produces a dimensional preferred orientation.

Extensive study has been made of preferred orientations produced experimentally in polycrystalline aggregates of crystals with the ability to glide. The cold working of metals is the best example but similar ductile behaviour is shown by ionic solids (NaCl, MgO, LiF_2, Al_2O_3), etc., and by minerals of geological importance such as calcite, dolomite, rock salt, galena, mica, etc., under appropriate conditions.

Work on metals (Barrett, 1952) has shown that even though a crystal may contain a number of possible glide systems, not all operate under all conditions. Movement is controlled by temperature, pressure, the presence of inclusions or dislocations, the amount of deformation, and also compositional factors which affect the stacking-fault energy and the ability of the metal to undergo cross-slip. As the preferred orientation depends on the relation between the atomic structure and the slip planes, the fabric pattern is also controlled by these factors.

In a face-centred cubic metal such as aluminium the dominant slip plane is that with closest packing, i.e. (111). Slip on (111) in the [110] direction is accompanied by rotation of [110] towards the axis of compression to give a lattice preferred orientation with [110] parallel to the axis of compression. However, if slip takes place on (112) as well, a "double texture" is formed with [111] and [100] both parallel. Aluminium in tension slips first on (111) [110] so that [110] rotates towards the plane of tension until duplex slip on two $\{111\}$ planes begins. The lattice then rotates so as to maintain the axis of compression in a plane symmetrical to the two acting slip planes, i.e. to keep equal stresses on the two active systems. Duplex slip and corresponding rotation continues until the two active slip directions lie in the same plane as the axis of compression and on opposite sides of it.

The mechanisms which produce preferred orientations are by no means simple even in metals. In single crystals of aluminium (110) rotates so as to be perpendicular to tension but in polycrystals (111) becomes parallel to tension.

Plastic deformation of a polycrystalline metal by cold-working produces a preferred orientation in which certain lattice directions are aligned with the principal directions of flow. Rolling of face-centred cubic metals produces a texture with $\{110\}$ parallel to the rolling plane and $\langle 112 \rangle$ parallel to the rolling direction; in a body-centred cubic metal $\{001\}$ is parallel to the rolling plane and [110] to the rolling direction, and in hexagonal close-packed, with basal slip, (0001) is parallel to the rolling plane.

Deformation twinning accompanies slip in some crystals and the two are closely related. Twinning may cause a marked change in orientation but does not allow such large strains as slip. Preferred orientations may be produced in calcite, etc., by twinning on $\{01\bar{1}2\}$ (Turner *et al.*, 1954, pp. 883–934; Griggs, Turner and Heard, 1960, pp. 81–87).

The textural effects of cold-working of materials with glide properties are:

(1) A slight to profound reduction in grain size as large crystals are broken into smaller ones.

(2) A change in grain shape with a marked dimensional preferred orientation. A close relationship exists between the shapes of the grains and of the total strain pattern.

(3) A lattice preferred orientation with a close relationship between the symmetry of the fabric and that of the total strain pattern.

Deformation of metals at a low temperature leads to *strain hardening* (an increase in strength). The dislocations which permit slip to occur initially become "piled up" at grain boundaries or tangled with each other and so slip becomes more difficult. If a metal is deformed at a sufficiently high temperature (*hot-working*) its microstructures may appear to be unaffected by the deformation, there is no strain hardening and slip lines are absent. The basic mechanism is still one of slip and twinning by the movement of dislocations, however at high temperatures, diffusion of interstitials, climb, and cross-slip prevent work hardening.

Cold-working leads to a stable preferred orientation which persists even with very large strains. Hot-working produces similar orientation patterns but with a much greater degree of scatter for a given amount of strain. The processes of cold-working pass into those of hot-working as the temperature rises. A higher temperature results in increased thermal movement of atoms, greater interatomic distances, greater ease of slip, greater freedom of dislocation-movement and greater mobility of grain boundaries. Hot-working leads to recrystallization which may strongly modify the microscopic texture and produce a preferred orientation which is transitional with those produced by pure crystallization processes.

Although gliding is an adequate explanation for preferred orientations in metal aggregates, it is not satisfactory when applied to rocks (except marble) because of the limited numbers of slip systems in most rock-forming minerals, their resistance to glide, and possibly a strong strain-hardening effect. Flinn (1965, p. 63) considered that the preferred orientation pattern in tectonites is simpler than that expected by translation glide alone.

Orienting Mechanisms during Recrystallization

The mechanical processes described above depend either on some dimensional inequality or on the presence of internal glide or twin systems, but a preferred orientation can be produced in a range of anisotropic (both gliding and non-gliding) crystals during recrystallization. Many strong preferred orientations of quartz occur in rocks which do not seem to be sufficiently deformed to have caused gliding or rotation of crystals.

Lattice preferred orientations may be produced during the recrystallization of an aggregate of randomly oriented crystals by a combination of:

(1) Increasing the proportion of crystals with desirable orientations.
(2) Reducing the proportion of crystals with undesirable orientations.
(3) Changing the orientations of crystals from undesirable to desirable orientations.

The preferred orientation could be produced during a period of increase in grain size, in which case, crystals of desirable orientation may grow more quickly than those with undesirable orientations. Alternatively all the old crystals may disappear at the expense of new crystals which tend to be nucleated in desirable orientations. In processes involving granulation and reduction of grain size, the preferred orientation is complex because a group of new small crystals commonly reflects the orientation of a single large crystal from which it has been derived.

A preferred orientation in an aggregate of non-gliding crystals may be due to:

(1) An original fabric (mimetic).
(2) Preferred nucleation.
(3) Preferred growth.
(4) Piezocrescence.

Mimetic crystallization is the process by which the position or orientation, either dimensional or lattice or both, of later crystals is influenced by that of earlier ones. The effect may be produced by simple replacement as in pseudomorphs, by older crystals influencing nucleation of newer ones, or by the shapes or orientations of older crystals controlling the growth of newer ones.

Structural discontinuities such as bedding or foliation may involve not only compositional differences, but also be marked by crystals (e.g. mica) with a preferred orientation of form and lattice and so may be preserved, or even emphasized, during recrystallization. Thermal metamorphism of schists, slates or phyllites may produce hornfelses, chiastolite-slates, etc., whose textures are dominated by the relict structure. Mimetic crystallization of parallel mica flakes in this way resembles a displacive transformation with low activation energy, high retention of old lattice elements, and minimum nucleation. Recrystallization of parallel micas to a randomly oriented aggregate is a reconstructive transformation with high activation energy, requiring maximum nucleation and involving minimum use of old lattice elements.

Preferred Nucleation

Recrystallization consists of the formation of nucleii (or sub-grain blocks) followed by their growth, and a preferred orientation fabric must result if the nucleii have preferred orientations (Burgers and Louwerse, 1931). Preferred orientations of nucleii may arise from: first, a process analogous to annealing of a cold-worked metal, e.g. thermal metamorphism of a mylonite or other deformed rock, or postectonic crystallization in regional metamorphism; or second, by crystallization during the application of stress, e.g. syntectonic crystallization.

(1) Nucleii produced during annealing tend to have an orientation which is related to that of the lattice of the host or adjacent crystal and so the crystals formed by annealing an oriented cold-worked metal may have a preferred orientation. As an example, the orientations of new crystals growing in cold-worked aluminium are related to that of the host because the process favours the growth of crystals which share a [111] axis with the host and whose lattices have a rotation of 30–45° around this axis (Beck, Sperry and Hu, 1950; Dillamore, 1964). The control may be strong enough to produce a preferred orientation during annealing and this fabric may be the same as the original (but weaker) or be dependent on the original orientation but with a different pattern.

During annealing, the original preferred orientation does not change up to a certain temperature, then randomness increases, then a *new* preferred orientation replaces the old one (Barrett, 1952, p. 508) e.g. the cold-worked texture (123) $[\bar{4}1\bar{2}]$ of Al–Mn alloy is replaced by a cube texture (parallelism of {100}) during annealing and the intensity of the cube texture increases with temperature (Koppenoak, Parthasarathi and Beck, 1960).

(2) Potential nucleii in a stress field differ in stability according to their orientation. MacDonald (1957, 1960), Kamb (1959), Brace (1960) and Schwerdtner (1964) have postulated that certain orientations of anisotropic crystals in a stress field are more stable than others, thus the total energy of a polycrystalline aggregate with an appropriate preferred orientation is lower than that of a random aggregate and, given the opportunity, a preferred

orientation might be expected to develop by crystallization while elastically strained.

The relationship between the most desirable orientation and the stress is not clear. Kamb (1959) considered that crystals should be aligned with their *weakest* axis parallel to the main compression but MacDonald (1960) considered that the *strongest* axis should be parallel to the compression.

As MacDonald (1960, p. 279) has pointed out, the chemical potential differences for the different orientations are very small, "on the order of calories per mole for stress differences of thousands of bars". The chemical potential difference at 1000 bars for quartz is 0·07 to 0·08 cal/mole and for calcite is 0·22 to 0·27 cal/mole (Kamb, 1959). These figures are at least four orders of magnitude less than the activation energies of most geological processes and in any case, the thermodynamic approach of Kamb, etc., does not explain the *mechanism* of reorientation to the stable position.

It is suggested that *crystals* in the less stable orientations do not necessarily disappear in favour of more stable orientations but that *nucelii* oriented in the preferred directions are more stable than others so that more nucleii form in more-stable orientations than in less-stable. The free energy of a nucleus is the sum of its internal free energy and its surface free energy, so that the favourably oriented nucleii could have a lower total free energy and thus a smaller critical size. The formation of such nucleii would be favoured if the activation energy of nucleation were to be affected by their orientation. If this mechanism were valid, then a mineral preferred orientation could be produced during crystallization within an elastically strained body and in this case no correlation would be expected between the degree of preferred orientation and the amount of permanent deformation.

It is difficult to assess the importance of preferred nucleation in the recrystallization of an aggregate in which nucleation is either easy or not required, e.g. the recrystallization of limestone to marble or sandstone to quartzite, because abundant nucleii must be available in a wide variety of orientations in addition to those which are thermodynamically preferred.

Preferred Growth

Both lattice and dimensional preferred orientations are produced by processes which favour the growth of some crystals, faster growth in certain directions, or the faster movement of some crystal boundaries in preference to others. The process operates in the static recrystallization of an aggregate with an existing preferred orientation (annealing), in the growth of strongly anisotropic crystals with their directions of fastest growth parallel to a direction of extension in the aggregate, in the preferred growth of crystals favourably oriented with respect to stresses, or in the preferred growth of crystals or movement of boundaries in directions controlled by structural inhomogeneities.

The grain boundary configuration of an annealing texture depends not only on the abundance and position of nucleation sites, but also on the ability of the nucleus to grow through adjacent grains and into symmetrical positions (Dillamore, 1964, p. 1006). A grain or nucleus with a certain orientation may grow by moving its boundary through adjacent grains of different orientations. The material in the parent grains (those which are destroyed) becomes rotated and joined to the daughter (that which is growing). The boundary moves by *strain-induced boundary migration* and the mobility of a grain boundary (i.e. its ability to move in a direction approximately perpendicular to itself) is strongly orientation-dependent. The mobility is greatest in high-energy boundaries of poor fit and apart from special coincidence types, decreases with decrease in the angle of misfit. Low-energy twin boundaries are stable and static; medium-angle twist boundaries are moderately mobile; high-angle (high-energy) misfit boundaries are most mobile. The mobility of all types of boundaries increases with temperature.

The resultant texture is the result of larger movements by more mobile boundaries and smaller movements by less mobile boundaries. The more mobile (higher energy) boundaries are reduced in area to the benefit of less mobile (lower energy) boundaries in the same way as the faster growing (high energy) faces of a crystal in solution are reduced in magnitude in favour of more slowly growing faces.

The *direction* of movement of a mobile boundary is important. A boundary which moves by virtue of *strain-induced boundary migration* moves from an unstrained region into a strained region (Fig. 55). If it is curved, it moves *away* from its centre of curvature and thus will tend to *increase* the total surface area of the grain boundaries. On the other hand, adjustment of grain boundaries, particularly in the last stages of annealing differ in that they take place in unstrained crystals; they are dominated by processes which tend to *reduce* the total surface energy by straightening the boundaries and adjusting the angles between them; grain boundaries will then migrate *towards* their centres of curvature.

Consider the three strained grains *A*, *B* and *C* in Fig. 55b (Beck and Sperry, 1950, p. 152, Fig. 4) at the beginning of annealing. The *AB* boundary moves from position 1 to 2 to leave an unstrained region *D*. The *BC* boundary is a static one between two strained lattices of differing orientation. The *AD* boundary is an abnormal one between two lattices with the same orientation, one lattice being strained and the other unstrained. The *DB* boundary is a mobile one between unstrained and strained material with different orientations. Beck and Sperry (1950, p. 150) stated with respect to strain-induced

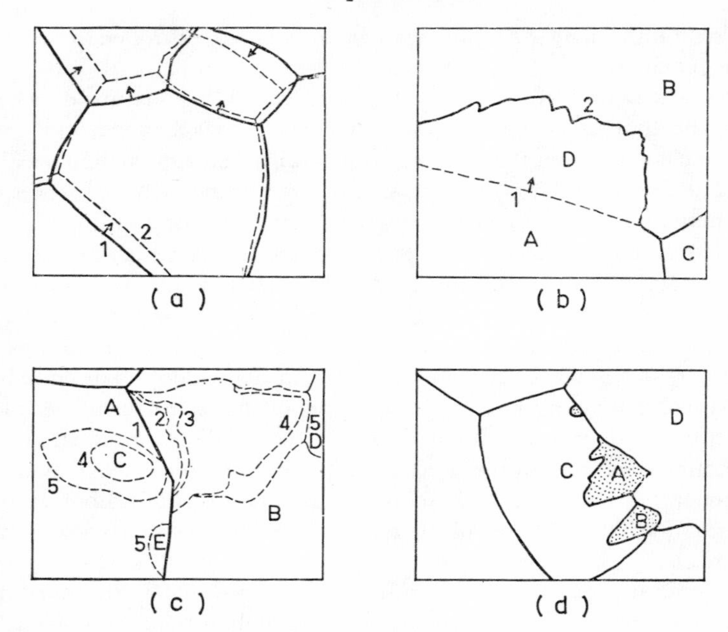

Fig. 55. Strain-induced boundary migration (after Beck and Sperry, 1950). (a) Grain boundaries in a cold-worked metal move from positions 1 to 2 during annealing. (b) Migration of the boundary between strained grains *A* and *B* from 1 to 2 leaving an unstrained region *D* with the same orientation as *A*. (c) Recrystallization takes place by grain boundary movement beginning at boundaries or within grains. Numbers show successive positions of the grain boundaries. (d) "Nucleation" of unstrained grains *A* and *B* at a *C–D* grain boundary

boundary migration during annealing, that "even though the grain boundary movements take place in a direction opposite to that required by surface-energy considerations, the total free-energy is decreased and equilibrium approached, as the strain-hardened grains of high free-energy content are replaced by low-energy annealed grains".

The grain shapes resulting from strain-induced growth are markedly different from those produced by normal recrystallization and the two processes are competitive (Burke and Shiau, 1948). Grain growth occurs at a rate depending only on the annealing temperature and present grain size but not on the strain. Strain promotes the formation of nucleii and the greater the strain, the shorter the incubation period. Recrystallization will dominate when the strain and grain size are sufficiently large to cause nucleation before the boundaries have moved distances comparable with the grain size.

Both lattice and dimensional preferred orientations are produced during annealing by rather complex mimetic processes but the textures can be accounted for by preferred growth with only some recourse to preferred nucleation (Beck, Sperry and Hu, 1950, p. 422).

The importance of preferred growth in the development of a dimensional and lattice preferred orientation in annealed aluminium single crystals was shown by Parthasarathi and Beck (1961). Nucleation was initiated by scratching one surface of a rolled single crystal which was then annealed. New crystals grew into the host crystal from randomly oriented nucleii on the scratched surface. The rate of growth of the new crystals depended on the orientation relation between each new grain and the host crystal. Certain orientations grew more quickly than others so that their fabric dominated that of the more slowly growing crystals to produce a preferred orientation.

Two opposed tendencies operate during recrystallization (Aust and Rutter, 1962). Nucleii with an orientation giving low-energy coincidence boundaries with the host are easier to form because they require a lower activation energy for nucleation hence a lower critical size and a shorter incubation time. However special orientations would also be favoured during growth because of the greater mobility of high-angle boundaries. Where coincident, twin or special low-energy boundaries are formed they remain immobile and persist in the final texture.

Griggs, Paterson, Heard and Turner (1960) produced annealing textures in calcite and stated that "this recrystallization of calcite is similar in all respects investigated so far to the recrystallization of metals under similar conditions". They produced recrystallization in deformed single crystals, in deformed marble (driving energy is strain-energy) and in fine-grained undeformed limestone (driving force is surface energy). Increased deformation or reduced grain size decreased the minimum critical temperature for recrystallization. True annealing preferred orientations were not produced and recrystallized grains had either no preferred orientation or a weaker one than the original material. An orientation relation between primary and secondary crystals was established, the *C*-axis of the recrystallized grain favouring an inclination of 25–30° to that of the host and in some examples, an *a* axis or $(r_1 : f_2)$ direction was shared. Immobile cleavage and lamellae boundaries controlled grain growth and produced a dimensional preferred orientation.

Voll (1960) considered that recrystallization of quartz in a rock with a preferred orientation should produce secondary grains which avoid the original orientation and develop *C*-axis fabric patterns with small-circle and great-circle girdles around the original concentrations.

Carter, Christie and Griggs (1964) and Raleigh (1965a) succeeded in producing a preferred orientation of quartz experimentally. A slight to strong dimensional preferred orientation normal to the compression as well as a lattice preferred orientation was found but unfortunately the lack of fully determined fabrics for the fine-grained experimental samples has so far hindered direct comparison with the predicted fabrics or with natural quartz fabrics (Raleigh, 1965a, p. 377). Much of the recrystallized quartz has a granoblastic-polygonal texture, is coarser than the original material and shows no sign of strain; these textures resemble those of annealing.

Preferred orientations will be produced by static recrystallization (annealing) only if the material is anisotropic and has some kind of fabric prior to recrystallization. The foliation, lineation and preferred orientation of most metamorphic rocks appear to be produced by syntectonic crystallization, i.e. accompanying movement, flow, strain or stress but there is very little experimental evidence relevant to syntectonic crystallization in the most important metamorphic rocks. Deformation of metals at high temperatures (hot-working) is essentially a continuation of the processes acting in cold-working and is dominated by gliding. Evidence from metals and from experimental deformation of calcite and marble is relevant only to fabrics of preferred orientation in marbles or other polycrystalline aggregates of material with the ability to glide freely.

The texture (grain size, fabric), etc., produced by hot-working depends on the initial texture of the metal, the total strain, the strain rate, the temperature and the composition. The correlation of higher strain rates with finer grain is probably of general application to regional metamorphic rocks.

Preferred growth during syntectonic crystallization is a possible process by which preferred orientations might be produced in metamorphic rocks. Assuming abundant nucleii with random orientations, it is possible that the rate of growth of those with certain favoured orientations might be greater so that they could dominate the fabric.

It has been suggested (Ramberg, 1952, p. 117) that those crystals in a randomly oriented aggregate growing in a stress field which have their lattice directions for fastest growth favourably oriented with respect to the operative stresses will grow most quickly. Thus those hornblende crystals in an amphibolite which happen to have their *C*-axes parallel to the direction of extension could grow most quickly and dominate the fabric.

Preferred growth will result if units are more easily or more quickly attached to a more-stable than a less-stable interface. The ease of attachment of new atoms depends on the grain boundary energy and some metallurgical results suggest that surface energies may be a controlling factor in some preferred orientations.

Piezocrescence

A preferred orientation results from preferred growth in which mutual grain boundaries move from crystals with more stable ("favourable") orientations into those with less stable ("unfavourable") orientations, thus reducing the proportion of the latter. This could take place by *piezocrescence* (Thomas and Wooster, 1951) which means "growth under pressure" or the growth of one crystallographic orientation out of another under the influence of stress. The simplest examples are the advance of the composition plane of a deformation twin, of a kink boundary or a martensitic boundary, and the process has been demonstrated experimentally in metals and ceramics (Burgers, 1963; Day and Stokes, 1966, p. 353) and in Dauphiné twins in quartz (Thomas and Wooster, 1951). The grain boundaries undergo *stress-induced boundary migration* and each moves by changing the lattice orientation of the crystal just ahead of the boundary into the orientation of that behind the boundary. The transformation is martensitic and its complexity depends on the complexity of the structure (it may be rapid, displacive and diffusionless in quartz Dauphiné twinning and in the deformation of high-symmetry metals where motion of dislocations allows restacking to the new orientation).

Piezocrescence in quartz appears to cause reorientation so that the crystal can store more elastic energy, i.e. present a direction of increased compressibility to the principal stress. The original example discussed by Thomas and Wooster (1951) dealt with Dauphiné twinning but there seems no reason why the principle should not be extended to include reorientations with subsidiary bond breakage. The adjustment to a more stable orientation is achieved by first compressing elastically some bonds and stretching others, then rotating ionic groupings such as silica tetrahedra, disconnecting some bonds by dislocation movement (possibly aided by silanol groups) then reattaching the ionic groupings in new orientations. The optical and lattice orientations thus change.

EFFECT OF INCLUSIONS ON DIMENSIONAL PREFERRED ORIENTATIONS

As the shape of a crystal in a polycrystalline aggregate is the sum of its boundaries with adjacent crystals, it depends on the relative mobilities of the various parts of its boundary. The mobility of grain boundaries can be strongly affected by inclusions and thus the shape of crystals in an aggregate may reflect the distribution of another phase.

The orientation of mica in a quartz–mica schist may be the controlling influence on the dimensional orientation of quartz which will be elongate parallel to the micas as (001) of mica controls the orientation of quartz–quartz boundaries in its vicinity (Hobbs, 1966, p. 704). Moving quartz–quartz boundaries will be pinned at mica flakes and statistically the mobility of quartz boundaries (and thus the length of the crystals) will be greater parallel to the mica trains than perpendicular to them (Plate XXVIIIa, b). The ability of inclusions to pin a moving boundary depends on their size and spacing relative to the extent of the boundary.

CHAPTER 9

Textures of Dynamic (and Shock) Metamorphism

MECHANICAL processes control the textures of rocks which have been deformed at comparatively low temperatures or at rapid rates. Recrystallization is at a minimum under these conditions although it may accompany or follow the deformation. *Dynamic* metamorphism is produced on a comparatively minor scale by load or tensional faulting, on a larger scale by thrusting, and on a regional scale by folding and there appear to be no clearly defined differences between the processes or products of dynamic and those of low-grade regional metamorphism although in the simplest cases the field occurrences differ and dynamic metamorphism effects may be limited to narrow zones.

Recent studies (Chao, 1967; Carter, 1965; Short, 1966) have recognized the new field of *shock metamorphism* which may be regarded as either a special type of dynamic metamorphism or as an extension of the conditions. Shock metamorphism consists of the very rapid application of high differential pressure as at a meteorite impact site or near an atomic explosion. The term *impact* metamorphism has been used more or less synonymously, but this has the direct implication that meteorite impact is responsible, whereas it has been argued that some shock metamorphics are not due to impact but to rapid tectonic pressures. Pressures rise to many megabars and the temperature may exceed 1500°C. for periods of a microsecond to a second.

DYNAMIC METAMORPHISM

The term has been applied to a group of processes which include brecciation, cataclasis, granulation, mylonitization, pressure solution, partial melting and slight recrystallization and which produce, according to conditions, rocks such as breccias, mylonites, phyllonites and slates.

Textures of dynamically metamorphosed rocks are characteristically complex and inhomogeneous and depend on the inter-relations of temperature, pressure, strain rate, presence of solvents, and on the mechanical properties of the rocks concerned. Some behave in a plastic fashion even at low temperatures, thus serpentinites, pelites, limestones, rock salt, etc., become folded, recrystallized and foliated whereas crystalline rocks under the same conditions behave in a more brittle fashion and undergo mylonitization or cataclasis.

In the least intense stage of dynamic metamorphism the rock is cut by rather closely spaced fractures from a few centimetres to a few millimetres apart. These slice through the grains and may have visible small displacements. Differential movement may take place on grain boundaries as well as tectonic surfaces or discontinuities including clean fractures with visible displacements and brecciated zones. The main effect of marginal granulation is a reduction in grain size. With increasing deformation the mineral grains themselves become affected. Crystals become bent (micas, amphiboles), show undulose extinction (quartz, feldspar, calcite, pyroxene, amphibole), contain deformation twins (feldspar, calcite), are fractured (garnet), have anomalous optical properties (quartz, mica, calcite), undergo inversion (orthoclase-microcline), or gliding (calcite, mica), are kinked (mica, pyroxene, kyanite), develop deformation lamellae (quartz, calcite), or show mineralogical changes such as exsolution feldspar, pyroxene, amphibole) or hydration (uralitization, chloritization, sericitization, serpentinization, etc.). Feldspar may become opaque.

Relict crystals or *porphyroclasts* are large, angular and irregular in less-deformed rocks, becoming smaller, more rounded, and even lenticular as deformation becomes more intense. The porphyroclasts are surrounded by a matrix which originally ranges from very fine-grained and optically irresolvable to finely crystalline but which may have been wholly or partly recrystallized later.

DYNAMICALLY METAMORPHOSED ROCKS

Table 8 shows what is possibly the most widely accepted nomenclature. The classification of these rocks is based on textural criteria but the terminology is complex and rather confused; it has been reviewed most recently by Hsu (1955) and Christie (1960, 1963).

Tectonic or *crush breccia* consists of large angular fragments produced by tectonic fragmentation of pre-existing rock. The original texture within the fragments is not disturbed and there is a powdered rock matrix. The term *breccia* is used for rocks containing fragments larger than about ½ cm, *fine breccia* for those with fragments from about 1 mm to ½ cm, and *microbreccia* for those with fragments less than 1 mm although the size ranges have not been rigidly defined.

TABLE 8. CLASSIFICATION OF DYNAMICALLY METAMORPHOSED ROCKS

Nature of matrix		Proportion of matrix			
Crushed		0–10%	10–50%	50–90%	90–100%
	foliated	crush (tectonic) breccia or conglomerate	protomylonite	mylonite	ultramylonite
	massive		protocataclasite	cataclasite	ultracataclasite
Recrystallized	minor	hartschiefer			
	major	blastomylonite			
Glassy		pseudotachylyte (hyalomylonite)			

Mylonite (Lapworth, 1885; Teall, 1918; Waters and Campbell, 1935) consists of strained porphyroclasts embedded in abundant (50 to 90%) fine-grained to cryptocrystalline matrix; a foliated structure is typical (Plate XX) and this distinguishes it from a *cataclasite* which has an unfoliated matrix (Hsu, 1955). The matrix of quartzose mylonites is typically finely crystalline and minor recrystallization may produce a somewhat granoblastic-polygonal texture. The matrix quartz may form elongate crystals or elongate aggregates and may or may not be undulose. Many strongly deformed mylonites lack porphyroclasts and have such a simple banded structure and fine, granular texture that they resemble comparatively undeformed, scarcely metamorphosed sediments. However, their origin is indicated by intrafolial folds between the planar foliation, and by a lattice preferred orientation of the quartz; this may be recognized by simply inserting a gypsum plate when viewing the slide under crossed polarizers (similarly oriented crystals display similar interference colours).

Repeated movements occur in some great thrust zones (e.g. the Moine Thrust) and Christie (1960) recognized *primary* and *secondary* mylonitic rocks. Primary mylonites are formed in one episode from a variety of pre-existing rocks but secondary mylonites contain evidence of several episodes and may be derived from primary mylonites.

Protomylonite is transitional between micro-breccia and mylonite and contains between approximately 10 and 50% of crushed matrix; protocataclasite is similar but with a non-foliated matrix.

Ultramylonite (Staub, 1915; Quensel, 1916; Sutton and Watson, 1950, p. 261) and *ultracataclasite* (Grubenman and Niggli, 1924; Tyrrell, 1926) are rocks in which few (less than 10%) porphyroclasts have survived and which consist of foliated or structureless rock powder respectively.

Hartschiefer (Holmquist, 1910; Quensel, 1916; Tyrrell, 1926) is a microbrecciated rock with rigid parallel banding and a matrix affected by minor post-tectonic recrystallization independent of the layering.

Blastomylonite (Sander, 1911; Knopf, 1931) is a mylonite in which late recrystallization is so pronounced that the original cataclastic nature can only be recognized with difficulty, if at all (Plate XXI). It is convenient to distinguish between hartschiefer in which only matrix has been recrystallized and blastomylonite in which both the matrix and lenticles or porphyroclasts have been recrystallized.

Augen schist as originally defined (Lapworth, 1885) is a schist containing augen of original rock (with cataclastic texture) or original minerals set in a schistose matrix of quartz and mica. It appears to differ from a mylonite and blastomylonite in the large proportion of coarse new syntectonic mica and quartz. The term *augen gneiss* covers a variety of rocks ranging from coarse augen schists to simple regional metamorphic gneisses with porphyroblasts of feldspar. *Mylonite gneiss* (Quensel, 1916) and *flaser rock* (Tyrrell, 1926) are somewhat similar.

Layered mylonite (Prinz and Poldervaart, 1964) is a mylonite with a compositional layering which has developed by metamorphic differentiation during deformation.

Hyalomylonite (Scott and Drever, 1954) is a mylonite containing glassy material produced by melting during deformation and is more or less synonymous with *pseudotachylyte*

The textures depend on the response of minerals with various properties (strength, elasticity, brittleness, ductility, tendency to recrystallize, chemical stability) which in turn depend on the interaction of the following controls:

(1) *Total strain.* The proportion of matrix to porphyroclasts increases with increasing deformation, e.g. from protomylonite to ultramylonite.
(2) *Confining pressure.* The degree of development of a foliation seems to increase with increasing confining pressure, e.g. from cataclasite to mylonite.
(3) *Temperature.* The amount of crystallization increases with temperature e.g. from mylonite through hartschiefer, blastomylonite, to augen schist or gneiss and ultimately perhaps to granulite (*sensu stricto*).
(4) *Strain rate.* This acts oppositely to temperature in that increased strain rate favours mechanical effects over crystallization, e.g. mylonite rather than blastomylonite.

Impact metamorphics and some pseudotachylytes are the result of very high strain rates. The degree of vitrification has been correlated with the degree of shock metamorphism, i.e. the amount of glass increases with increase of strain rate, total strain and temperature. Dynamic metamorphism grades into shock metamorphism with an increase of strain rate and into regional metamorphism with a decrease.

Phyllonite (Sander, 1911, p. 301; Knopf, 1931) or "phyllite-mylonite" is a rock of phyllitic appearance formed from a coarser-grained rock by mylonitic retrograde metamorphism (Plate XXII). The rock has a characteristically lenticular structure and the dominant foliation is at least partly due to folding and transposition of an older foliation.

It is difficult, and in some instances impossible, to decide whether the texture of a highly deformed rock with strained porphyroclasts in a mylonitized or schistose matrix is a mortar or a blastopsephitic texture.

Cement texture indicates angular mineral fragments, many of which show matching outlines, cemented by another mineral. It is a microscopic texture characteristic of brittle minerals such as pyrite. A rather irregular texture due to partial replacement (*pseudo-cataclastic* texture of Anderson, 1934) superficially resembles some cataclastic textures and is equivalent to *ice cake* texture (rounded or corroded remnants surrounded by a secondary mineral).

Slate

Low-grade regional metamorphism with slight recrystallization produces much the same results as dynamic metamorphism and it is of little importance whether slates should be included among the products of dynamic or of regional metamorphism.

Fine-grained, clay- and mica-rich pelitic rocks are structurally weak and tend to flow in a plastic fashion when deformed, with the production of various kinds of cleavage and the conversion of pelite, argillite, shale or mudstone into *slate* (a fine-grained argillaceous rock with a cleavage, foliation or direction of splitting which is independent of the bedding). Brecciation, mylonitization and cataclasis are rare in fine-grained pelitic rocks and recrystallization is extensive although the crystals are very small. A *shale* will split along the bedding because of tiny clastic flakes of clay and mica lying on the bedding planes (a depositional, preferred orientation fabric). Deformation of the shale produces a *slaty* or *flow* cleavage due to the mechanical rotation of these clastic particles together with the recrystallization and growth of new illite, sericite or chlorite flakes in a preferred direction. In thin section, slates consist of an extremely fine-grained aggregate of nearly parallel flakes of a white mica which is referred to as "sericite" or "illite", round to elongate grains of quartz, and perhaps chlorite, feldspar, hematite, and an indefinite carbonaceous material which may be amorphous carbon in some rocks and graphite in others.

Fracture, strain slip and crenulation cleavages commonly develop in foliated rocks such as slate, phyllite or schist later in their tectonic history.

They appear to require the pre-existence of a closely spaced weakness which is generally a slaty cleavage or schistosity but which may (rarely) be bedding. The rock splits along a series of fractures or planes of weakness which have a small but finite separation in contrast to slaty cleavage in which the potential splitting planes are due to the parallel orientation of flaky minerals and which have an almost infinitely small separation.

Sandstones

Folding of interbedded sandstones and argillites may produce various kinds of cleavage in the unmetamorphosed sediments. Textural changes accompanying such deformation are related to those of dynamic or low-grade regional metamorphism and may be briefly considered here.

The mechanical behaviour of a sandstone and its resultant texture depends on the amount of deformation and on the relative proportions of sand size clastic grains and clay- or mica-rich matrix. Folding of a quartzose-sandstone with a closed framework results in undulose extinction, shattering and similar strain effects within the grains. There is little tendency to produce a foliation although plastic deformation and pressure solution at grain boundaries may produce some grain elongation. Grains become sliced up and granulated.

Fellows (1943) has traced the textural evolution of mortar structure and preferred orientation from sandstone to quartzite and recognized five textural types of quartz:

(1) Original clastic grains.
(2) Peripheral growth quartz (diagenetic overgrowths).
(3) Crush quartz (fine-grained matrix produced by "crushing").
(4) Needle quartz (elongate quartz formed by "polygonization" of large undulose grains).
(5) Recrystallized quartz (an even-grained aggregate of elongate quartz produced by growth of the matrix).

Rocks with greater amounts of matrix and an open framework (such as greywackes) develop different textures because movements are mainly concentrated in the weak matrix.

Small deformations leave the larger clastic grains strained and develop a fracture cleavage in the matrix. This foliation consists of rather closely spaced, irregular, anastomosing and intersecting dark lines when viewed in thin section. It tends to divide the rock into rhombic fragments and wraps around the larger clastic grains. Clastic micas are kinked and rotated into parallelism with the foliation. The foliation consists approximately of two surfaces at about 30° to each other with the acute bisectrix approximately parallel to the axial surfaces of folds.

Where folding has been tight (as in the cores of rather open folds), the sandstones become more strongly deformed and recrystallization takes place in the matrix. Parallel flakes of new sericite grow in the matrix or in "tails" or pressure shadows adjacent to large quartz grains to give a slaty cleavage which is approximately parallel to the fracture cleavage. New mica may grow along the fracture cleavage where it is parallel to the slaty cleavage or may lie obliquely where the fracture cleavage occurs away from the slaty cleavage direction.

Original clastic grains become undulose but it is not generally possible to determine what proportion of the strain is due to deformation of the rock itself and what was present in the grains in their original clastic state. Dimensional preferred orientation of elongate clastic grains may be strong and a lattice preferred orientation may, or may not, be present. The presence of a pelitic matrix appears to act against the development of a lattice preferred orientation of clastic grains because movements are taken up in the matrix rather than the grains.

The large clastic grains may be sliced into elongate fragments, smashed into irregular fragments, or recrystallized into aggregates of small crystals with blurred grain boundaries. Peripheral replacement by sericite commonly gives crenulated margins. Recrystallization of both quartz and mica takes place in many folded, but unmetamorphosed sandstones. The quartz grains retain their clastic size but lose their undulose extinction. Mica (sericite and chlorite) forms thin layers and lenses of parallel flakes along the anastomosing fracture cleavage so that the rocks resemble *semi-schists* produced by regional metamorphism.

Calcareous Rocks

The behaviour of limestone or marble varies widely, depending on the nature of the original rock and the physical conditions (temperature, confining pressure and deformation rate). Dynamic metamorphism of very impure fine-grained limestones may give slaty rocks. Limestones shatter, undergo pressure solution and recrystallize readily during dynamic metamorphism even at low confining pressures.

Coarse-grained calcitic rocks (clastics, or those which have been coarsened by diagenesis or metamorphism) behave somewhat similarly to the quartzites described above, but differ because of the strong gliding and twinning ability of calcite. Mortar texture may be produced with large, lenticular to strongly elongate calcite crystals (with undulose extinction, several sets of deformation twins, deformation lamellae and glide planes) set in a finely crystalline matrix (Plate XXVIIIc). Recrystallization may produce calc-schists or schistose marbles similar to those produced by regional metamorphism.

Crystalline Rocks

Cataclasis affects the different silicates in coarse polymineralic rocks in different ways (Beavis, 1961). The minerals in granites, pegmatites, porphyries, gabbros, gneisses and granulites (even coarse-grained arkoses might be included even though they are not crystalline) may be divided into three groups:

(1) Brittle minerals such as feldspar.
(2) Somewhat ductile minerals with some glide properties such as mica, amphibole, pyroxene and olivine.
(3) Minerals which recrystallize easily such as quartz and calcite.

Cataclasis causes granulation (particularly marginal) of all minerals but affects the brittle ones most. Feldspar tends to remain as large porphyroclasts with undulose extinction, deformation twinning, fractures, twisted twin planes and exsolution perthite lamellae. Orthoclase inverts to microcline. All feldspars tend to be sericitized or made opaque by tiny inclusions and plagioclase may be saussuritized. Pyroxene, amphibole and biotite are first bent, twisted, twinned and frayed, then converted into lenticular or ribbon-like aggregates of fine-grained parallel flakes, fibres or prisms.

Cataclasis in quartzite leads to a *mortar texture* in which large, ragged, lenticular quartz porphyroclasts with undulose extinction (or various related kinds of deformation substructure) and deformation lamellae, are set in a fine-grained aggregate of quartz (Plate XXX). The resistance of feldspar to granulation and deformation is in direct contrast to the behaviour of quartz. Large feldspar porphyroclasts persist as well formed, apparently unstrained, fresh crystals and thus may very closely resemble post-tectonic porphyroblasts (Hsu, 1955, p. 55). Feldspar can be recognized as probably pre-tectonic if it forms "trains" or groups of fragments, is undulose, contains deformation twins, is wrapped around by the foliation, is xenoblastic and rounded, contains compositional zoning (especially zoning oblique to the crystal boundaries) or is perthitic. Feldspar is probably post-tectonic if it is helicitic, idioblastic, spongy with crenulate margins, untwinned, if the matrix foliation abuts against its boundaries, if it is markedly elongate athwart the foliation, or if zoning is complete in each fragment and is parallel to the crystal boundaries.

Deformation of a polycrystalline aggregate at temperatures below which there is either simultaneous recrystallization or widespread slip, causes the grain boundary regions to be highly strained. Inhomogeneous grain boundary slip may produce corrugations on the boundaries and then mechanical crushing to produce large crystals in a crushed matrix. Mortar texture however appears to involve some recrystallization and experiments on metals and ceramics (particularly of magnesia, Day and Stokes, 1966) suggest that the strained boundary regions polygonize to produce rather fine-grained interlocking crystals which still reflect the shape and orientation of the host crystal (e.g. Plate XXXc). Boundary polygonization at higher temperatures leads to "cellular" or granular recrystallized regions between large strained porphyroclasts.

The small crystals between the porphyroclasts are generally not undulose, do not contain deformation lamellae or substructure, and some have boundaries meeting at 120° triple-points. Generally both large and small quartz grains are flat and elongate and thus impart a foliation and lineation to the rock.

A dimensional preferred orientation of the porphyroclasts appears to develop by the initial formation of undulose extinction, then accentuation into deformation bands (bounded by walls of dislocations perpendicular to the basal glide planes) and then disruption along these sub-grain boundaries (not along the glide planes) to give elongate slices, lenticles or ribbons which are mechanically rotated into parallelism by movements of the matrix.

Petrofabric analysis shows a variety of relationships. There may be a lattice preferred orientation in the large porphyroclasts but not in the small matrix grains, or in the small but not the large; there may be patterns of preferred orientation which differ between large and small or there may be no preferred orientation at all.

Quartz forms fine-grained, recrystallized aggregates in the matrix where it may be accompanied by feldspar, mica and amphibole as foliated layers swirling around large relict porphyroclasts. Coarse, crystalline rocks thus are first converted into irregularly textured schistose rocks such as augen schists or augen gneisses. *Auge* is German for "eye" (*augen* is plural) and refers to the lenticular shape of relict porphyroclasts or mineral aggregates. The term *flaser* (*Ger.* streak, lenticle) has been similarly applied, e.g. flaser gneiss, flaser gabbro.

Strong deformation produces mylonites. Large, original grains of quartz are strained or are broken down into tiny granules with intensely crenulated margins and in places the quartz appears to have flowed in curving streaks between the feldspars. The feldspars (with enclosed unstrained quartz and mica) become rounded but may retain their original crystal shapes to a remarkable degree. Large crystals become disrupted into lines of small oval granules. As the amount of matrix increases, a kind of irregular flow pattern is traced out by lines of minute biotite flakes and iron oxides which swirl around the relict, lenticular porphyroclasts. The foliated texture of the mylonite, which is prominent in thin section, is different from the schistosity of schists and is not generally associated with any strong mineral orientation or macroscopic fissility. The black colour of mylonite in hand specimen is due mainly to strong light-absorption by the very fine-grained mineral aggregate with an additional effect by finely divided, disseminated biotite or iron oxides.

Dynamic metamorphism may convert an originally massive rock into a *layered mylonite* by metamorphic differentiation which is the process whereby the mineralogical homogeneity of a rock is reduced by metamorphic processes, i.e. contrasted mineral assemblages develop from rock in which the minerals are more or less uniformly distributed. It may produce layering in an originally massive homogeneous rock or form large mineral segregations. The term was first used by Stilwell (1918) and Eskola (1932b) but the first well-documented description appears to have been of layered schists formed from massive, unbedded greywackes (Turner, 1941). A general discussion and examples of the process are given by Ramberg (1952), Bennington (1956) and Turner and Verhoogen (1960, p. 581).

Metamorphic differentiation appears to act contrary to most metamorphic processes which tend to flatten out chemical gradients; it involves an increase of order from more disordered, homogeneous material and thus a decrease in entropy. This kind of process always occurs to some extent in metamorphism even in the formation of a porphyroblast from disseminated components. The coarse, parallel layering in some banded gneisses is probably due more to metamorphic differentiation than to original bedding, igneous flow structure, partial melting or lit-par-lit.

The two kinds of metamorphic differentation which are most important are:

(1) Segregation under low to high pressure, but no directed pressure, to produce aggregates rich in one or more minerals, e.g. concretions in sediments, monomineralic segregations in hornfelses and granulites. The chemical concentration results from gradients in chemical potential due to pressure differences (Ramberg, 1952, p. 215), to some original chemical discontinuity (for example fragments of wood or fossils causing a local change in pH or eH), to differences in grain size or shape (i.e. in surface energy), to differences in ionic migration and to other lesser-known factors.

(2) The formation of a compositional layering parallel to a metamorphic foliation. The layering is due to segregation of light-coloured minerals such as quartz and feldspar on the one hand, and of dark-coloured minerals such as biotite, hornblende, pyroxene, epidote, garnet, etc., on the other. This layering occurs in mylonites (Schmidt, 1932; Wenk, 1934; Prinz and

Poldervaart, 1964; Sclar, 1965) and is very common in schists. Massive igeneous rocks pass into layered greenschists and amphibolites (Van Diver, 1967).

Alternate layers of different mineralogical composition are found in some mylonites formed from rocks which were essentially massive but which contained minerals with different mechanical properties. They have been derived from non-layered granite–gneiss, biotite–hornblende–cummingtonite gneiss and dolerite. The original crystalline rocks consisted of various mixtures of quartz, feldspars, micas, pyroxenes and amphiboles, whereas the layered mylonites have alternations of felsic (quartz, feldspar) and femic (amphibole, mica) minerals. Inosilicates have been separated from phyllosilicates. Original feldspar tends to persist, quartz recrystallizes, and pyroxene is granulated and altered to amphibole which gives a dark matrix swirling around hornblende or feldspar porphyroclasts.

Layering in mylonites is best explained by the hypothesis of Schmidt (1932) who considered that it was the result of different mechanical behaviour of the various minerals. More ductile minerals such as amphibole and mica adjust by gliding and recrystallization and are smeared out into intensely sheared layers while the more brittle minerals such as feldspar and quartz are rotated and segregated into layers between the shears.

Sander (1930), Eskola (1932b) and Ramberg (1962) outlined an alternative hypothesis involving an interaction of mechanical and chemical factors. This is, as yet, poorly understood but may be applicable to banding in schists and gneisses. Eskola considered that the relatively high solubility of quartz, albite and calcite is the essential property which causes segregation.

Cataclasis of acid porphyritic rocks (porphyries, rhyolites, acid tuffs, etc.) is a common phenomenon. The fine-grained or glassy matrix is transformed very easily into a fine-grained schistose aggregate of quartz, sericite, chlorite, paragonite and albite. The schistosity wraps around porphyroclasts of quartz and feldspar which were originally phenocrysts. These crystals may retain typical igneous characteristics such as embryo-like embayments, subhedral or euhedral forms of plagioclase or high-temperature quartz, igneous-type twins (Carlsbad, etc.) or igneous mineralogy (sanidine, high-temperature perthite). Such rocks have been called *porphyroids*.

Acid rocks can be converted into layered mylonites in which the thicker, lighter coloured bands have a sparkling appearance in hand specimen and contain abundant original rock material in contrast to the black layers of mylonite which are streaky and irregular, black, dense and dull (Read, 1951).

Feldspar is the first mineral to show the effects of deformation in the dynamic metamorphism of gabbro consisting of a coarse granular interlocking aggregate of pyroxene and labradorite with possible minor iron oxides, olivine, biotite and mesostasis. Lattice curvature causes undulose extinction; there are irregular, wedge-shaped deformation twins, and microfractures (like tiny faults) displace the twin planes. The formation of these textures in feldspar is followed by marginal granulation and mortar texture. Mineralogical breakdown may occur and the plagioclase may become saussuritized and riddled with sericite, zoisite, epidote or calcite or transformed to albite. Pyroxene may show purely mechanical effects such as undulose extinction, twinning, kinking and exsolution-lamellae but it commonly undergoes uralitization to give aggregates of fine, fibrous actinolite and chlorite. True mylonites of gabbroic composition appear to be rare and dislocation metamorphism appears to be generally accompanied by some recrystallization (Read, 1951, p. 245).

Pseudotachylyte

Pseudotachylyte (glassy mylonite, hyalomylonite, flinty-crush rock, *puree-parfait*, *gang-mylonit*, dyke-mylonite, *friksjonslgass*, trap-shotten gneiss, pseudo-eruptive mylonite) is thought to result from extreme dynamic metamorphism of crystalline rock. The term *pseudotachylyte* was first used by Shand (1916) to refer to a black rock resembling tachylyte cutting the streaky gneiss of the Vredefort granite in Orange Free State.

Pseudotachylyte generally occurs as very small, discontinuous lenticular bodies or branching, anastomosing dykes or veins across granite, quartzite, amphibolite and gneiss and has the composition of an acid igneous rock. In hand specimen it is black, fine-grained to glassy and structureless (Plate XXIIIc, d). In thin section it consists of sparse porphyroclasts of quartz and feldspar in a dense base. Quartz fragments may be embayed and surrounded by pale glass and the feldspar may be rimmed with microlites. The base is generally dense, dark brown to black and almost opaque but somewhat distinct varieties may be distinguished. It may be very dark coloured and isotropic but not opaque, with banding which wraps around porphyroclasts and some spherulites, vesicles and abundant tiny opaque (magnetite?) granules, or completely opaque in patches but somewhat clearer in others. Devitrification and crystallization cause the base to be cloudy and feebly anisotropic with mantles around some relict porphyroclasts, and streaky patches of very fine-grained biotite or hornblende, or to be dark and weakly anisotropic. Devitrification or recrystallization of the base produces crystallites, interlocking spherulites (which may be so close as to be polygonal) or vague crystalline patches.

The structural relations of the intersecting and "blind" veins imply that some type of fluid phase was present and three origins have been suggested:

(1) Frictional fusion at high temperature in the presence of abundant gas produces the type with glass, vesicles and microlites of high-temperature feldspar (Philpotts, 1964).
(2) Extreme mylonitization and tectonic injection produces a finely comminuted intrusive type of cataclasite which may lack evidence of having been at a high temperature (Philpotts, 1964).
(3) An igneous process produces a tuffisite by comminution of brecciated country rock in an entrained fluidized system without cataclasis or mylonitization (Reynolds, 1954; Roberts, 1966).

Clough, Maufe and Bailey (1909), Park (1961) and Shand (1916, p. 212) considered that there is a transition from mylonite into "fritted mylonite" or flinty crush rock, then into "fused mylonite" or pseudotachylyte, thence into the partly recrystallized types of pseudotachylyte. The various gradational types have not been clearly defined and the "transition" is so abrupt in some examples as to make its validity questionable.

Violent brecciation and heat generated by friction along faults (Bhattacharjee, 1964) have been postulated. Sutton and Watson (1950, p. 261) suggested that folding was followed by mylonitization in bands parallel to the axial surfaces and thence by the formation of flinty crush rock. It seems possible that heating might have been due to a very rapid conversion of mechanical to thermal energy by some shock process, and Philpotts suggested general heating prior to frictional heating.

Doubt has been cast on the origin of the original pseudotachylyte at Parijs (Willemse, 1937; Reynolds, 1954; Christie, 1960) and it has been suggested that the general term *hyalomylonite* be used for other occurrences. However this has a genetic implication and although it might be applicable to most pseudotachylyte occurrences (Armstrong, 1941; Drever, 1961; Park, 1961; Bhattacharjee, 1964), there is still considerable mystery about the source of the liquid phase.

There may be some relationship between the vitreous base of pseudotachylytes and of glass-bearing impactites. Mechanical effects dominate in the alteration of rocks at high strain rates and it is instructive to consider a series of groups of physical conditions ranging from cataclasis, through experimental deformation, to the extremely high but transient directed pressures at sites of nuclear explosion or meteorite impact. While such alteration is rare on the Earth's surface, it may well be common on the Moon's.

Texturally these rocks resemble mylonites and cataclasites with relict porphyroclasts in a very fine-grained matrix and mineralogically they are typified by the presence of high pressure polymorphs (coesite, stisbovite), glass and Ni-Fe spherules. Quartz is converted entirely to silica glass (*theto-morphic silica*) in moderately altered types with no evidence of melting. Strongly affected varieties contain coesite or stishovite embedded in a silica-glass base with flow structure indicating actual melting.

PRESSURE FRINGES AND PRESSURE SHADOWS

The dynamic metamorphism of an inhomogeneous rock in which there are large crystals more rigid than the matrix may result in the formation of minor textural inhomogeneities adjacent to the rigid crystals. These inhomogeneities have been called *pressure fringes*, *pressure shadows*, *eyed structure*, *halos d'étirement*, *Streckungshöfe*, etc. (Mugge, 1930; Pabst, 1931; Bain, 1933; Magnée, 1935; Miles, 1945; Fairbairn, 1950; Frankel, 1957; Hills, 1963; Joplin, 1968, Figs. 20, 68). These terms have not been defined rigidly and it is customary to use them interchangeably but it appears that there are two rather distinct textures and therefore the terms *pressure fringe* and *pressure shadow* are here redefined to cover these two.

A *pressure fringe* is an approximately ellipsoidal mineral growth adjacent to a central rigid crystal. The orientations of the crystals within the fringe are related to those of the boundaries of the rigid crystal and the foliation in the host rock abuts against the boundary of the fringe.

A *pressure shadow* is an approximately ellipsoidal region adjacent to a central rigid crystal; the texture within the shadow differs from that of the host rock and the foliation wraps around the central crystal plus its shadow.

Pressure fringes are distinguished from pressure shadows in that they contain oriented mineral growths completely distinct from the rock texture away from the fringes.

Pressure Fringes

A pressure fringe is generally composed of fibrous quartz, calcite, chlorite or muscovite; the central crystal is generally of magnetite or pyrite and the host rock is a low-grade regional or dynamic metamorphic rock (slate, phyllite or phyllonite). The fringe as a whole is approximately ellipsoidal and its external shape is related to that of the rigid internal crystal about which it formed. The *fringe boundary* (Fig. 56a) is discordant with the foliation in the enclosing matrix. The internal structure may be simple or complex and many fringes are composed of a number of *segments* which differ in the orientation or composition of the constituent fibres or flakes, and which are separated by *segment junctions*. The *fibre axis* is defined by the orientation of the long axes of fibres or flakes; it may be straight or curved. The total structure may be symmetrical or asymmetrical.

Pressure fringes are classified:

1. Straight fibre (non-rotational)
   (a) Simple (symmetrical or asymmetrical)
   (b) Compound (symmetrical or asymmetrical)
2. Curved fibre (rotational); simple or compound.

STRAIGHT FIBRE PRESSURE FRINGES

The simplest variety has axial to orthorhombic symmetry with coincidence between the symmetry planes of the matrix, fringe envelope, internal fringe structure and the shape of the internal rigid crystal. The shape of the fringe envelope is closely related to that of the crystal boundary, e.g. there is a match between *AB* and *AF* in Fig. 56a and between *AB* and *EF* in Fig. 56d.

The correspondence in detail of the size and shape of the boundary of fringe and crystal may be very close (Magnée, 1935, Plate II, Fig. 1a, b; Hills, 1963, Fig. 2a) or there may be slight departures (Mugge, 1930, Figs. 1, 3, 4 and 5; Magnée, 1935, Plate IV, No. 1) and this implies strongly that the fringe boundary was originally in contact with a face of the rigid crystal and was subsequently detached.

The internal structure of each segment is related to a face of the rigid crystal. Thus fibres of quartz in the fringe are perpendicular to (111) or (100) of magnetite or pyrite, and the basal plane (001) of chlorite or muscovite is parallel to (111) or (100) of magnetite or pyrite. These relations cause the formation of junctions between segments with fibre axes in different directions.

This fixed relation between fibre axis and crystal shape (but not necessarily between fibre axis and fringe boundary) implies that the fibre orientation depends on the crystal face and hence that nucleation took place there.

Mica flakes in the fringes have the normal elongation parallel to (001) but the quartz fibres are not elongated parallel to C or to any other crystallographic direction. The quartz forms *mutual impingement* texture and Pabst (1931) demonstrated that although there is a dimensional preferred orientation of the fibres, there is no preferred orientation of optic axes.

The idioblastic rigid crystals themselves do not normally have a preferred orientation (Mugge, 1930, p. 475) although Ramberg and Ekstrom (1964) described an unusual example of a symmetrical preferred orientation of pyrite crystals.

Mugge (1930, p. 476) suggested that chlorite was nucleated on magnetite because of a structural similarity between (001) chlorite and (111) magnetite (a "*Topochemische Reaktion*", i.e. an epitaxial relation). This seems

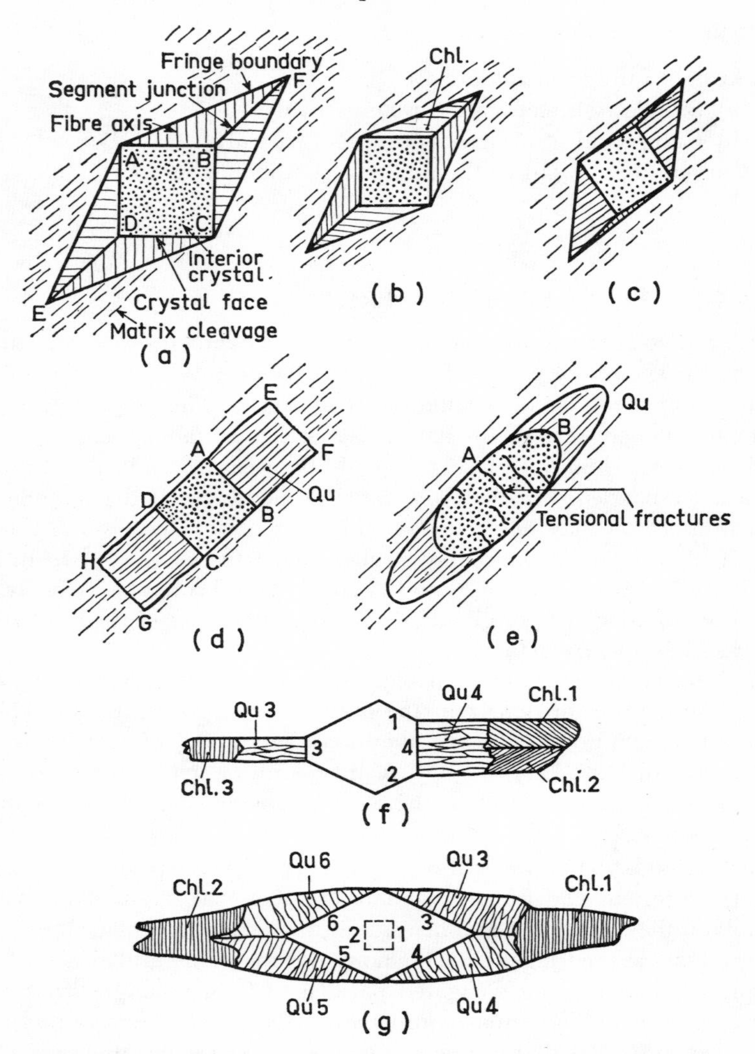

FIG. 56. Straight fibre pressure fringes. (a) Nomenclature of the simple symmetrical fringe containing quartz. (b) Simple fringe with chlorite. (c) Asymmetrical fringe produced by a non-symmetrical relation of crystal faces and cleavage. (d) Simple pressure fringe; *EF* and *HG* match faces *AB* and *DC* respectively and were once attached to them. (e) Ellipsoidal fringe around pyrite which is striated at *A* but not at *B*. (f) Compound, symmetrical pressure fringe. The chlorite and quartz in each segment is numbered to correspond to the face on the central crystal on which it was nucleated. (g) Compound, symmetrical fringe; chlorite in segments 1 and 2 are related to faces 1 and 2 which have been obliterated by later growth of the central crystal. The porphyroblasts are probably pretectonic and the fringes syntectonic in examples (a) to (e), but both porphyroblast and fringe are largely syntectonic in (f) and (g)

unlikely because chlorite forms parallel to both (100) and (111) and it seems that the magnetite merely acts as a catalytic surface for nucleation. The fringe minerals are nucleated at the interior crystal face and then are detached ("forced away") whereupon further growth occurs at the interface. The parts of the fringe *farthest* from the rigid crystal face were formed *first* and those now in contact with the face were formed *last.*

The crystal faces and fringe boundaries are generally discordant with the matrix cleavage, implying that both the rigid crystal and the fringe were formed after the cleavage. If the cleavage formed after the central rigid crystal, it should be deflected around the crystal plus fringe instead of abutting against it. However, there is a little evidence suggesting that some crystals plus their fringes grew at the same time as the cleavage:

(a) The central crystals in Figs. 56f, g grew simultaneously with at least the later parts of the fringe.

(b) The amount of strain represented by elongate fringes such as in Figs. 56f, g and 57a, b, is probably comparable in magnitude with that required to produce the cleavage.

The fringe is due to an inhomogeneous strain pattern around the crystal; extension has taken place along the long axis of the fringe accompanying, and possibly due to, shortening perpendicular to it. The crystal–matrix interface was detached and pulled away and crystallization took place in the low-pressure region at the interface. Thus most porphyroblasts are older than the movements which produced the fringe (i.e. are pretectonic) and the crystals in the fringe are of the same age as the movements (i.e. are snytectonic). Both may be later than the cleavage although fringe and cleavage may form simultaneously if the crystal was present before the cleavage was formed. The cleavage will then be concordant with the fringe boundary and wrap around both the crystal and its fringe. Pyrite with pressure fringes in sheared igneous rock from the Witwatersand (Frankel, 1957) in Fig. 56e is polished at *A* but not at *B* and the corners are rounded. Striations and grooves at *A* on the pyrite surface are parallel to slickensides in the matrix, and the pyrite is broken by tensional cracks perpendicular to the long axis of the fringe. That the pyrite existed before the fringe was formed is also shown by the shattered pyrite with fringes figured by Magnée (1935).

Asymmetrical fringes with straight fibre axes (Fig. 56c) are formed around crystals whose faces were not symmetrical with respect to the cleavage or were formed during movements where compression was not normal to the cleavage. This is the general case and although the highly symmetrical example discussed in detail above is a special case, the same principles apply. Asymmetry can also be caused by the fibre axes in different segments being related to faces belonging to different crystal *forms* on the porphyroblast.

Fringes composed of more than one mineral or of one mineral which crystallized at distinctly different times are termed *compound.* The minerals in different segments were formed at different times in Fig. 56f; the quartz in segment *Q*3 and in *Q*4 is related to faces 3 and 4 but the chlorite, which is now further from the crystal than the quartz, was formed before the quartz and detached from the crystal faces, 1 2, and 3 when the crystal was smaller.

A complication in this chronology is shown in Fig. 56g where some of the outer, older chlorite has an orientation which depends on crystal faces not now present on the porphyroblast. This implies that the porphyroblast is not older than the fringe but was growing during the movements and during the crystallization of the fringe minerals so that early crystal faces were obliterated by growth of the porphyroblast. The mismatch in size and shape between the ends of fringes (oldest part) and the present porphyroblast shape also implies this. *Q*3, *Q*4, *Q*5 and *Q*6 fibre axes are related to the existing faces of 3, 4, 5 and 6, but the orientations of the chlorite in *C*1 and *C*2 (as well as the shape of the ends of the envelope) are related to faces 1 and 2 which are no longer present and which have been obliterated by later growth of faces 3, 4, 5 and 6.

CURVED-FIBRE PRESSURE FRINGES

Asymmetrical pressure fringes characterized by curved fibre axes have a symmetry of monoclinic or lower, curved fringe boundaries and segment junctions, and an envelope whose shape is generally not closely similar to that of the porphyroblast (Fig. 57). The mechanism of formation is similar to that of the symmetrical type except that the curved fibre axes show that the porphyroblast was rotating as the crystal fibres grew against its surfaces.

The quartz fibres are curved, not because they have been "bent" but because the orientation of the older, outer parts remained constant as they were moved away from the porphyroblast while new units were added to the inner end with a dimensional preferred orientation perpendicular to the porphyroblast crystal face. This orientation probably results from quartz–quartz grain boundaries tending to be perpendicular to the quartz–magnetite or quartz–pyrite interface.

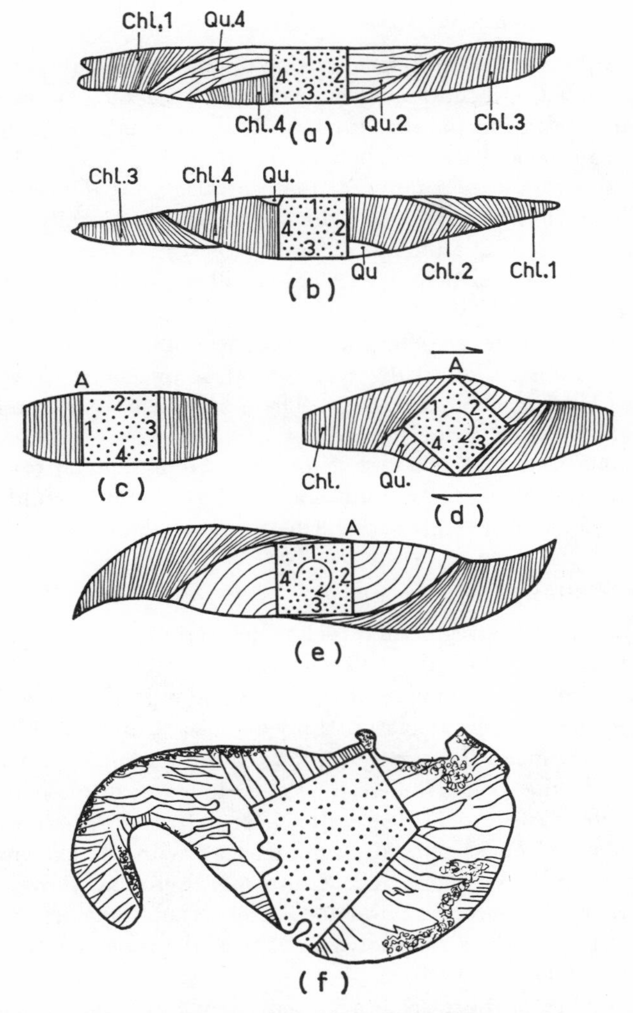

FIG. 57. Curve-fibre pressure fringes. (a), (b) Compound, asymmetrical, curved fibre, rotational pressure fringes. Minerals in segments numbered according to the porphyroblast face which controlled them. (c), (d) and (e) Stages in growth of a pressure fringe around a rotating porphyroblast. (f) Rotational, quartz, pressure fringe (after Pabst, 1931). The fringes are syntectonic and the porphyroblast largely pretectonic although a narrowing of the fringe ends may imply some syntectonic growth of the porphyroblast

Relations of Fringe Structure and Strain

In general, straight fibre fringes are due to non-rotational movements and can be correlated with flattening movements (shortening parallel to *AC* in Fig. 56a and extension parallel to *BD*). This may indicate flattening perpendicular to the cleavage (Fairbairn, 1950, Fig. 3).

Curved fibres indicate rotation of the central crystal with a recognizable sense of shear (Fairbairn, 1950). The strain, at least locally, is more like simple than pure shear although Collette (1959) and Ramsay (1962) have shown that rigid idioblastic porphyroblasts will rotate during "flattening" of the matrix.

Some pressure fringes are extremely irregular in shape, they may have very little or no symmetry, and fringes may even be one-sided (Magnée, 1935, Plate I, 3a; Harker, 1939, Fig. 67B). These may be due to complex strains or repeated strains in various directions.

It is emphasized that pressure fringes may represent only a small and insignificant part of the total strain and extreme caution should be used in extending deductions from pressure fringes to the whole rock.

Pressure Shadows

A *pressure shadow* is a region in a rock which is protected from deformation by the presence of a relatively rigid object.

Shadows occur where a foliation wraps around a porphyroblast in a schist or gneiss, around a porphyroclast in a mylonite, or around a pebble or fossil in a sheared sediment. They may contain coarse granular quartz where all other quartz in the rock is fine-grained and elongate; or may contain relics of old foliation surfaces which are elsewhere destroyed. Shadows are elliptical with pointed ends, generally symmetrical, with the external foliation concordant with the envelope. The central porphyroblast shows such strain effects as fractures, undulose extinction, deformation lamellae and deformation twins, etc. Shadows are shown in Plates Id (upper part) and XIIId, and Figs. 58b, 63d, e and 64b.

They resemble pressure fringes in that they form approximately ellipsoidal envelopes around porphyroblasts (or similar rigid objects) but differ in that they do not contain the internal structure of fringes. Some fringes have an imperfect internal structure and may be indistinguishable from shadows but in general the matrix foliation abuts against a fringe but wraps around a shadow.

Pressure shadows are formed:

(1) By differential compaction of the matrix of sediments adjacent to rigid objects such as fossils, pebbles or large crystals.

(2) By flow in porphyritic igneous rocks, particularly rhyolite (Hills, 1963, Figs. XII-4A, 4B).
(3) By various kinds of deformation in metamorphic rocks.

The porphyroblast is older than the foliation, i.e. is pretectonic with respect to the deformation which produced the foliation but the significance of such crystals has commonly been misconstrued and the bowing of the foliation has been said to be due to the crystal having pushed aside the foliation by its "force of crystallization" (Harker, 1939, p. 219).

Harker figured a chloritoid crystal (ibid, Fig. 175B) in Manx slate and stated that it had "visibly thrust aside the surrounding matrix". Inspection of the original slide (Cambridge collection No. 11069), however, shows that the chloritoid is *older* than the cleavage because the chloritoid is fragmented with matching separated fragments, S_i is discordant with S_e, and pressure shadows are present around the porphyroblasts. The textural evolution of the Manx slate has been discussed by Simpson (1964).

Harker (ibid., Fig. 177B) also figured a pressure shadow in Manx slate adjacent to garnet. Inspection of the original slide shows that the garnet contains S_i which are less plicated than S_e, and that chlorite flakes occur parallel to crystal faces of garnet within the pressure shadows. The garnet is thus post-tectonic to an early deformational phase which produced the cleavage but is pre-tectonic to the phase which contorted the cleavage. The chlorite formed in pressure shadows during this later phase.

The time relations of many pressure fringes and shadows are similarly complex. Many idioblastic crystals with pressure fringes in foliated rocks are post-tectonic to the cleavage whereas the minerals in the *fringe* are syntectonic to a later very minor movement which caused the *fringe*. The crystals in a pressure *shadow* may be syntectonic with respect to the foliation or may be, like the porphyroblast, pre-tectonic to the foliation.

SHOCK METAMORPHISM

This is best regarded as a type of dynamic metamorphism in which the strain rate is very high. The confining and directed pressures are transiently very high and the temperature ranges from low to very high.

Typical products range from somewhat shattered rocks, through breccias ("suevite") to glassy rocks. The original texture is dominant in the less altered varieties but becomes progressively destroyed with crushing and glass formation. Many such rocks are associated either with known meteorite craters (Bunch and Cohen, 1964; Chao, 1967) or with the circular fracture systems of "cryptovolcanic structures" (Currie, 1967) or "astroblemes"

(Deitz, 1963). Experimental work relevant to an understanding of the textures has been discussed by De Carli and Jamieson (1959); Wackerle (1962); Ahrens and Gregson (1964); Christie *et al.* (1964); and the products of "metamorphism" due to man-made explosions by Short (1966). Little is known of the petrology of shock and impact metamorphics but some details have been given by Currie (1967), Carter (1965), Crook and Crook (1966) and Chao (1967).

The effects of shock metamorphism are:

(1) Mechanical and plastic deformation including fragmentation, fracturing and bending of crystals.
(2) Partial to complete vitrification of minerals such as quartz, feldspar and ferromagnesians without melting.
(3) Mineralogical change, mainly phase transformation to high-pressure polymorphs (quartz to coesite or stishovite).
(4) Breakdown or melting of some minerals.

The changes in quartz are illustrative. In slightly shocked specimens the quartz is cut by intersecting sets of closely spaced fractures which are parallel to such directions as $\{10\bar{1}3\}$, $\{01\bar{1}3\}$ and $\{11\bar{2}2\}$ as well as the more common $\{10\bar{1}1\}$, $\{01\bar{1}1\}$, (0001), $\{10\bar{1}0\}$ and $\{11\bar{2}0\}$. The occurrence of two or more sets of planar fractures, particularly parallel to (0001) and $\{10\bar{1}3\}$, is considered to be unique to shock metamorphics produced by meteorite impact, i.e. *impactites* (Short, 1966; Carter, 1965). These fractures differ from the common deformation lamellae and occur in crystals lacking undulose extinction (but having a fine mottled extinction).

A greater degree of metamorphism results in the formation of very narrow (a few microns) bands which have similar orientations to the fractures mentioned above but which consist of isotropic material of lower refractive index than the host crystal. Increased deformation changes the whole crystal to this low refractive index, isotropic vitreous phase but it should be emphasized that the change from crystal to glass is a solid state transformation taking place by the loss of long-range order without melting. Moderate metamorphism may convert a quartzite to silica glass ("thetomorphic silica") with R.I. of 1·46 without losing the identity of the original grains. The glass lacks vesicles and flow structure.

Intense metamorphism gives tiny crystals of coesite or stishovite embedded in silica glass which may show evidence of melting and flow. Quartz has been experimentally converted to silica glass (De Carli and Jamieson, 1959) by explosion impact with pressures exceeding 600 kb but temperatures of less than 1400°C.

Plagioclase behaves similarly to quartz in that it develops fractures and a vitreous phase but high-pressure phases such as jadeite are not found,

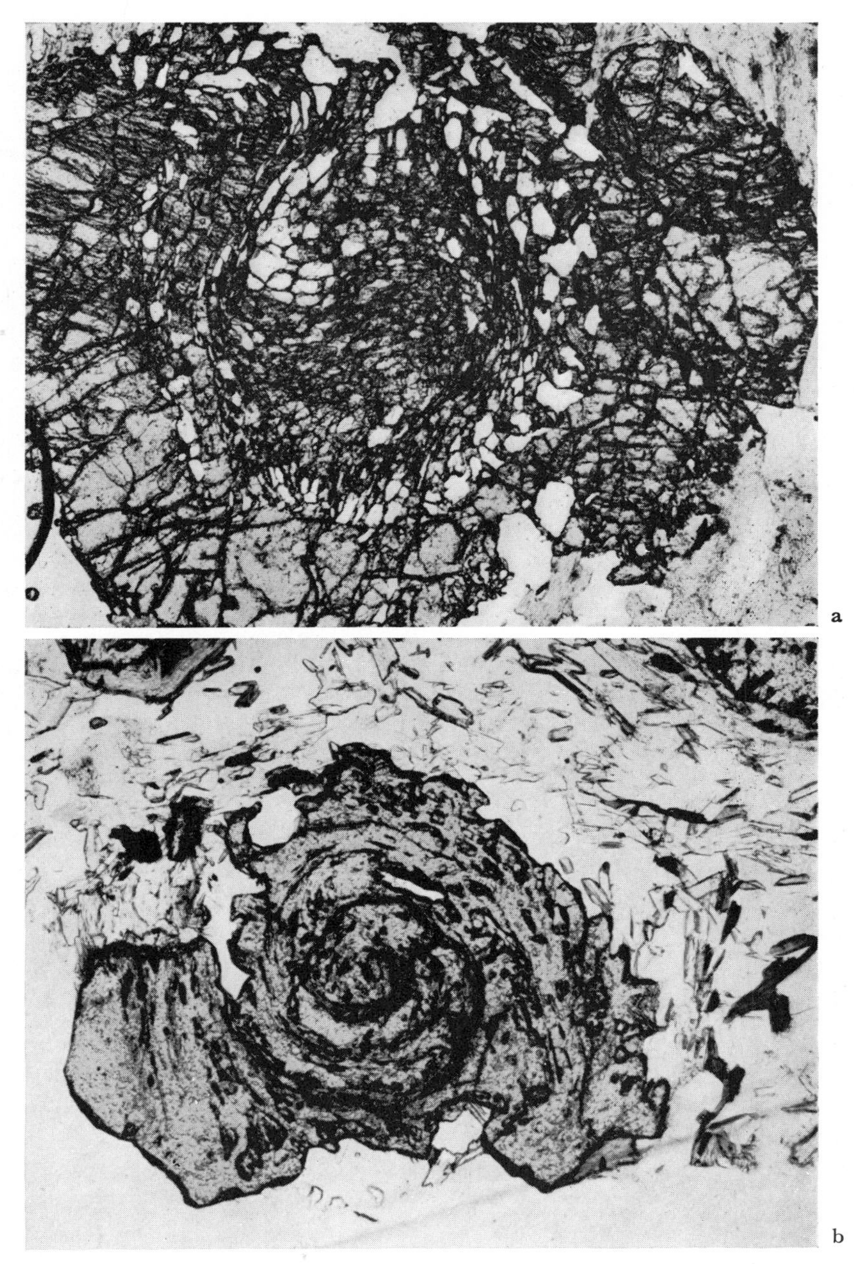

PLATE XXV. Syntectonic textures. (a) Garnet with rotational core and massive post-tectonic rim, Raglan Range (×25). (b) Snowball garnet with 540° rotation, Carn Chuinneag. This crystal (U.K. Geological Survey slide No. 11801) was figured by Flett (1912). Photomicrograph by courtesy of Dr. Sabine, Crown Copyright reserved

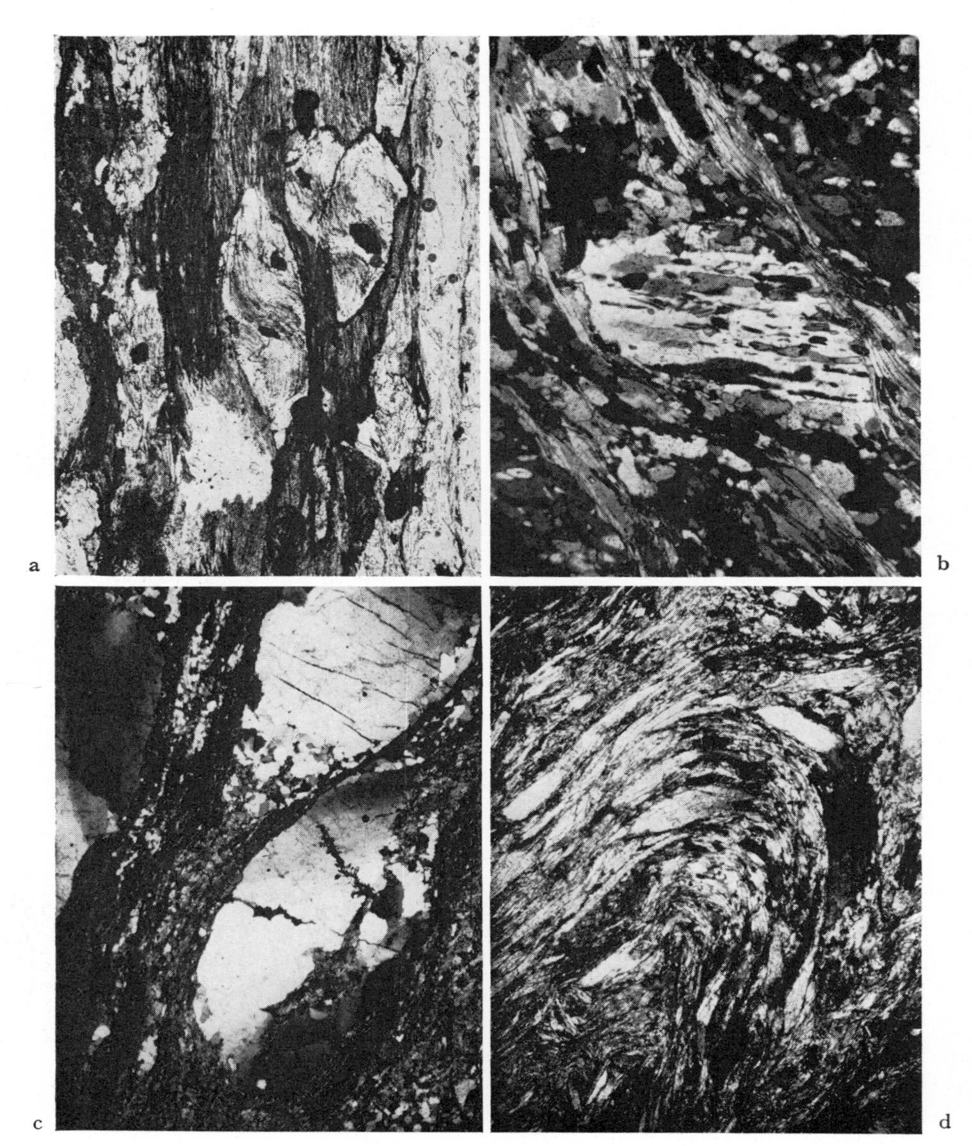

PLATE XXVI. Post-tectonic textures. (a) Helicitic albites containing contorted S_i. Governor River Phyllite (×30). (b) Helicitic albite containing S_1 as straight lines of inclusions; wrapped around by S_2. Albite schist, Raglan Range (×38). (c) Large quartz crystals in schist which has been later thermally metamorphosed. Lenticular, undulose quartz crystals are partly polygonized into unstrained aggregates. Trial Harbour (×10). (d) Polygonized muscovite flakes around fold in glaucophane schist, Berkeley Hills (×20)

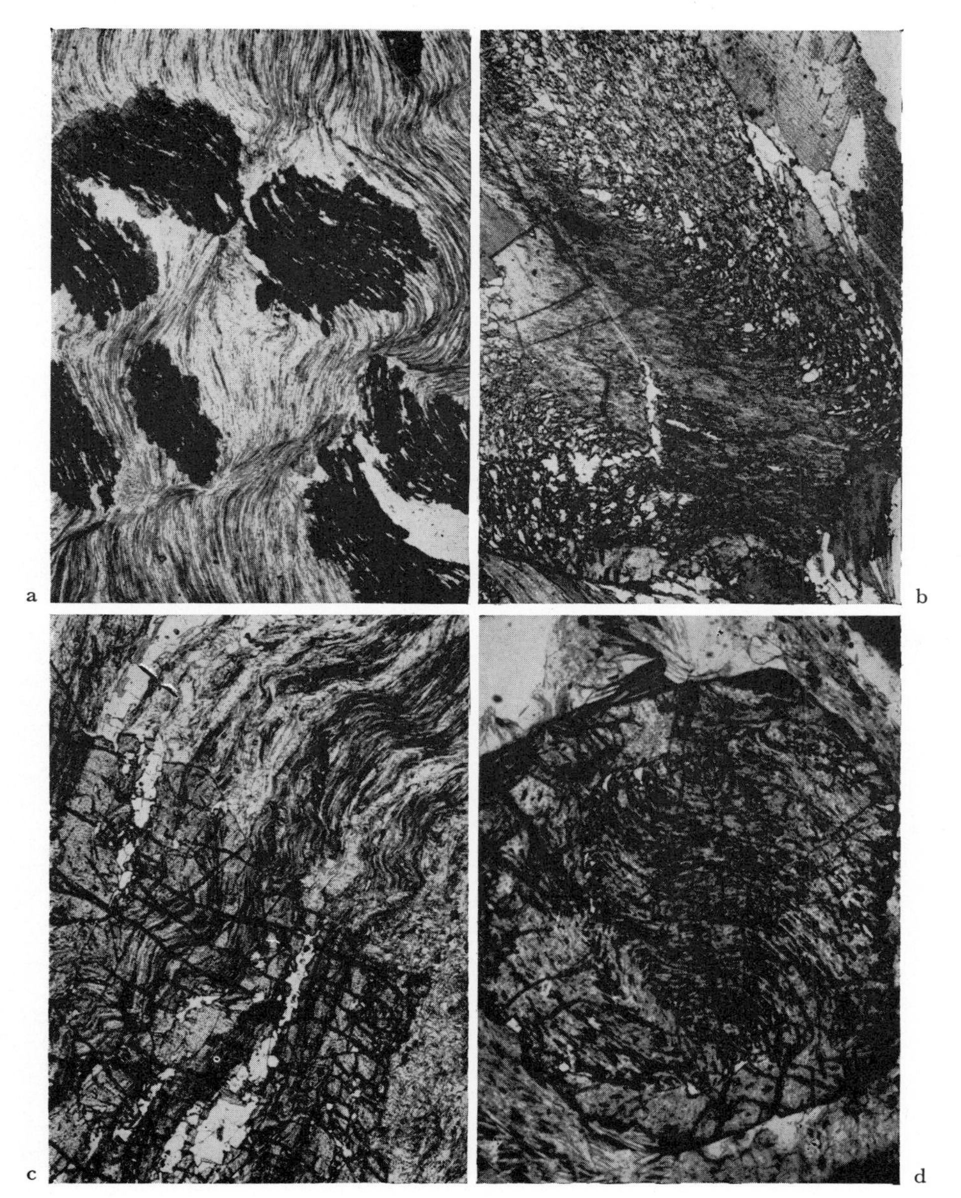

PLATE XXVII. Post-tectonic textures. (a) Helicitic chloritoid crystals in schist, Val Somnix, Switzerland. The "spiral" internal structure is due to recrystallization over folds, not to syntectonic rotation (×10). (b) Helicitic staurolite which has grown over a fold, Spain (×10). (c) Helicitic garnet. Crumpled S_e in the upper part of the photograph pass concordantly into S_i in the garnet (in the lower part) (×10). (d) Contorted S_i in helicitic garnet (×18)

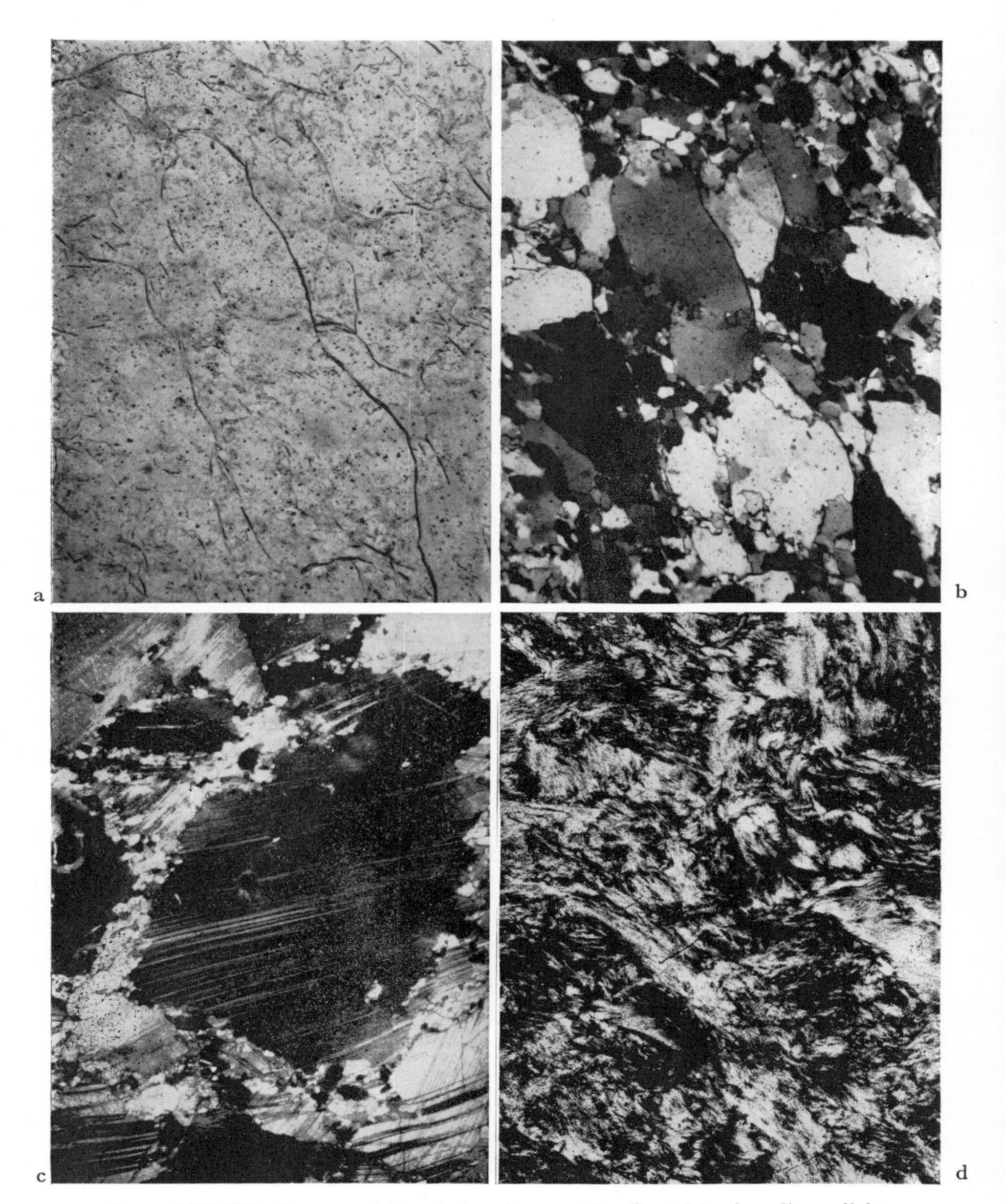

PLATE XXVIII. Quartz–sericite relations in quartzite, Goat Island; ordinary light (a) and crossed polarizers (b) of the same field. Mica forms continuous shear surfaces which prevent the lateral migration of quartz grain boundaries and thus control its shape. (c) Mortar texture and deformational twinning in calcite of marble (×18). (d) Matted fibrolite (×25). Contortion is probably due to growth, not deformation.

PLATE XXIX. (a) Simple schistosity in quartz-muscovite schist ($\times 25$). (b) Crenulated calc-schist in which an older foliation is cut by a younger, widely spaced foliation formed along the limbs of asymmetrical folds. Calcite (high relief grains) has been displaced from the high-pressure regions along the new foliation on the appressed limbs to adjacent low-pressure areas ($\times 10$). (c) Phyllonite with S_0 (bedding), S_1 (foliation parallel to bedding in upper part) and S_2 (strain slip foliation with compositional layering formed along crumples in S_1, in lower part) ($\times 12$). (d) Foliations in micaceous marble. The main foliation in a north-westerly direction is marked by lines of parallel muscovite flakes. A foliation is also present in a north–south direction due to cleavage in the calcite ($\times 40$)

PLATE XXX. Textures of quartzites. (a) Non-metamorphosed orthoquartzite with secondary silica cement on well-rounded clastic grains, Rocky Cape (×12). (b) Slight dynamic metamorphism of (a) produces a dimensional (but not lattice) preferred orientation by plastic deformation of quartz grains (×12). (c) Mortar texture. Large, old, strained grains surrounded by fine, recrystallized new quartz; Mary Group (×25). (d) Dimensional preferred orientation in crystalline regionally metamorphosed quartzite. The quartz also has a lattice preferred orientation (×30). Goat Island. (e) Annealed texture. Post-tectonic recrystallization of regionally metamorphosed quartzite destroys much of the dimensional preferred orientation of (d) and tends to form a granoblastic-polygonal network (×20). (f) Ribbon texture in mylonitic quartzite from the Moine Thrust. Texture due to crushing and intense plastic deformation. (Photograph by J. Christie, also Plate 9 of Christie, 1963

PLATE XXXI. Granulites. (a) True granulite from Rohrsdorf, Saxony. Typical ribbon structure in quartz (×15). (b) Moine "ganulite" with sparse parallel biotite flakes in a granoblastic quartz-feldspar matrix (×15)

apparently due to lack of time. Ferromagnesian minerals such as biotite and hornblende are less easily altered than quartz and plagioclase but finally break down into glass plus iron oxides. Refractory minerals resist alteration but rounded "droplets" of ilmenite, rutile, pseudobrookite and baddeleyite indicate temperatures exceeding 1500°C. The presence of Ni-Fe spherules is taken to support an impact origin for some shocked specimens.

The term "shock" metamorphism is used here in preference to "impact" metamorphism because it is not certain that all occurrences are due to impact. However, the shock features are unique and most are not found in rocks from normal geological environments so that an impact origin seems most probable for most. The occurrence of glass in hyalomylonites (pseudotachylytes) attributed to dynamic metamorphism points to a link between dynamic and shock metamorphism.

CHAPTER 10

Textures of Regional Metamorphism

THE general term *dynamo-thermal* has been applied to rocks which have been deformed and recrystallized at approximately the same time, but is little used at present and the term *regional* metamorphism is applied to rocks which have been affected in this way over large areas. The simultaneous action of stress and heat, however, may arise in a number of ways, not only by the common process of the deformation of solid sediments during regional heating but also by *protoclasis* (the movement of magma during solidification) and by recrystallization in the aureole of a forcibly intruded igneous body. Regional metamorphic rocks are more common and as textures due to regional, protoclastic and contact dynamo-thermal metamorphism are all very similar, discussion will be mainly concerned with the first.

Regional metamorphic rocks differ texturally from those formed by thermal metamorphism because deformation produces some kind of foliation, lineation or banding. No significant mineralogical differences between thermal and regional rocks have been recognized and the division of minerals into *stress* and *anti-stress* (i.e. those requiring stress for their formation and those which do not formed in a stressed environment) is not valid.

Directed pressure and heat tend to act in opposite senses in the formation of metamorphic textures in that deformation tends to break the rock down, by bending and breaking crystals, but crystallization builds the rock up.

Directed pressure ("stress") is important thermodynamically in that it adds energy to the system, i.e. lattice strain energy, surface energy due to the formation of extra surface area on broken crystals, and mechanical energy which is converted to frictional heat. It is important kinetically in that many solid state reactions are accelerated by accompanying deformation. It aids recrystallization by assisting nucleation of new crystals at points of high strain, e.g. dislocation concentrations, regions of high lattice curvature, intersecting deformation twins, etc.

In general, the fine-grained crushed and mylonitized textures of dynamic metamorphism and the random, polygonal or decussate textures of thermal metamorphism are replaced by a preferred orientation of minerals in regional metamorphism. Rocks produced by low-grade regional metamorphism are mineralogically and texturally indistinguishable from those produced by dynamic metamorphism. Some rocks produced by low-stress regional metamorphism are mineralogically and texturally indistinguishable from those produced by thermal metamorphism.

Original textures tend to be destroyed during regional metamorphism so relict textures are much less common than in contact rocks. Textures involving large compositional differences or large discontinuities may be preserved even in rocks which have been strongly deformed but delicate sedimentary or igneous structures are only preserved in weakly deformed rocks, e.g. in the centres of dykes, in "protected" parts of structures, or in a low-stress regional environment. Terms such as *blastoporphyritic* (Plate Id), *blasto-amygdaloidal* (Plate XXIVd), *blasto-granitic*, *blastopsephitic*, *blasto-foliate*, etc., are applied in the same way as in thermal metamorphism although the textures are different because of the deformation.

Typomorphic textures in regional metamorphic rocks range from granoblastic-polygonal to some kind of preferred orientation such as *foliation*, *cleavage*, *schistosity*, etc.; compositional *layering* on a small or large scale is common. *Porphyroblastic* and *poikiloblastic* textures occur.

The detailed understanding of regional metamorphic textures involves the use of a genetic terminology depending on the interpretation of various time relations in the textural evolution of the rock. Whereas thermal or dynamic metamorphism may be comparatively brief in duration, regional metamorphism is a long-continued process and the old concept (due chiefly to Becke and Harker) of a *crystalloblastic* texture due to the simultaneous crystallization of all minerals has been found to be an oversimplication when applied to most regional metamorphic rocks. Criteria for recognizing the *order of crystallization* of minerals are given below. Some minerals were formed at the same time as tectonic movements which produced the schistosity (e.g. parallel micas and elongate quartz) and crystallization of these is said to be *syntectonic*; some minerals were formed before the movements (e.g. shattered cordierite) and crystallization of these is said to be *pre-tectonic*; some minerals were formed after the movements (e.g. cross-cutting helicitic porphyroblasts of albite, chloritoid, chiastolite) and crystallization of these is said to be *post-tectonic*.

Pre-tectonic Crystallization

Elevation of temperature before the onset of tectonic movements may result in some crystallization, and as these crystals precede the deformation they show the same strain effects as the original crystals in the rock. Defor-

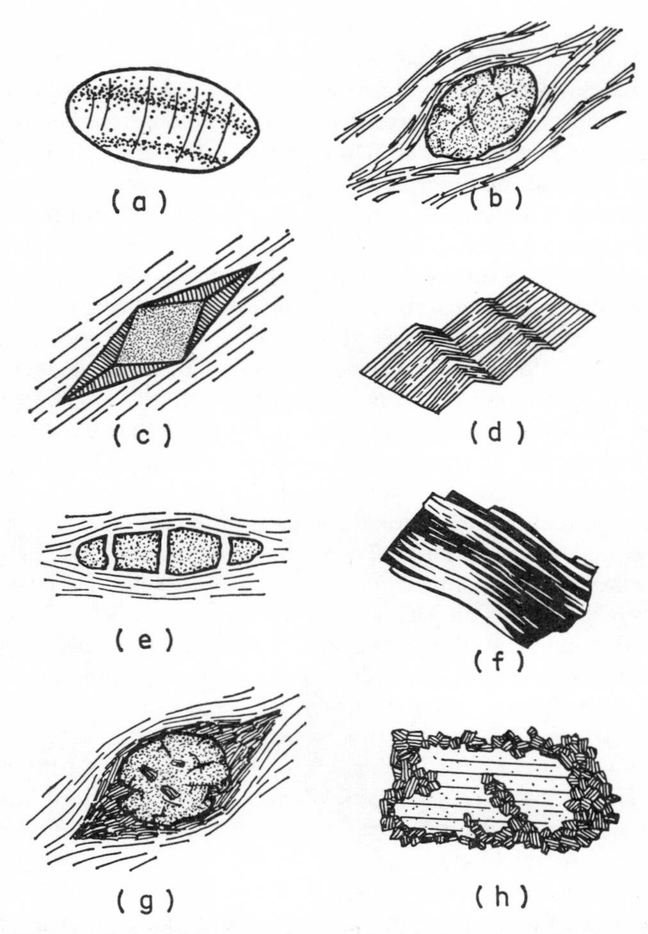

FIG. 58. Characteristics of pre-tectonic crystals. (a) Undulose extinction and deformation lamellae in quartz. (b) Cracked garnet wrapped around by the foliation. (c) Pressure fringes around pyrite. (d) Kinked biotite. (e) Fragmented garnet. (f) Plagioclase with deformation twins. (g) Garnet with a sheath of chlorite along the foliation. (h) Large amphibole crystal breaking down to an aggregate of small crystals (mortar texture)

mation and crystallization tend to obliterate early minerals but they may be preserved as relics distinguished by various strain effects (undulose extinction, deformation lamellae, mechanical twins, fractures, kinks, curved cleavage or twins, anomalous optical properties and exsolution textures (Fig. 58). The foliation is formed later and wraps around pretectonic crystals which generally have pressure fringes or pressure shadows (Grout, 1932, p. 382).

Syntectonic Crystallization

Most minerals in a regional metamorphic rock are thought to have formed *during* tectonic movements although this is difficult to prove. Preferred orientations such as the alignment of hornblende or elongate quartz parallel to a fold axis or of mica flakes parallel to the axial surface is probably due to the crystallization of these minerals during deformation; however, a foliation may be mineralogically complex and contain pre-tectonic crystals which have been mechanically rotated and post-tectonic crystals which have grown mimetically along a pre-existing foliation.

Rotational (snowball, pinwheel) structure may be preserved in certain minerals (garnet and albite) which are rotated by tectonic movements *at the same time as they grow* (Spry, 1963b, Plate XXV, Figs. 59 and 60). "Rotational" in this sense is not synonymous with "rotated" (Niggli, 1954, p. 302) meaning rotated *after* formation.

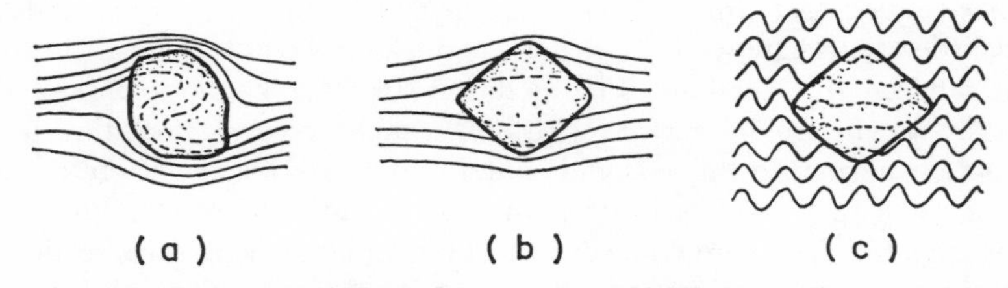

Fig. 59. Characteristics of syntectonic crystals. (a) Snowball garnet. (b) Andalusite porphyroblast which grew during flattening of a pre-existing foliation. (c) Porphyroblast which grew during the crumpling of a pre-existing foliation (b and c after Zwart, 1961, 1963)

The crystals contain a set of discontinuous internal *S*-surfaces (S_i) which have been called "trails" or "trends" and which are composed of round or elongate blebs of quartz, or wisps and grains of graphite or magnetite. The crystal itself may be approximately round, sickle-shaped or web-like and skeletal. The defining criteria are that in sections cut parallel to the single plane of symmetry, the internal structure has a symmetrical spiral form. The structure is generally called *rotational* if the apparent rotation of the crystal is less than 90° and *snowball* if it is greater.

Post-tectonic, helicitic crystals which happen to have been nucleated on tiny folds resemble rotational structures and have been commonly confused with them. Singewald (1932, p. 459) and Tobi (1959, p. 201) drew attention to errors of this kind by Niggli (1912), Gansser (1937) and Sarrot-Reynaud (1958). Clough (1897) and others have correctly interpreted the helicitic

Cowal albites as post-tectonic but Bailey (1923) regarded them as syntectonic. The illustration of Goguel (1962, Fig. 15) is a post-tectonic crystal but is taken to indicate "rotation of the crystal during the course of its growth" (ibid., p. 52).

It may be difficult to decide on the origin of a single crystal with an S-shaped inclusion pattern (Plate XXVIIa) but examination of a number of crystals from the same slide should indicate their origin. A group of true snowball crystals all contain spirals of the same type and sense of rotation whereas the internal structures in a group of helicitic crystals will be varied and range from full S-shapes to half-S, W, straight, and irregular or asymmetrical forms.

A rotational crystal has a homogeneous crystal structure and although this cannot be seen optically in isotropic garnet, anisotropic albite crystals are not undulose and cleavage planes are straight. This distinguishes them from pre-tectonic crystals which may have been bent into S-shapes but not from post-tectonic crystals which also are not bent or undulose.

The exact geometrical form of the inclusion pattern in rotational garnets has not yet been determined because of the lack of suitable large crystals and what little knowledge we have depends on the shape of the spirals in somewhat random thin sections. The inclusion structure can be considered as basically a cylindroidal surface with an S-shaped cross-section, thus a thin section cut normal to the axis will contain an S or Ƨ form, but oblique sections show)(, //, (), ŒD forms depending on the angle of section (Powell and Treagus, 1967). The true shape can be obtained by viewing thick sections on the universal stage. The sections are rotated until the rotation axis is vertical whereupon the pattern becomes clear and symmetrical and the boundaries of most quartz inclusions become vertical.

In the simplest examples, the S_i pass out into the matrix where they are concordant with S_e, the internal structure is cylindroidal, the *S*-axes of the crystals are parallel, each *S* faces the same way (i.e. has the same sense of rotation), and petrofabric analyses of quartz in inclusions and matrix give related patterns.

Relationships may, however, be complex. Repeated deformation may cause the rotational axes of different crystals to be non-parallel, S_i to be discordant with S_e and the internal and external quartz fabrics to be different. A complex process of growth and rotation causes the internal spiral surface to be perhaps hyperbolic (longitudinal S_i are curved, not straight), or compound.

Concordance of S_i and S_e (the foliation in the matrix curving continuously into the centre of the crystal and out the other side) is taken to imply that the central S_i was once parallel to (or part of) S_e and that it has been rotated during growth of the crystal. The growth of a garnet from Raglan Range

(Spry, 1963b, Fig. 4a) is traced in detail in Fig. 60. However, many rotational crystals look as though they did not simply envelop the foliation but grew preferentially along it and absorbed elongate crystals of quartz from the matrix. It appears that the crystal grew in a medium undergoing simple

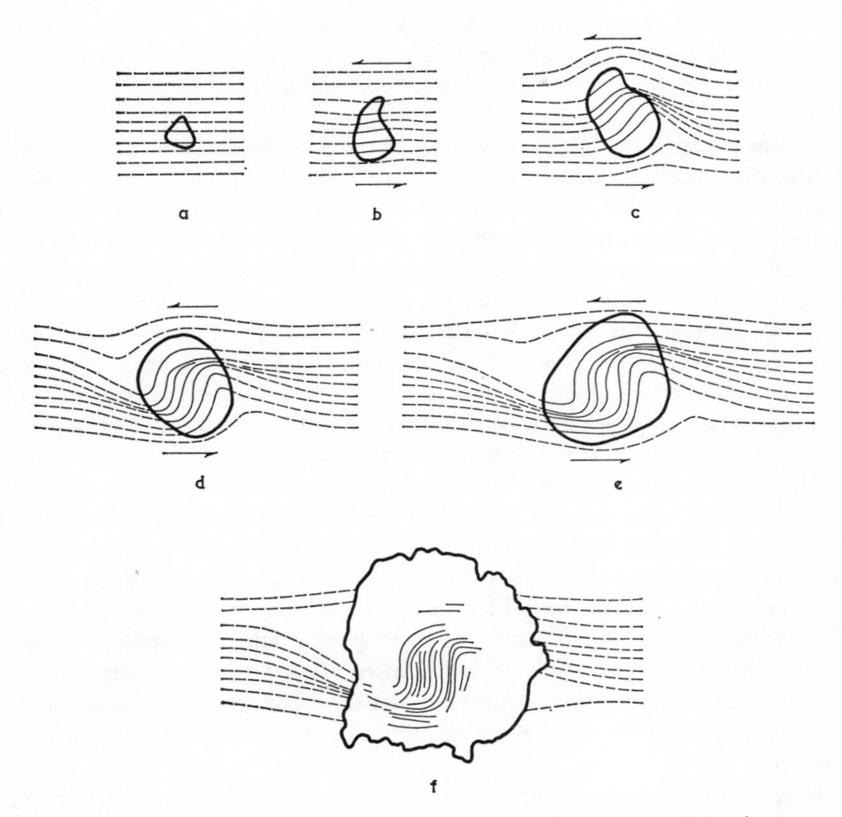

Fig. 60. Snowball garnet. Growth and simultaneous rotation of a syntectonic garnet crystal from the Raglan Range. The shapes of successive stages of the crystal are obtained from Spry (1962d, Fig. 4a). The S-shaped inclusion trends indicate anticlockwise rotation of 95°; the total angle through which the S_i curve is 110°, the difference being due to distortion of S_e adjacent to the crystal. Rotation between stages (a) to (e) is followed by static post-tectonic growth in (f)

shear or laminar flow along the foliation, and that the crystal lattice was insensitive to the movements. Successive units which were added to the crystal as it rotated retained the same lattice structure as the centre. An S-shape indicates anticlockwise rotation and a Ƨ-shape a clockwise direction of rotation.

Ramsay (1962) proposed that rotational crystals might form under a compressive regime (pure shear) rather than simple shear and has supported this by experimental work (as yet unpublished). Ramsay's explanation only applies to crystals which have been rotated by less than 90° but this constitutes the majority because the few measurements available suggest that most rotational crystals have been rotated about 50–60°. This explanation could not reasonably explain crystals which have been rotated by angles of 120° (Spry, 1963b), 130° (Rast, 1958), 173° (Schmidt, 1918), or 540° (Flett, 1912).

If it is assumed that the orientation of the S_i is the same as that of the S_e at the time that the S_i was formed, then it follows that all S_i which are parallel were formed simultaneously if growth were simple. Changes in shape and orientation of the crystal during growth can then be determined by outlining successive concentric zones containing parallel S_i (Spry, 1963b, p. 216). The growing crystal can be regarded as analogous to a snowball picking up snow as it rolls down a snow-covered slope. The point of contact between snowball and slope becomes enclosed within the next layer of snow and the locus of the point of contact is a spiral. The distance the snowball has rolled is equal to the length of the spiral. By analogy, the growing garnet crystal has two points of contact and the internal structure is a double spiral whose shape depends on the relative rates of growth and rotation. An estimate of the strain in the enclosing rock can be made by estimating the amount of slip of the foliation surfaces bounding the crystal as this depends on the length of the internal spiral. The accuracy of the method is severely limited because of the assumptions which must be made (Spry, 1963b).

It is clear that although accepting these assumptions involves a gross over-simplification, it seems likely that the differential slip between layers in the immediate vicinity of the crystal (or of the layer containing the crystal) is at least of the order of the length of the internal spiral measured in cross-section.

Non-rotational syntectonic crystals have been described by Zwart (1960a) and are shown in Fig. 59. The crystal in Fig. 59c contains S_i which are straight in the centre but which become progressively more crumpled towards the rim until they pass out into crumpled S_e. The crystal grew during deformation which was buckling a pre-existing foliation. The central non-crumpled S_i was enveloped before appreciable buckling and was protected from further deformation by the surrounding crystal. Continuous growth gradually enveloped parts of S_e which had been progressively more buckled.

A group of syntectonic porphyroblasts may be nucleated successively during a tectonic episode which crumples an existing foliation but does not produce a new one. Early-formed porphyroblasts will contain non-crumpled foliation as S_i, whereas late crystals will contain crumpled S_i.

Post-tectonic Crystallization

Crystallization after movements have ceased is extensive in some rocks and the formation of minerals such as kyanite, chiastolite and chloritoid are almost entirely restricted to this stage (Fig. 61). Post-tectonic crystallization may result in the formation of randomly oriented crystals replacing a

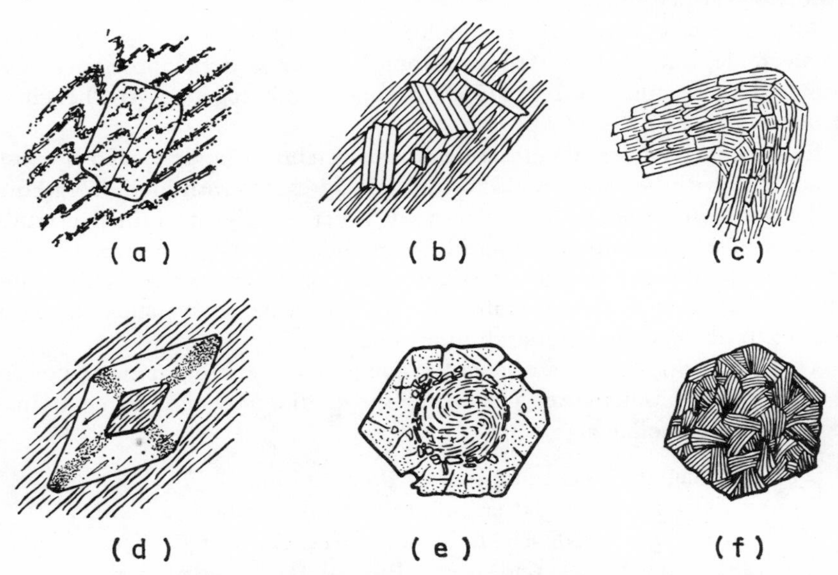

FIG. 61. Characteristics of post-tectonic crystals. (a) Helicite structure in albite, S_i (S-surface within the crystal) concordant with S_e (S-surface outside of the crystal). (b) Cross-micas. (c) Polygonized micas in a fold. (d) Chiastolite with central helicitic and outer cross structure and boundaries discordant with the foliation. (e) Garnet with syntectonic rotational core and post-tectonic idioblastic rim. (f) Random aggregate of chlorite as a multi-crystal pseudomorph after garnet

preferred orientation, the growth of porphyroblasts with *helicitic* structure (Plates XXVIa, b; XXVII), the polygonization of bent crystals (Plate XXVIc, d), the emphasis of a foliation by mimetic growth, the replacement of early minerals to give pseudomorphs, and the formation of oriented overgrowths to give zoned or idioblastic crystals. Post-tectonic porphyroblasts grow across the foliation which abuts against them without deflection as though the crystal outline had been punched out with a biscuit cutter (Plate XXVIIc). Deflection of adjacent cleavage due to crystal growth is extremely rare.

Helicitic texture consists of contorted lines of inclusions (S_i) within a poikiloblastic crystal (Figs. 61, 63). The S_i represent a folded S-surface which has been enveloped without appreciable distortion by the growing crystal.

Helicitic crystals are post-tectonic (Fig. 64). The inclusion patterns may trace out folds, parts of folds, straight or irregular surfaces. The texture is best shown by albite, andalusite, staurolite (Plate XXVIIb), garnet (Plate XXVIIc, d) and chloritoid (Plate XXVIIa) and abundant examples are figured by Turner and Hutton, 1941; Rast, 1958; 1965, plate I; Zwart, 1960, 1963; Spry, 1963a.

The normally accepted terminology distinguishes between rotational (syntectonic) and helicitic (post-tectonic) textures. Some observers have confused these and Collette (1959) grouped both together as "helicitic" structure.

Evidence of syntectonic crystallization will tend to be destroyed by post-tectonic growth because crystals or crystal aggregates with high energy due to lattice strain tend to be replaced by lower energy unstrained crystals. Interfacial energy will be reduced by an increase in grain size. General coarsening, mimetic growth, formation of porphyroblasts, growth twins, polygonization, etc., can be explained by virtue of the close analogy between post-tectonic crystallization and annealing.

The effect of post-tectonic crystallization on a pre-existing preferred orientation, however, is not clear. It has been shown that annealing of metals may cause one of the following:

(1) No change of an existing preferred orientation.
(2) A change from one pattern of preferred orientation to another.
(3) A loss of preferred orientation or a "randomizing" effect. Sahama (1936) and Harris and Rast (1960b) have suggested that quartz fabrics tend to be disoriented in the post-tectonic phase.

Large post-tectonic muscovite flakes in many schists tend to follow the general direction of micas in a foliation layer although the parallelism may not be as pronounced. Chlorite on the other hand commonly grows across the foliation ("cross-micas").

Strained or bent crystals commonly become polygonized in the post-tectonic stage (Plate XXVIc, d). An undulose quartz crystal or the mica flakes bent around the apex of a fold undergo incipient recrystallization and become resolved into crystals containing unstrained segments with low-angle sub-boundaries.

Microscopic folds of mica-rich layers in many schists contain some muscovite flakes which are bent and others which are unstrained. This structure has been repeatedly (Sander, 1930, p. 246; Knopf and Ingerson, 1938, p. 109; Hills, 1963, p. 411) quoted as an example of syntectonic crystallization of mica during the formation of the fold. It has been supposed that mica formed continuously, the early-formed crystals being bent but the later crystals being less affected, or unaffected, by the folding. In the many examples seen by the author, it is clear that the bent micas were formed along a foliation S_1 during F_1, then S_1, was deformed to give a fold composed of bent F_1 mica flakes during F_2, then some micas were polygonized and others completely recrystallized in the post-tectonic phase to give unstrained flakes. Deformation and crystallization can be said to be concurrent only in the most general way.

Isolated post-tectonic porphyroblasts of chloritoid, kyanite or staurolite in a matrix of differing composition are commonly randomly oriented although kyanite in a kyanite-rich rock tends to be roughly aligned along the layering.

Epitaxial nucleation of isolated post-tectonic porphyroblasts on parallel crystals of another phase (e.g. garnet or albite on muscovite) can lead to a mimetic preferred orientation even of isotropic garnet (Powell, 1966).

Low- and High-stress Metamorphism

Regional metamorphism covers an immense range of physical conditions even if the field were to be taken to exclude the related high-pressure, low- to moderate-temperature glaucophane schists and the high-pressure, high-temperature eclogites and granulites. In addition to a possible division of regional metamorphics into low-pressure (hydrostatic) and high-pressure types on mineralogical grounds, it seems advisable to make a distinction into low-stress and high-stress types on textural grounds.

Low-stress regional metamorphic rocks outcrop over broad regions but have many textural characteristics of contact rocks. Preferred orientations are weak and are largely confined to dimensional rather than lattice orientations; mimetic crystallization along bedding is common. Foliation and lineation in hand specimens are weak to absent. Quartz-rich rocks tend to be granular and the quartz has a weak to absent lattice preferred orientation. Fossils and sedimentary structures are well preserved and show comparatively slight deformation in some parts. There is a complete range of low to high temperature and confining pressure. Post-tectonic crystallization tends to be well developed. The rocks of the Buchan region (Aberdeenshire) are typical; other examples are the Kanmantoo Group of South Australia (Joplin, 1968, Figs. 26–30) and the "granulites" and associated rocks of Cloncurry, Queensland.

High-stress, regional metamorphic rocks are widely recognized and appear to be more common than the low-stress types although it is not easy to be sure from published descriptions because of the failure to recognize definitive criteria. Criteria for the recognition of temperature effects are well established qualitatively even if not quantitatively. The grade of metamorphism is estimated from the grain size and presence of individual "index" minerals or groups of minerals formed from a given bulk composition. The effects of confining (hydrostatic, or non-directed) pressure are not so clearly known but it appears that the presence of such minerals as kyanite, jadeite–omphacite, glaucophane, lawsonite and certain garnet compositions indicate rather high pressures, whereas a mineral such as andalusite indicates rather low pressures for a given composition and temperature.

The effects of directed pressure, "deformation", or "stress" are less understood and the amount of deformation must be evaluated on the basis of the strain of objects of known shape (fossils, pebbles, etc.) and in a very vague way, on the degree of development of a preferred orientation, folds or other textural characteristics. There is no evidence that any metamorphic mineral will only form in the presence of stress. It is possible that the stability fields of some minerals might be enlarged or reduced by stress but this is not certain. The position is confused by an association of directed and non-directed pressures in that some regions of low-pressure metamorphism are also low-stress regions; however this is not always the case.

Further complications are found in many regional metamorphic areas where an early phase (F_1) may be for instance a high-temperature, high-stress, low-pressure metamorphism (producing, for example, sillimanite–garnet schist) followed by an interkinematic, non-stress, period with growth of andalusite or cordierite, followed by a second phase with low-stress, high-temperature, high-pressure metamorphism (with kyanite).

Textural Classification of Regional Metamorphic Rocks

The most prominent macroscopic structure in regional metamorphics is a foliation. This is commonly parallel to a compositional layering and even in coarse gneisses, granulites and charnockites is tacitly assumed to be bedding by many workers although in some examples, at least, it can be shown to be a tectonic feature due to segregation and metamorphic differentiation. It may develop obliquely to the bedding which may be entirely obliterated, preserved as vague relicts or disrupted and dragged around into parallelism with the layering. In some examples it appears that an early tectonic foliation (with or without layering) forms along the bedding, particularly in regions with nappes and recumbent isoclinal folds. The bedding anisotropy may have a considerable influence on the formation of such a foliation but there appears to be little support for the Daly (1917, pp. 400–6) concept of *load metamorphism* producing a horizontal schistosity in horizontal metamorphosed sediments by a vertical load.

The study of textures of regional metamorphic rocks is largely concerned with the sizes, shapes, orientations and order of crystallization of the minerals. Simple equilibrium textures are recognizable in some high-grade types

TABLE 9

	Original rock	Regional			Plutonic
		Low	Medium	High grade	
ORIGINAL TEXTURE DOMINANT	Qu. sandst.		Quartzite		
	Greywacke	Semi-schist			
	Conglom.	Deformed conglomerate		Conglomerate gneiss	
	Calcareous sediment	Schistose limestone			
	Pelite	Slate, Slaty siltstone			
	Acid lava, porphyry	Schistose porphyry, Porphyroid			
	Acid to Int. plutonic	Gneissic granite, Gneissic diorite, Granulite, etc.			
	Basic lava	Metabasalt			
	Basic intrusive	Metadolerite metagabbro	Laxford and Buchan amphibolite, Coronite, Granulite		
	Ultrabasic	Serpentinite			
METAMORPHIC TEXTURE DOMINANT	Qu. sandst.	Quartz schist, Quartzite			Quartzite
	Greywacke	Semi-schist	Schist	Gneiss	Gneiss, Granulite
	Conglom.	Deformed-conglomerate		Conglomerate-gneiss, Granulite	
	Calc. sed. Pure	Marble	Marble		Marble
	Calc. sed. Impure	Calc-schist	Calc-silicate gneiss		Granulite
	Pelite to semi-pelite	Phyllite	Schist	Gneiss	Granitic gneiss Granulite
	Acid to Int. igneous	Schist to Gneiss		Gneiss, Granulite	
	Basic to ultrabasic	Greenschist	Normal amphibolite		Basic granulite, Charnockite, Eclogite

(particularly *plutonic* metamorphics) but lack of textural equilibrium is seen almost everywhere in low- to medium-grade rocks.

In simple terms the average grain size may be a function of the temperature and length of time of crystallization although the grain size may be reduced initially in normal recrystallization, and crystal growth will always be opposed by any granulation. In a multi-component rock, the relative grain size of various constituents will be controlled by their concentration, ease of nucleation (surface energy) and ease of growth. The shape of a crystal will depend on its surface energy but also on its chronological position; pre-tectonic crystals tend to be broken, syntectonic crystals to be irregular and xenoblastic, and only post-tectonic crystals have the opportunity to develop crystal faces. Thus the relation of shape to order of crystallization is the reverse of that in igneous rocks where the earliest crystals tend to be euhedral and the latest anhedral.

Regional metamorphism involves a widespread rise in temperature (due to heat of unknown origin) forming a kind of domal "thermal structure" (outlined by isotherms), accompanied by intermittent deformation of large magnitude. Regional metamorphism thus is analogous with, and apparently indistinguishable from, dynamo-thermal metamorphism. Few valid examples of dynamo-thermal metamorphism are extant, the best known example being in the vicinity of Donegal intrusives (Pitcher and Read, 1960, 1963; Pitcher, 1965) where the combination of lateral stresses imparted by forcible intrusion of granitic magma plus the heat of the magma gave "contact schists". Repeated intrusion produced polymetamorphic rocks which are petrographically indistinguishable from normal polymetamorphic regional metamorphics.

The prefix *ortho-* is used to signify that a metamorphic rock was originally igneous (e.g. ortho-gneiss) and *para-* to indicate that it was originally sedimentary (e.g. para-gneiss). Even though this is a difficult decision to make in some instances it is a natural division. A second subdivision can be usefully made to distinguish between the products of normal regional metamorphism and plutonic metamorphism. A third subdivision is made on general textural grounds according to whether the texture is dominated by the original or the metamorphic texture (Table 9), e.g. a rock may be classed as a meta-dolerite when its igneous texture predominates, and as an amphibolite when the metamorphic texture is dominant.

THE TEXTURES OF REGIONAL-METAMORPHIC ROCKS

Quartzites and Semi-schists

A pure quartz–sandstone is regionally metamorphosed to a quartzite (Plate XXX). Under conditions of low temperature and high stress the rock may be granulated or crushed to produce initially a mortar texture and finally a fine-grained aggregate of elongate grains. At moderate deformations and temperatures the original quartz grain fabric may be essentially retained with the grains interlocking to give a crystalline granoblastic texture due to boundary pressure solution, plastic flow and minor recrystallization; the quartz grains are generally somewhat elongate.

Most metamorphic quartzites develop *granoblastic elongate* texture with no sign of original grains and this is presumably due to the nucleation and growth of entirely new quartz crystals at the expense of the old grains (Plate XXXd). Pre-metamorphic lattice strain and grain boundary energy probably provide part of the driving energy for recrystallization and the texture is controlled mainly by the orientation, relative magnitudes and rates of application of the various stresses. The quartz crystals tend to be flat and elongate with their long and mean dimensions giving a foliation which is parallel to the axial surfaces of folds and their long dimensions parallel to the fold axes. This dimensional preferred orientation is commonly accompanied by a lattice preferred orientation, a girdle of quartz axes normal to the lineation being typical.

The grain boundaries may be straight and may meet in triple-points. The interfacial angles may be equal (Hobbs, 1966) but generally are not because of the inequant shape of the grains. A considerable amount of grain boundary slip is probable at elevated temperatures and as the grain boundaries move by stress-induced boundary migration, there is little opportunity for them to achieve a low-energy state.

Recrystallization under low stresses, or abundant post-tectonic crystallization may however produce a *granoblastic-polygonal* quartzite. Annealing of the early quartz fabric may produce a new pattern of preferred orientation, or may destroy the old one (Plate XXXe). Heating may cause coalescence of fine-grained elongate quartz to give large crystals with a vague mosaic texture (Crampton, 1963) but it is not possible at present to distinguish between such a crystal and a large porphyroclast which has been deformed and polygonized.

Many grain boundaries are not straight but irregular (Fig. 1). They may be curved about inclusions, very irregular, or *sutured* (i.e. with irregularities like the sutures in a skull) (e.g. Plate VIId). The origin of sutured boundaries is not clearly understood; in some rocks they appear to be an insignificant pressure solution effect (somewhat like stylolites) due to minor late stresses. In other rocks they appear to be typical of mobile quartz boundaries and probably represent a "frozen" image of a non-equilibrium boundary produced during metamorphic recrystallization. Voll (1960, p. 521) considered that there may be some crystallographic control of such boundaries, the serrated parts representing interpenetrating rhombohedral planes of both grains.

Various mineral phases may occur in quartzites. Regional metamorphism under dry conditions recrystallizes both quartz and feldspar but the feldspar in many rocks tends to be broken into fragments while quartz is recrystallized; this results in shattered or undulose pre-tectonic porphyroclasts in an elongate-granoblastic aggregate of quartz. Feldspar is altered to sericite or epidote, etc, and an arkose may be converted to a quartz–mica schist where H_2O is available at low to moderate temperatures.

Quartz–mica schist (schistose quartzite, quartz schist) is also produced from sandstones with a pelitic impurity. The behaviour during deformation of a quartz–mica aggregate depends on the relative proportions of the two minerals. Where a little mica is present it may be enclosed within the quartz but generally lies along the quartz grain boundaries and the micas will tend to be parallel and give a foliation because the quartz is elongate. The presence of mica on quartz grain boundaries increases the ability of the grain boundary to slide, because of the gliding ability of the mica. When the mica reaches such a proportion that continuous mica-covered surfaces are formed, shearing movements will take place preferentially along these. The rock should then be regarded as a series of parallel mica layers separated by layers of quartz. Grain boundary migration of quartz will be inhibited by all micas parallel to the foliation and thus a dimensional preferred orientation of quartz grains may result from preferred growth parallel to the foliation. Quartz–quartz boundaries commonly meet (001) of mica perpendicularly so that the lenticular shape of the quartz may be considerably modified by the mica (Plate XXVIIIa, b).

The fabric evolution of greywackes into *semi-schists* and schists at low grades has been outlined in detail by Hutton and Turner (1936), Turner (1938) and Ghent (1965). Four subzones, based on texture, were established for "chlorite zone" meta-greywackes in Southern New Zealand. They are numbered 1 to 4 in order of increasing intensity of deformation.

Subzone 1. The greywackes retain their original clastic texture, slight cataclasis is evident and reconstitution of the matrix has commenced but there is no schistosity.

Subzone 2. Greywackes are converted to *semi-schists* in which the clastic texture has been partly obliterated. The grain size has been reduced, a definite schistosity has appeared as irregular, intersecting surfaces of shear. Recrystallization is advanced but not complete.

Subzone 3. The original clastic texture has been destroyed but a few relict clastic grains remain. Recrystallization is almost complete. The grain size is small and the schistosity formed by parallel layers of mica is well developed and accentuated by a poorly defined compositional layering.

Subzone 4. The rock has been converted into a coarse schist with a strong foliation. A coarse, irregular but well-defined layering is outlined by alternations of quartz plus albite, and mica.

Like most coarse-grained rocks, greywackes are initially reduced in grain size during metamorphism; recrystallization at moderate to high grades involves (*inter alia*) the growth of these small crystals. The textures of meta-greywackes share characteristics with both pure sandstones and with pelites. The sedimentary inhomogeneity is important in the early stages as micaceous schistose layers develop from matrix and enclose lenses of the disrupted coarser-grained material. At moderate grades all clastic structures are destroyed, except perhaps for large clastic fragments, to give medium-grained, somewhat inhomogeneous schists which do not differ much from pelitic schists in high-stress environments. Reconstitution at high grades produces quartzo-feldspathic schists and granitic gneisses with predominant quartz, feldspar and mica, and subsidiary garnet and perhaps sillimanite.

Meta-rudites (Deformed Conglomerate, Sheared Grit and Conglomerate Gneiss)

Metamorphosed coarse clastic sediments (rudites or psephites) tend to retain their fragmental structure despite strong deformation and recrystallization. The degree of retention of recognizable relict structure depends on the size of the fragments and their compositions (relative to each other and to the matrix). Deformed rudites are particularly interesting structurally as the strain of the individual particles gives information about the amount and kind of deformation which the rock as a whole has undergone.

Rudites contain clastic particles ranging in size from 2 mm to 20 cm and upwards in a sparse to abundant matrix. The particles differ in composition and physical properties from the matrix which is commonly pelitic and mechanically much weaker. The behaviour of a conglomerate during deformation is thus inhomogeneous and the matrix is more strongly deformed than the pebbles. The deformation of pebbles and of pebbly rocks takes place under a very wide range of conditions ranging from low-temperature dynamic metamorphism to high-temperature, high-pressure regional metamorphism.

Regional metamorphism at low grade of a fine-grained conglomerate or greywacke may produce a *semi-schist* as discussed earlier if the matrix is sparse and the fragments are closely packed. If the rock has a disrupted or open framework with isolated large fragments in a fine matrix the behaviour is different. The matrix is converted into foliated quartz–sericite (with perhaps chlorite and biotite depending on composition and grade) resembling a slate or phyllite. The foliation wraps around the clastic particles which are undeformed and form small augen (eyes) with pressure fringes or shadows. Fine-grained rocks have been called *schistose grits* (Harker, 1939, p. 241).

Stronger deformation (particularly at higher temperatures) causes deformation of the clastic fragments as well as the matrix. The behaviour of each clastic particle in a single rock depends on its size and composition:

(1) Small fragments are deformed more than large ones.
(2) Fragments of slate are deformed more than those of limestone which are in turn deformed more than greywacke, then quartzite, and least are feldspar crystals, porphyry and granite fragments. The texture of each rock type responds to metamorphism in much the same way in a fragment as in the bulk rock.

The distinction between fragments and matrix tends to be lost if they are similar in composition. Fossils, fossil fragments or limestone fragments in a calcite matrix tend to blend together to give a homogeneous marble. Quartzite fragments in a siliceous matrix retain their identity at first but then tend to merge with the matrix.

There is a strong resemblance between the mortar texture produced by cataclasis of a coarse quartzite and the blasto-psephitic texture of a deformed, siliceous, poorly sorted sediment. Both contain large undulose, ragged, lenticular and deformed fragments with pressure shadows in a fine-grained, foliated matrix consisting of elongate quartz and perhaps mica. The blasto-psephitic texture may be recognized if some relict clastic fragments are polycrystalline aggregates (not single crystals) with differing grain sizes or compositions, or if the matrix is distinctly different from the clastic fragments.

The cataclastic-disruptive texture might be recognized if it is possible to fit now-separate fragments together although even then they might have been large clastic fragments which were disrupted. It may be necessary to rely on macroscopic field evidence to make the distinction but even then it is impossible to decide whether some deformed fragmental rocks are true meta-psephites (stretched pebble conglomerates) or pseudo-conglomerates.

Feldspathic rock fragments (e.g. granite) or feldspar crystals tend to be resistant to deformation and remain as relict clasts wrapped around by recrystallized quartz. However, if regional metamorphism is accompanied by accessible H_2O, the potash feldspar is converted to sericite or muscovite at low to medium grades. Altered feldspathic particles are merged with the mica-schist matrix.

Coarse clastic sediments tend to be richer in the minor "heavy" clastic minerals, particularly iron oxides. These may crystallize to hematite or magnetite in the matrix or react to give syntectonic or post-tectonic chlorite or randomly oriented post-tectonic chloritoid in the matrix or in clastic fragments.

Amphibolite Facies conglomerates from Ontario and Rhode Island (Walton, Hills and Hansen, 1964) contain deformed pebbles with concentric, mineralogically distinct zones. It has been suggested (ibid., p. 25) that the conglomeratic nature might be obliterated by diffusion and homogenization in the granulite facies and the rocks converted to feldspathic gneisses.

Marble, Calc-Silicate Rocks and Calc-Silicate Gneisses

The regional metamorphism of a pure limestone gives a pure marble. Predominantly mechanical effects similar to those of dynamic metamorphism are favoured by low temperatures and rapid deformation. Crystallization is accompanied and followed by deformation and the calcite crystals show textures typical of pre-tectonic crystallization (or more aptly, of post-crystalline deformation). The texture is crystalline, with interlocking elongate crystals giving a dimensional preferred orientation (*granoblastic-elongate* texture). The rock may be foliated due to the long flat crystals but may split in a number of directions because of parallelism of the three $\{10\bar{1}1\}$ cleavages. The calcite crystals have bent cleavages, deformation twins, deformation lamellae and a lattice preferred orientation (Plate XXVIIIc).

High temperatures and low deformation rates favour recrystallization and the marble is then *granoblastic-polygonal*. A lattice preferred orientation, however, is commonly present even if a dimensional one is lacking. There may be little difference between the microscopic appearance of marbles produced by moderate to high temperature thermal, regional or plutonic metamorphism.

Recrystallization of a dolomite gives a dolomite-marble which resembles a calcite-marble. The presence of dolomite and calcite together does not lead to the formation of periclase or brucite and the two crystallize side by side, the dolomite tending to be idioblastic towards the calcite.

A siliceous impurity in a dolomitic limestone may simply recrystallize as isolated quartz grains or there may be extensive reaction to give talc. A clay impurity may give muscovite (and/or phlogopite) to produce schistose marbles or calc-schists (Plate XXIXb, d).

The presence of both muscovite and carbonate may give a strongly foliated rock. The mica will occur on calcite grain boundaries if it is sparse, but if abundant will form lenses or continuous layers parallel to the foliation. The relation of calcite to mica is much like that of quartz to mica discussed previously but calcite tends to become strongly elongate and to develop intergranular boundaries which are rational lattice planes. Post-tectonic crystallization of calcite appears to take place readily and patches of coarse granoblastic and polygonal carbonate develop throughout the oriented texture to give an annealed texture.

Impure limestones contain, in addition to calcite, various amounts of clays, quartz, dolomite, iron oxides and feldspars. Regional metamorphism forms minerals such as pyroxene (commonly diopsidic), garnet (rich in the grossular molecule), amphiboles, the epidote group, phlogopite or other micas in a matrix (or with intergranular masses) of calcite. The calc-schists of low grade with strong foliation give way to poorly foliated, granular or perhaps irregularly textured rocks. If calcite is abundant the rock is called a marble but where the silicates dominate, the general term *rock* is used with an appropriate prefix (e.g. "diopside-grossular rock"). High-grade calc-silicate rocks are commonly granoblastic, even polygonal, with the shape of the grain-boundaries controlled by grain boundary energy rather than by mechanical processes.

The presence of calcite in a rock appears to promote ionic mobility. Calc-silicate rocks tend to become very coarse and irregularly textured (almost pegmatitic) and metamorphic differentiation produces large monomineralic masses (of pyroxene, amphibole, garnet, etc.) or a strongly defined (but irregular and lenticular) segregation banding. Coarse-grained, banded varieties are commonly called *calc-silicate gneisses* even though they are granoblastic and not particularly well foliated.

The characteristic lack of a dimensional preferred orientation of calcite or of ferromagnesian minerals such as hornblende, pyroxene and epidote was noted by Harker (1939, p. 255) who attributed it to the "relatively yielding nature of a rock composed mainly of calcite which makes any very high measure of shearing stress impossible". Whatever the reason, many regional metamorphic calc-silicate rocks are unfoliated and are not unlike calc-silicate hornfelses.

Aluminous and siliceous impurities may lead to the formation of abundant calcic plagioclase and an impure limestone may be converted into a granular (dioritic) rock consisting of plagioclase and pyroxene or amphibole. Abundant iron and magnesia in addition leads to the formation of *para-amphibolites* consisting of hornblende, plagioclase and garnet which are difficult to distinguish from *ortho-amphibolites* derived from basic igneous rock (Walker *et al.*, 1959). High-grade (Granulite Facies) metamorphism may differentiate an impure calc-silicate rock into layers of pure marble and emphibolite.

Scapolite is a common minor mineral in calc-silicate rocks of all grades from Greenschist to Granulite Facies and is abundant in some calcareous rocks which have undergone regional chlorine metasomatism (Edwards and Baker, 1954; White, 1959) in low-stress provinces. Scapolite occurs with diopside and/or hornblende and grossular at all grades, *diopside-scapolite rock* being characteristic (Joplin, 1968, Figs. 64, 65). Scapolite resembles quartz in crystallization behaviour and forms granoblastic and polygonal aggregates with the diopside. The texture is characteristically evenly grained with the scapolite a little larger than the diopside which is located on scapolite boundaries or at triple-points. The rather small grain size appears to be due to diopside pinning the scapolite boundaries; where a grain boundary is freed, the single scapolite crystals grow to a large size and envelop the diopside poikiloblastically. A compositional banding which appears to be relict bedding is well preserved in some rocks which consist of alternate thin layers of scapolite plus diopside and scapolite plus hornblende. Scapolite shows little tendency towards a dimensional preferred orientation and these rocks tend to be unfoliated, resembling hornfelses; however, a lattice preferred orientation may develop.

A distinctive textural rock type (*Garbenschiefer*) may be considered here with convenience although it is not only derived from impure calcareous sediments but also from tuffs or from intermediate to basic igneous rocks.

The German terms for somewhat schistose rocks (both hornfelsed slates or phyllites, and regionally metamorphosed rocks with strong post-tectonic crystallization) with spots or knots are:

fleckschiefer—with spots or flecks of indeterminate mineral,
fruchtschiefer—spots the size of grains of wheat,
garbenschiefer—small porphyroblasts the size of caraway seeds,
knotenschiefer—conspicuous porphyroblastic knots.

Garbenschiefer are characterized by the presence of poikiloblastic, subidioblastic porphyroblasts of amphibole, commonly hornblende (*Hornblendegarbenschiefer*). Most of the amphibole prisms are parallel to the foliation but are randomly oriented or form stellate, radiating aggregates along the foliation. The texture is due to pronounced post-tectonic mimetic crystallization of the amphibole. The large size and stellate arrangement are the result of nucleation difficulties, the random orientation is due to the lack of stress coupled with ease of growth along the foliation, and the poikiloblastic texture can be attributed to rapid growth.

Slates, Phyllites and Schists

Dynamic or very low-grade metamorphism converts pelitic rocks into slates in which the foliation may be a widely spaced fracture-, false-, crenulation-, or strain slip-cleavage at one extreme or a flow- or slaty-cleavage due to tiny parallel illite or sericite flakes and elongate grains or aggregates of tiny quartz crystals at the other. Pressure fringes may be present around scattered large clastic grains or around pre-tectonic pyrite or magnetite crystals which were either authigenic or introduced into the pelite before the cleavage was formed.

Three compositional terms which have some textural significance have been widely used in connection with Scottish rocks but not so much elsewhere. A *pelite* (pelitic-schist, etc.) is very rich in clay minerals and becomes a mica-rich schist on metamorphism. A *semi-pelite* is composed of quartz and clay (perhaps a siliceous mudstone or siltstone); a *psammite* is a sandstone. Semipelite and psammite give rise to quartz-rich metamorphic rocks such as micaceous quartzites or rather massive granular schists (the "Moine granulites").

A *slate* is a very fine-grained rock in which the grains are scarcely visible under the microscope. A *phyllite* is a slightly coarser-grained (higher grade) rock in which the mica flakes have become sufficiently large to be easily recognizable under the microscope, but not in hand specimen where their presence gives a sheen or glossy lustre to the foliation surfaces. There is no clearly defined boundary between slate and phyllite nor between phyllite and schist and all gradations occur. *Schist* is a foliated rock rich in mica (or possibly amphibole); it is coarser than a phyllite from which it can be distinguished by having macroscopically recognizable mica flakes. Schists may be finely banded but are generally rather homogeneous. There is a transition into *gneiss* which is more coarsely grained (quartz, feldspar and micas more than a few millimetres across) and characterized by a richness in feldspar and a general inhomogeneity (Wenk, 1963).

A simple pelitic schist consists of interlocking, elongate quartz crystals (somewhat similar to those described previously in quartzites) as layers or lenses alternating with layers of interleaved, parallel muscovite and biotite flakes forming the foliation. The quartz–mica relations have been described in mica–quartzites. Quartz, muscovite and biotite may have all crystallized syntectonically and simultaneously.

Plagioclase may be present as:

(1) Pre-tectonic crystals; shattered, undulose with deformation twins (Plate XIc), pressure shadows and wrapped around by the cleavage.

(2) Syntectonic crystals; commonly clear, elongate to granoblastic, untwinned and interspersed with quartz from which they are practically indistinguishable although some have sigmoidal spiral trains of inclusions.

(3) Post-tectonic crystals; large, subidioblastic porphyroblasts which cut across the cleavage without disturbing it and possibly contain helicitic structure (Plate XXVIa, b). Twinning is due to growth and simple twins are most common. Such porphyroblasts may be due to soda metasomatism.

Garnet may similarly occur as pre-tectonic, syntectonic or post-tectonic crystals. Pre-tectonic crystals form shattered, possibly partially chloritized, porphyroblasts with pressure shadows and the cleavage wrapping around. Syntectonic crystals may be rounded, sickle-shaped or skeletal with the characteristic S-shaped spiral structure (snowball or rotational structure) (e.g. Plate XXV). Post-tectonic crystals of garnet are commonly homogenous (or with zonally arranged inclusions), idioblastic and do not disturb the foliation; helicitic structure occurs but is not common (Plate XXVIId). Some garnets have a syntectonic core and a post-tectonic idioblastic rim (Plate XXVa).

Two other characteristic textures form in post-tectonic garnet. Skeletal *fish-net* garnets are large web-like crystals containing so many quartz inclusions that the quartz predominates. The grain size and fabric of the quartz inclusions is similar to that in the matrix and it appears that the garnet has crystallized along the intergranular boundaries of the quartz aggregate. This is a high surface energy form and would appear to be less stable than a single homogeneous idioblast of garnet. It is due to formation of garnet along grain boundaries perhaps due to crystallization from an intergranular fluid. A second, rather unusual form of garnet is the *single-crystal aggregate-pseudomorph* which consists of a single, large skeletal crystal which has replaced an aggregate of chlorite or biotite laths (Plate XIIb).

Andalusite, chiastolite, staurolite and kyanite in schists affected by one metamorphic phase are almost invariably post-tectonic (pre-tectonic crystals occur in multi-phase schists). Syntectonic kyanite is very rare, if indeed, it occurs at all. The aluminium silicates tend to form large, idioblastic porphyroblasts which cut across but do not bend the cleavage. Helicitic structure is common in andalusite and staurolite (Plate XXVIIb). Chiastolite is a special variety of andalusite which is rich in inclusions which form a growth-controlled pattern (Plates XIIIc and XVI). Kyanite forms subidioblastic prisms which tend to be randomly oriented. Bent kyanite (demonstrating its excellent glide properties) occurs as interkinematic crystals in polymetamorphic rocks. The time relations of the crystallization of kyanite show that its formation cannot be explained by "tectonic overpressure".

The most common habits of sillimanite in schists and gneisses are:

(1) Relatively large idioblastic, elongate, prismatic crystals with typical square cross-section (Plate XVc).

(2) Finely felted, fibrous, aggregates of acicular "fibrolite" which forms lenses, schlieren and twisted layers (Plate XXVIIId). Some fibrolite is intergrown with biotite and replaces it.

(3) In "*faserkiesel*" (quartz–fibrolite veins and knots).

(4) As isolated groups of small idioblastic needles in fascicular or rosette aggregates (Plate XVb).

(5) As pods of fibrolite which appear to represent large kyanite crystals which have been replaced.

(6) As discrete trains of needles enclosed helicitically by later muscovite.

Experimental and petrographic evidence suggests that all of the aluminium silicates are rather difficult to nucleate and sillimanite is no exception. Even in polymetamorphic rocks containing kyanite, sillimanite shows a strong preference to form within or upon biotite and this is best regarded as nucleation on the mica rather than simple alteration to sillimanite (Chinner, 1961).

The relations between fibrolite and coarse idioblastic sillimanite are not easily recognized, but in some examples at least (Chinner, 1961, at Glen Clova) the fine fibrolite has changed to the coarse variety by grain-growth. However, in other rocks the two varieties appear to have formed simultaneously. The high-density nucleation of fibrolite (considering that sillimanite is a mineral which normally does not nucleate easily) would appear to require special circumstances, perhaps strong supersaturation. The common association of fibrolite with metasomatism (Pitcher, 1965) suggests that perhaps the presence of a fluid phase might be one requirement.

Most fibrolite masses have a complex internal structure (Plate XXVIIId) as though they have been deformed but it seems likely that much of this "contortion" is not due to deformation but to irregular growth following seeded nucleation. Twisted and non-twisted fibrolite occur closely together, and contorted fibrolite curves around older crystals of biotite, etc., which show no signs of deformation (Binns, 1964, p. 304).

Andalusite and staurolite appear to be stable together but although two or even three of the group andalusite–sillimanite–kyanite have been found together (Hietanen, 1956) such assemblages occur in polymetamorphic rocks and there is no evidence that they are in equilibrium (the lack of obvious disequilibrium textures such as reaction rims cannot be taken to indicate equilibrium). Kyanite forms pesudomorphs after andalusite where an early, lower pressure, metamorphic phase is followed by a higher pressure phase.

Cordierite in schists is colourless, rather fine-grained, of low relief and is practically indistinguishable from plagioclase. It is commonly pre-tectonic and thus occurs as shattered crystals with typical curved and tapered polysynthetic deformation twins similar to those in plagioclase. It is somewhat prone to alteration to a yellowish isotropic material or a white mica ("pinite") which together with pleochroic haloes, helps to distinguish it from feldspar. Syntectonic cordierite is rare in schists (Zwart, 1960b). Cordierite is most common as large, xenoblastic, poikiloblastic, post-tectonic crystals.

Chloritoid occurs in low- to medium-grade pelitic schists which are relatively rich in FeO and Al_2O_3 but poor in K_2O. Harker (1939, p. 150) included it among his "stress minerals" and this has been supported by Read (1934). However, the mineral typically occurs as randomly oriented, subidioblastic prisms which cut across the cleavage and have all the characteristics of post-tectonic crystals. Niggli (1912) considered it to be syntectonic in the Alps but this has been denied by Tobi (1959) whose views are supported by the excellent helicitic structure shown in Plate XXVIIa, where Alpine chloritoid has crystallized over a contorted cleavage. Syntectonic chloritoid appears to be unknown and thus there is no evidence of any relation between the crystallization of chloritoid and directed stress.

REGIONALLY METAMORPHOSED IGNEOUS ROCKS

Classification on a textural basis is made into two groups according to whether the original igneous texture or the new metamorphic texture predominates. Those in which the metamorphic texture is well developed can be divided into two groups according to whether their texture is schistose (mainly low- to medium-grade rocks) or granoblastic (mainly high-grade rocks). Each class can be usefully subdivided on bulk chemical composition into acid plus intermediate, basic, and ultrabasic groups.

Although this classification is fundamentally descriptive it is also genetic and a given rock type which is defined on compositional grounds may belong to several categories; thus amphibolites have a wide range of composition and origin. Three kinds of metamorphosed dolerites recognized by Sutton and Watson (1951) are the Laxford and Buchan type (low to medium grade, basic, original texture dominant), the normal or South-West Highlands type, studied in detail by Wiseman in 1934 (low- to medium-grade, basic, new schistose texture dominant) and the pyroxene–granulite variant (high-grade, granoblastic, basic, new texture dominant). Amphibolites may be derived from either sedimentary or igneous rocks during either thermal or regional metamorphism. They may be derived directly from basic, ultrabasic or some intermediate igneous rocks, or indirectly by retrograde metamorphism of eclogite, basic-granulite or a higher-grade amphibolite. They range in composition from rocks composed entirely of amphibole through those containing garnet, epidote or zoisite, plagioclase, etc., to those rich in plagioclase and even with a little quartz.

Original Texture Dominant

Preservation of a palimpsest igneous texture and failure to achieve an equilibrium texture is caused by low ionic mobility which is favoured by low stress and an absence of intergranular fluid. It is also favoured by low temperature and short duration of metamorphism but these, perhaps surprisingly, do not appear to be very important because relict textures are preserved in many rocks which have been plutonically metamorphosed and belong to the Granulite Facies. Relict textures may be strong in some low-grade rocks, e.g. the schistose porphyries or porphyroids, but the deformation accompanying low-grade regional metamorphism generally destroys any original texture. Relict textures are most likely to be visible in thin sections cut *parallel* to the foliation.

The textural changes in dolerite at Laxford described by Sutton and Watson

(1951, p. 27) are typical of medium-grade, low-stress, regional metamorphism.

(1) The least altered dolerite retains its original texture although the plagioclase is somewhat cloudy and bent; secondary hornblende forms rims on pyroxene, and garnet occurs in some reaction rims between feldspar and pyroxene.

(2) The next stage in the metamorphism is the replacement of pyroxene by multi-phase, multi-crystal, hornblende–quartz pseudomorphs; the feldspar is cloudy and somewhat sericitized but has fresh rims. The multi-crystal pseudomorphs become reorganized to single-crystal pseudomorphs (large poikiloblastic hornblende crystals with quartz inclusions and relics of the rim hornblende from the previous stage).

(3) The next stage sees the partial loss of the igneous texture as plagioclase laths become recrystallized to granular aggregates.

(4) Wholesale recrystallization of both hornblende and plagioclase destroys the original texture and a foliated aggregate of hornblende, with granoblastic plagioclase, quartz, garnet and sphene develops.

(5) The greatest alteration appears to be shown by amphibolite in which metamorphic pyroxene is generated and the texture becomes granoblastic.

The best examples of the preservation of igneous textures in high grade metamorphic regions are shown by the *coronites* and related rocks. Igneous assemblages including gabbros, anorthosites, diorites, etc., which belong to the upper Amphibolite and Granulite Facies, appear to have been metamorphosed under "dry" conditions with low stress such that igneous textures are only slightly modified. Deformation effects appear to be restricted to slight intergranular crushing, bending of feldspars, deformation twinning in plagioclase and exsolution in feldspar and pyroxene. Recrystallization is limited to the formation of reaction rims and coronas which may be complex. It is difficult to be certain which minerals belong to the igneous phase and which to the metamorphic and consequently there has been extensive debate as to whether the metamorphism took place when the material was partly molten (protoclasis) or is entirely metamorphic.

Some gneissic granites are fundamentally granitic rocks with an igneous texture which has been modified by a weak preferred orientation (chiefly of mica) due to deformation at low to medium grades; gneissic texture can be produced without the rock undergoing complete reconstitution at high grade.

Metamorphic Texture Dominant

Deformation accompanied by crystallization and recrystallization replaces the original features by a new texture and mineralogical assemblage. This is favoured by directed pressures, elevated temperature and the presence of an intergranular fluid, and is regarded as the "normal" process which tends towards achieving mineralogical and textural equilibrium.

Three fundamentally different textures are recognized:

(1) *Foliation:* Due to the presence of minerals which are platy or fibrous, e.g. micas, chlorite, amphibole, talc, serpentine, etc., and thus most pronounced in low grade rocks.

(2) *Layering:* Formed in originally massive igneous rocks by metamorphic differentiation.

(3) *Granoblastic texture:* Produced by the formation of minerals of equant habit, particularly proxene and feldspar, and more common in high-grade rocks. It results from recrystallization during deficient stress, the formation at medium grades of equant minerals in rocks of special composition (diopside scapolite, calcite hornblende), or from extensive post-tectonic crystallization.

METAMORPHISM OF BASIC IGNEOUS ROCKS

Basic igneous rocks are converted by regional metamorphism into greenschists at low temperatures and low to moderate pressures, to glaucophane schists at low temperatures and high pressures, to amphibolites at medium temperatures and pressures, to pyroxene–granulites (basic granulites, basic charnockites) at high temperatures and pressures, and to eclogites at very high pressures and moderate to high temperatures.

Low- to medium-grade greenschists and amphibolites are derived either directly from igneous rocks or indirectly by retrograde processes from high-grade amphibolites, eclogites or basic granulites. Some high-grade rocks may have been formed by progressive metamorphism but rocks with an equilibrium texture may contain no evidence of the path of crystallization. The proportion of high-grade rocks containing relict igneous textures is sufficiently high to suggest that in many cases there has not been a progressive change from fine-grained basalt to fine-grained strongly foliated greenschist to medium grained, moderately foliated amphibolite to coarse-grained granoblastic granulite, but that metamorphism normally proceeds directly from basalt to greenschist, basalt to amphibolite, or basalt to granulite. Alternative views have been advanced by Binns (1964) and Atherton (1965) and the controlling factor depends on the relative rates of heating and crystallization.

Wiseman (1934), Harker (1939, pp. 278–87) and Sutton and Watson (1951) have traced what might be regarded as the "normal" textural evolution of regionally metamorphosed basic rocks. The following composite picture of the textures at various grades is not meant to imply necessarily progressive metamorphism.

The original rock may be taken as a fine-grained massive or amygdaloidal basalt, an intergranular, intersertal or ophitic dolerite, or a granular gabbro. The primary minerals are pyroxene (ortho- and clino) and labradorite with various amounts of olivine, hornblende, calcite, zeolite, feldspathoid, ilmenite, magnetite, etc.

At low grades the rock consists of albite, epidote, actinolite, chlorite, serpentine, carbonate and a little quartz. Traces of the original texture may be preserved (amygdales, phenocrysts or coarse gabbroic or ophitic texture), and relict pre-tectonic igneous crystals or structures may be deformed and wrapped around by a foliation due to parallel crystals and layers of elongate actinolite and prisms or trains of granules of epidote and flakes of chlorite or possibly biotite.

Distorted pseudomorphs (cloudy saussuritized plagioclase, uralitized pyroxene) may be present but the texture is dominated by ragged and fibrous crystals of actinolite and chlorite and aggregates of small, irregular to possibly polygonal albite, epidote, quartz or calcite. The rock is a *greenschist*, i.e. fine-grained, foliated, green-coloured schist with an approximately basic igneous chemical composition and containing abundant amphibole and/or chlorite. Special varieties rich in a certain mineral may be called chlorite-schist, actinolite-schist etc. The terms Greenshist and Amphibolite *Facies* are moderately well defined (Turner and Verhoogen, 1960, p. 533; Francis, 1956) but the difference between the *rocks* greenschist and amphibolite is very vague and appears to be mainly on the basis of texture. The grain size boundary seems to be placed at about 1–2 mm for amphibole but the distinction is somewhat arbitrary. Greenschists are more foliated and amphibolites more granular. Transitional rocks have been called *hornblende schist* and this term has also been used synonymously with *schistose amphibolite*. *Hornblende-gneiss* is approximately equivalent to *banded amphibolite*. The greenschists of the chlorite and biotite zone (or the lower Greenschist Facies) pass gradationally into the amphibolites of the garnet zone (upper Greenschist and Almandine Amphibolite Facies).

Original textures are preserved in few medium- to high-grade amphibolites and then probably only because of low stresses associated with metamorphism or protection within the inner parts of massive igneous bodies. Preservation of original features may be considered on various scales. Metamorphism may preserve even microscopic details such as the size and shape of crystals, intergrowths, amygdales, etc. On the other hand, small-scale textures might be destroyed and only large-scale compositional banding preserved. The whole body may be completely reconstituted on the microscopic and mesoscopic scale but still retain an original gravity differentiation sequence such as that shown by the Insch mass.

A typical, medium-grade amphibolite consists of hornblende and plagioclase with various amounts of garnet, epidote–zoisite, and quartz. The texture depends on:

(1) The degree of deformation.
(2) The relative proportions of felsic to femic constitutents.
(3) The amount of post-tectonic crystallization.
(4) The presence of porphyroblastic species.
(5) The possible influence of earlier textures, either igneous or metamorphic.

Layered amphibolites contain layers alternately rich in felsic (light-coloured) and femic (dark-coloured) constituents. Layering is due to metamorphic differentiation and the layers are parallel to a tectonic foliation. They are very thin and lenticular in greenschist but become thicker and more regular in amphibolite. The darker layers consist mainly of ragged, fibrous actinolite or well-crystallized parallel hornblende (possibly with garnet) and the lighter-coloured layers of plagioclase, quartz and epidote which may be elongate and lenticular or rather coarse and granoblastic.

High-grade examples (sillimanite zone) have been called "striped amphibolite" by Evans and Leake (1960). The layers range from 1 mm to 2 cm in thickness, the darker consisting of an even-grained foliated aggregate of elongate xenoblastic hornblende and the lighter consisting of a granoblastic aggregate of plagioclase and pyroxene.

The final texture of an amphibolite depends also on the time-relations of its constituent crystals. Pre-tectonic crystals (perhaps igneous feldspars) are broken and wrapped around by the foliation which is outlined mainly by syntectonic crystals (elongate, feathery, ragged subidioblastic amphiboles, parallel prisms of epidote or zoisite and elongate or lenticular crystals or aggregates of plagioclase and quartz). Syntectonic garnet may have spiral structure with S_e concordant with S_i. Post-tectonic crystallization tends to give the crystals a more-regular form and even produces idioblastic garnet, amphibole and epidote. The preferred orientation is broken up by the growth of cross-cutting randomly oriented hornblende and epidote. Granoblastic-polygonal aggregates of quartz, plagioclase and amphibole, and large porphyroblasts of garnet, plagioclase or amphibole (perhaps with helicitic structure) are formed. The grain size generally increases with post-tectonic crystallization and schistose amphibolite may be converted into a coarse granoblastic rock similar to a contact-metamorphic amphibolite. Binns (1964, pp. 296–302) traced the textural changes in high-grade amphibolites at Broken Hill in some detail. He showed that in progressively higher grades there is an approach to textural equilibrium controlled by intergranular energies with:

(1) A general increase in grain size, particularly of felsic minerals.

(2) A tendency towards equivalence in grain size. Hornblende in low-grade rocks may be twice as big as quartz or feldspar, but all species tend to be of the same size at high grades.

(3) A change from ragged to smooth and straight grain boundaries.

(4) A decrease in the number of inclusions (i.e. in poikiloblastic texture).

Textures in amphibolites depend to a certain extent on the relative proportions of amphibole and plagioclase. Metamorphism of more mafic types produces an amphibole-rich rock with the plagioclase dispersed as regular "round-grained" crystals giving a rock of very characteristic spotted appearance. It seems likely that the plagioclase originally had an elongate, lenticular form and that it was converted to a new, spherical, low surface energy shape during post-tectonic crystallization. Metasomatic post-tectonic feldspar in amphibolites commonly has the "round grain" form.

Equal proportions of hornblende and plagioclase may result in either a banded rock consisting of almost monomineralic layers of elongate hornblende and of plagioclase or in a homogeneous diablastic intergrowth of hornblende prisms and xenoblastic plagioclase. Metamorphosed intermediate rocks containing more plagioclase (perhaps with quartz) than hornblende tend to become gneissic and consist of an aggregate of elongate plagioclase and quartz with streaks of parallel hornblende. High-grade regional metamorphism results in the formation of pyroxene rather than amphibole and this is accompanied by a granoblastic texture.

The discussion above has assumed simple one-phase metamorphism, but the textures of most amphibolites are more complex due to repeated metamorphism. It is emphasized that even the amphibolites from Laxford and Buchan are polymetamorphic rocks.

The regional metamorphism of ultrabasic rocks has not been extensively studied. Original peridotites and pyroxenites consist mainly of olivine and pyroxene but some have been wholly or partially converted to a massive or schistose serpentinite prior to regional metamorphism.

Low-grade metamorphism converts the rocks into greenschists rich in such minerals as serpentine, chlorite, talc and magnetite. The rocks may be strongly foliated and also thinly banded because of metamorphic differentiation. Coarse crystalline ultrabasic rocks tend to be rather resistant to recrystallization and unaltered relict masses within a metamorphosed body are common. The simplest igneous schistose or massive serpentinite may be converted into a simple serpentine schist with a change of serpentine from chrysotile to antigorite. Shattered relics of spinel or magnetite and streaky aggregates of talc or chlorite are common. Post-tectonic crystallization produces octahedra of magnetite and spherulitic aggregates, fascicular growths or cross-cutting flakes of chlorite, etc. Some ultrabasic rocks rich in magnesite change to coarse, irregularly textured carbonate-rich schists. Parallel syntectonic needles, feathery crystals and post-tectonic, randomly oriented, cross-cutting, idioblastic porphyroblasts of amphibole (tremolite or anthophyllite) are common at low to medium grades. Talc schists derived from ultrabasic rocks (Vesasalo, 1962) have irregular textures. Talc forms pseudomorphs after amphibole; oriented overgrowths of colourless cummingtonite form rims on hornblende cores.

A medium grade of metamorphism of an ultrabasic rock produces massive or weakly layered amphibolite rather similar to a basic amphibolite, or to the more basic layers in a striped basic amphibolite, the difference being mainly in the lack of plagioclase and the persistence of chlorite.

Cordierite–anthophyllite rocks are formed by regional metamorphism of ultrabasics over a wide range of conditions from rather low- to high-grade with possibly small directed pressures. The rocks range from fine-grained and schistose to coarse-grained and granoblastic. Cordierite–anthophyllite rocks form directly from ultrabasic rocks or by the iron–magnesia metasomatism of gneisses, granulites, etc. (Turner and Verhoogen, 1960, p. 573).

Many apparently unmetamorphosed ultrabasic igneous rocks have deformation textures plus a dimensional or lattice preferred orientation of olivine or pyroxene and a considerable controversy has arisen over whether the textures are due to crystal settling and movement in a dominantly igneous environment, to flow during the intrusion of a crystal mush, or to the deformation of a solid body. The matter is too complex and the literature too extensive to pursue here.

Gneisses

A gneiss is typically light-coloured, coarse-grained, rich in feldspar, somewhat inhomogeneous and possessing some kind of foliation. It differs from a schist in being coarser (grain size about 2 mm), richer in feldspar and poorer in mica, but all gradations between the two are possible. Williams, Turner and Gilbert (1954) discarded the term gneiss and used "quartzo-feldspathic schist" but although "gneiss" is a broad imperfect term, it has a firm place in geological nomenclature. The terms "amphibolite–gneiss" or "basic gneiss" have been used for coarse-grained, dark-coloured, amphibole-rich rocks but the term gneiss is generally used for light-coloured rocks equivalent in composition to acid or intermediate igneous rock varieties.

The term *gneissic texture* is wide in its scope. It involves textural units such as parallel flakes of mica or prisms of amphibole, elongate crystals or ribbons of quartz, a wavy lenticular structure, aggregates of mica and feldspar, and a layering which may be partly mimetic after bedding or may be entirely tectonic and emphasized by metamorphic differentiation.

Approximately six, somewhat distinct, textural categories of gneiss have been recognized:

1. *Granite (diorite, etc.) gneiss:* A massive, homogeneous, light-coloured, granoblastic to porphyroblastic quartz–feldspar or feldspar rock with separate parallel flakes of mica, small micaceous streaks (schlieren) or prisms of amphibole.

2. *Banded gneiss:* A megascopically banded rock composed of alternate granoblastic (quartz and feldspar) and schistose (mica) layers.

3. *Augen gneiss:* An inhomogeneous, rather light-coloured rock with irregular lenticular banding and foliation due to layers, lenses and patches of differing grain size and composition, elongate quartz, parallel mica flakes and aggregates of flakes, and large *augen* of feldspar. The term "phacoidal" has been applied to flaky lenticles or augen-shaped mineral aggregates in schist and gneiss.

4. *Pencil gneiss:* A rock (related to the granulites) whose structure is dominated by a lineation rather than a planar foliation and which breaks into pencil-like fragments: the lineation is due either to the intersection of two or more coaxial foliations or to the presence of elongate aggregates of quartz and other minerals.

5. *Conglomerate gneiss:* A moderately to strongly foliated rock composed of elongate or flat fragments which are closely packed and separated by a schistose matrix. This metamorphosed conglomerate would generally be called *meta-conglomerate* but at high grades the relict clastic texture tends to be blurred by recrystallization and the rock is called a *conglomerate gneiss*.

6. *Non-foliated gneiss:* Very inhomogeneous, irregularly textured, contorted rock with a moderate to dark grey colour mainly because of fine-grained patches. It is composed of high-grade minerals such as sillimanite and garnet in addition to quartz, feldspar and mica, is texturally complex and typically, but not necessarily, polymetamorphic. The name *rock*, with suitable qualifiers (e.g. *sillimanite–garnet–quartz–microcline rock*), might be better than gneiss.

Gneisses are formed by a wide variety of processes. Those derived from igneous rocks are called *ortho-gneisses* and those from sediments are called *para-gneisses*. Although gneisses are generally regarded as high-grade, regional-metamorphic rocks this is not necessarily so; the name is given on petrographic and not genetic criteria.

Some special petrographic names are given to gneisses. A mineral prefix may be added to indicate some abundant or significant mineral, e.g. *sillimanite gneiss*. A *granitic gneiss* or *granite gneiss* is a gneiss of granite composition with a variety of possible origins. A *gneissic granite*, however, is fundamentally a granite with a superimposed gneissic texture.

The main genetic varieties are as follows:

(1) Ortho-gneisses formed by low-grade regional or dynamic metamorphism of coarse crystalline rocks.
(2) Ortho-gneisses formed by medium- to high-grade regional metamorphism of coarse crystalline rocks.
(3) Ortho-gneisses formed by protoclasis, e.g. syntectonic intrusion of magma with movement accompanying crystallization.
(4) Para-gneisses formed by high-grade regional metamorphism of sediments.
(5) Migmatites formed by high-grade regional metamorphism.

Many gneisses are low-grade regional (or dynamic) metamorphosed igneous rocks related to mylonites. The coarse grain and the mineralogy belong to an original igneous (or perhaps high-grade regional or plutonic) rock, and the foliation is due to limited deformation. The minerals show strain effects (bending, twinning, exsolution, alteration, fracturing). Recrystallization is minor and limited to some of the quartz and mica. The texture is due mainly to mechanical processes and there is no sign of equilibrium. Retrograde metamorphic changes are well developed.

Such gneisses belong to basement complexes which have been folded together with a thin cover of overlying sediments in orogenic zones. Many of the Alpine massifs contain such rocks and they occur also in the Lewisian basement of Scotland. Basement regions may undergo a considerable rise in temperature during orogeny and may become regionally metamorphosed along with the superstructure rocks. The "mylonitic" low-grade, non-equilibrium gneisses may thus grade into, or be replaced by, high-grade gneisses. The degree of approach to equilibrium and the consequent textures produced depend not only on the temperature attained but also on the rate of deformation and the presence of H_2O.

The regional metamorphism of acid to intermediate plutonic igneous rocks may produce gneisses or alternatively a series of schists rather similar to those produced from greywackes of similar chemical composition. At low grades there is pronounced granulation and reduction of grain size to produce phyllonites and schists. Potash feldspar tends to be sericitized and plagioclase to be sericitized or saussuritized; the mica initially forms pseudomorphs but these are destroyed by deformation and drawn out into streaks. Quartz is recrystallized to elongate crystals; biotite and hornblende are bent, kinked, fragmented, chloritized or recrystallized into schistose aggregates. The feldspar may only be partly altered (to a kaolinitic material) and remain as large shattered or strained augen with pressure shadows. Garnet may be present. The rock is a medium to coarse quartz–mica or quartzo-feldspathic schist at medium grades with a strong foliation.

The high-grade metamorphism of a granite may be progressive (i.e. it may have been reduced first to quartz–mica schist and then the micas changed back to feldspar at high grade) or it may be metamorphosed directly. In the latter case there is little driving force for mineralogical change because original igneous minerals such as quartz, orthoclase, plagioclase, biotite and hornblende are stable under the new metamorphic conditions. However deformation causes textural changes, chiefly to quartz and micas which recrystallize to give a dimensional and lattice preferred orientation. Feldspars may undergo slight changes such as exsolution, mechanical twinning or transformation from orthoclase to microcline. There may be slight chemical rearrangement to give replacement textures such as graphic or granophyric intergrowths in the marginal zones of feldspars or replacement perthites and antiperthites. Biotite may alter to garnet plus sillimanite as radial acicular aggregates. New hornblende or biotite sprout from iron oxides. Relict crystals of igneous plagioclase are recognized by their zoning, but the original igneous textures tend to be destroyed and it is rarely possible to distinguish on textural grounds between gneisses derived directly from granitic rocks, from sediments, or from various rocks by metamorphism plus metasomatism.

Nepheline gneisses range from coarse rocks of igneous appearance consisting of perthite, nepheline and biotite to banded and foliated varieties which are irregularly granular with parallel layers, lenses and clots of biotite flakes. The main minerals are microperthite, biotite and nepheline with subsidiary albite, ilmenite, magnetite, pyrite, zircon, green hornblende and rare corundum and secondary calcite, sericite, muscovite, cancrinite and zeolites. Nepheline tends to form lobate crystals and to be poikiloblastic towards perthite and biotite; it appears to have replaced perthite in places. Some rocks are granoblastic and others show abundant evidence of crushing, such as undulose extinction, deformation twins, bent lamellae, marginal granulation and mortar texture.

Bloomfield (1959), Sturt (1961) and Sturt and Ramsay (1965) considered that such rocks were formed by metasomatism of biotite–microperthite gneiss but there seems to be no *textural* way of distinguishing these from metamorphosed (but unmetasomatised) alkali syenites.

Ortho-gneisses formed by Protoclasis

Regional, dynamo-thermal metamorphism is produced in solid rocks by deformation plus crystallization. A similar combination is possible in igneous rocks which flow when they have partly crystallized or are deformed tectonically as they crystallize. The concept has arisen of normal or *subsequent* igneous bodies which are intruded after deformation and of *synchronous* bodies which are introduced into the crust during its deformation or even during an episode of regional metamorphism.

The term *protoclastic* (Brögger, 1890, p. 105) refers to structures produced by the grinding together of crystals in an igneous rock during its crystallization. Examples include:

1. Flow of partially crystalline magma past an igneous contact (wall of a dyke or pluton) to produce a protoclastic structure in granitic or porphyritic rock.
2. The intrusion of magma into an area which undergoes deformation and/or regional metamorphism while the magma is crystallizing.
3. The forcible injection of a partially crystallized magma, e.g. the hypothetical ultrabasic crystal "mush" lubricated with an intergranular fluid phase.

The second process has been widely postulated to explain the origin of granitic gneisses in high-grade metamorphic complexes. Gradations may be found from unmetamorphosed sediments through successive zones of regional metamorphism into schists, banded or veined gneisses thence into gneissic granite and finally massive granite. Arguments have been given at length as to whether this arrangement represents simultaneous magmatic intrusion and metamorphism or whether the granite is simply the end product of metamorphism plus metasomatism. It is emphasized that metamorphic convergence takes place at the high temperatures and pressures of regional metamorphism and little difference is possible in mineralogy or texture between the different hypothetical types. Arguments are largely based on gradations of various kinds from one rock to another but the interpretation of such gradations is highly subjective and liable to serious error. There is little reason to suppose that, once the dominant process has proceeded very far towards equilibrium, it is possible to distinguish *texturally* between an igneous granite which has been deformed while it crystallized, a granite which was regionally metamorphosed at high grade after it had solidified, and a sediment which has been regionally metamorphosed and granitized. The distinction between such rocks may be made if megascopic field criteria are available.

Protoclasis was once considered to be widespread (Miller, 1916; Balk, 1931) but the concept has become progressively less popular (Buddington, 1939; Pitcher and Read, 1963; Binns, 1966). One recently described example (Waters and Krauskopf, 1941) is a protoclastic gneiss which forms a strip $\frac{1}{4}$ mile wide of the border of the Colville batholith. The granitic rock has been converted into augen-gneiss, irregular gneiss with "swirled" foliation and mylonite (Waters and Krauskopf, 1941). The deformation is considered to have been due to a period of mechanical movement which interrupted the course of consolidation and partially granulated the early solidified components. The streaky matrix around the feldspar augen is not composed of a mylonitic powder but is crystalline and lacks cataclastic structure. It consists of a quartz–feldspar–biotite aggregate with long bladed crystals of quartz. A potash–feldspar-rich mesostasis binds crystals and fills cracks.

The textures are thus similar to those of some regional metamorphic gneisses or to some blastomylonites, and are due to both deformation and crystallization. As the rocks are part of an igneous body the crystallization could be attributed to igneous heat. The deformation may be tectonic and it is open to debate whether or not it is due to magmatic movements of a partially crystallized magma.

Metasomatic Para-gneisses

Homogeneous schists develop feldspar porphyroblasts and are converted into augen-gneisses, irregularly banded gneisses, granitic gneiss and finally granite when high-grade regional metamorphism is accompanied by alkali metasomatism. Extensive discussions and literature surveys have been given by Grout (1941), Read (1957) and Goodspeed (1959). The textural evolution of such rocks is very difficult to recognize and many studies of granitization and migmatization are so subjective as to be unconvincing and of doubtful value.

The terminology of these rocks is extensive. A *migmatite* is a rock which is believed to be of *mixed* composition, i.e. igneous and metamorphic material, and most are gneisses with banded or lenticular structure. Migmatites have been subdivided into *diadysites* (with cross-cutting veins), *embrechites* (with concordant veins and layers), *anatexites* and *nebulites* (nebular confused structure), and *agmatites* (migmatitic breccia with angular fragments and a crystalline "granitized" matrix).

These gneisses consist essentially of quartz, K-feldspar, plagioclase (near oligoclase), and biotite with possible accessory garnet, sillimanite, cordierite, etc. The textures tend to be complex (many are polymetamorphic) and have been interpreted as indicating feldspathization approximately simultaneously with deformation. In some gneisses of this type the feldspars form augen which are wrapped around by the foliation, are shattered and contain deformation twins. The feldspar therefore is pre-tectonic and it is almost impossible to prove unequivocally whether it has been introduced or not. The textures produced by syntectonic feldspathization are difficult to predict and the petrographic evidence given in the literature is conflicting. Even post-tectonic feldspathization producing cross-cutting, helicitic porphyroblasts is difficult to confirm as can be seen by following the discussion of the Cowal albite (Clough, 1897; Cunningham-Craig, 1904; Bailey, 1923; McCallien, 1929; Bailey and McCallien, 1934; Reynolds, 1942; Jones, 1961; and Bowes and Convery, 1966).

Feldspathization may take place syntectonically and be accompanied by metamorphic differentiation such that dark minerals tend to be concentrated in foliae with elongate quartz and feldspar layers between. A second possibility is that while the schist is held at high temperatures and pressures (at the same time as or later than the movements) feldspathic or quartzo-feldspathic material is introduced as a granitic fluid, as hydrothermal solutions or as diffusing ions. Movement of these materials is easiest parallel to the foliation and quartzo-feldspathic material crystallizes so as to give the impression of a "*lit-par-lit*" (bed by bed) injection. A third possibility is that the rock is held within the lower part of its melting range so that differential melting takes place and produces a melt which crystallizes to the quartzo-feldspathic layers. A fourth possibility is that the injection of fluid, diffusion, partial melting, etc., takes place post-tectonically.

These four origins may be considered to be distinct, but the processes which take place at the crystal level (and hence which affect the texture) are very similar in each case. There is convergence between the mechanisms of melting, solution and solid diffusion, and between recrystallization in the solid and crystallization of a melt, taking regard of the ultimate temperature, pressure and rate of cooling. In any case, post-tectonic recrystallization will be so extensive that most of the earlier textures will be destroyed. No textural way is known of recognizing whether a foliation due to parallel flakes or strings of mica in a gneiss is relict bedding, an early low-temperature mylonitic foliation, a layer formed by protoclastic flow of a partially crystallized magma, or a layer formed by deformation of a solid quartzo-feldspathic aggregate. An additional complication is that late stresses, perhaps completely unrelated to the main high-grade process, may superimpose cataclastic textures on the gneisses.

Textures are only a useful aid in the interpretation of a metasomatic rock if the replacement is simple, partial, and not confused by deformation. Even simple replacement textures such as rim feldspathization of quartz or compositional change in feldspar (rim alteration or replacement perthite) are open to misinterpretation. Many "basement" gneisses have a very complex history despite their superficial structural simplicity; e.g. the polymetamorphic Lewisian gneisses (Sutton and Watson, 1950; Ramsay, 1963) and the Carn Chuinneag and Inchbae gneisses (Harker, 1962).

Glaucophane Schists

The Glaucophane Schist Facies of Turner and Verhoogen (1960, p. 541) is defined by a mineralogical association of glaucophane with lawsonite, jadeite–quartz, aegerine and pumpellyite. The facies includes the derivatives of sediments (pelites, cherts, calcareous rocks and greywackes) and basic lavas, tuffs and intrusives (Coleman and Lee, 1963). They occur in the more highly deformed portions of post-Palaeozoic eugeosynclines where they are associated with serpentinites and feebly metamorphosed pillow lavas, spilites, keratophyres, radiolarian cherts and greywackes (Ernst, 1963). Metasomatic introduction of soda is common but not essential in the formation of glaucophane schists which belong to high-pressure, low- to moderate-temperature conditions.

As far as is known, textures of most glaucophane schists do not differ from those of the more common regional metamorphics, e.g. the transition from greywacke through semi-schist to quartz-mica-lawsonite schist belonging to the Glaucophane Schist Facies (Ghent, 1965). The different mineral species do not all crystallize simultaneously and some show paragentic sequences and evidence of polymetamorphism.

Some varieties, however, are most unusual in that an original sediment or igneous rock passes abruptly into a non-foliated or weakly foliated rock which is composed of new metamorphic (high pressure) minerals and thus has apparently been statically recrystallized but not deformed. Jadeite-bearing metagreywackes (Bloxam, 1956, 1960) retain their original clastic texture and bedding even though detrital albite has been replaced by jadeite which occurs as large crystals, aggregates of small crystals and as radiating aggregates. Associated dolerites retain an undeformed igneous texture even though the feldspar has been completely replaced by chlorite, lawsonite and jadeite. The relationships between these rocks and normal foliated glaucophane schists are not known.

Metamorphosed Ores

The origin of ore bodies is a subject of considerable debate and one which does not concern us here. However, it is clear that some were formed during metamorphism, that some have been metamorphosed along with the enclosing sediments and that some have been formed after the metamorphism of their host rocks.

It is emphasized that sulphides recrystallize very easily and even low-temperature deformation such as folding or faulting may be accompanied by recrystallization so that the textures of dynamically altered ores must be considered along with regionally metamorphosed rocks (Ramdohr, 1950;

Bastin, 1950, p. 73; Edwards, 1960, p. 32). Sulphides such as galena, sphalerite, chalcopyrite, etc., have low hardness, weak bonding, excellent glide properties, low melting points and relatively high solubilities in intergranular water. They undergo plastic deformation at low temperatures (Buerger, 1928; Lyall and Paterson, 1966) to give textures similar to those of cold-worked metals. The individual crystals are shattered, bent or twinned; *schistose ores* or *sulphide schists* have a dimensional and lattice preferred orientation (Uglow, 1917; Newhouse and Flaherty, 1930; Osborne and Adams, 1931). Minerals differ in their ability to glide, thus gneissic or layered ores may develop by metamorphic differentiation. A brittle mineral such as pyrite may be fractured and the broken fragments strewn along the foliation of elongate galena. These mineral features (typical of pre-tectonic crystals) may be strongly developed in a mass of sulphide even though the deformation has scarcely left any imprint on the host rock.

A sulphide ore body will be altered along with the enclosing rocks during regional metamorphism, but few studies (e.g. Stanton, 1960, 1964; Kalliokoski, 1965; Richards, 1966) have attempted to integrate the history of the ore and that of the country rocks. The readiness of the sulphides to recrystallize as they are deformed produces textures similar to hot-worked metal ("steely-galena"). Richards (1966, p. 19) proposed a two-fold division into *sulphide breccias* and *sulphide schists*. The gradation from breccia to schist (i.e. from brittle to plastic behaviour) can be correlated with the kind and intensity of deformation and also with the sulphides present. Sphalerite-rich rocks tend to be more brittle and galena-rich rocks more plastic. A dimensional and lattice preferred orientation may be produced. The gangue minerals will have a mineralogy appropriate to the grade although the behaviour of quartz, etc., when enclosed in a matrix of plastic galena is quite different from quartz in a quartzite, schist or gneiss. The quartz may have a fabric which is different from that in the country rocks or may have no preferred orientation at all.

A sulphide may have shared the successive tectonic phases of polymetamorphism but may retain no mineralogical or textural evidence of them. Criteria for recognizing polymetamorphism are well preserved in few rocks and only the last episode may be represented in the texture of the ore due to its readiness to recrystallize. There appears to be no great scope for the critical interpretation of the textures of polymetamorphic (regional) ores chronologically because complex crystallization deformation histories may only represent a very late episode. Post-tectonic crystallization is strongly developed in most ores and a granoblastic polygonal texture tends to replace the older directed textures. Stanton (1964) has discussed at length the equilibrium textures and grain boundary relations in recrystallized ores and casts considerable doubt on the validity of "paragenetic" criteria. As he pointed out, many texturally simple ores may have had a complex history and need a much more critical interpretation than they have been given in the past.

The sulphide minerals show a range of textures because of their different responses to conditions. Many galena textures result from deformation and annealing (Edwards, 1960, p. 39; Stanton, 1960, 1964, p. 71). Deformation of coarse aggregates produces megascopic contortion and bent crystals with slip bands and twins. Fine aggregates show pronounced flattening of grains and a schistose structure. Foliae of galena sweep around crystals of garnet and sphalerite or fragmented crystals of pyrite. Galena is easily polygonized, recrystallized and "annealed" to give a massive, apparently undeformed, aggregate with straight grain boundaries and equal-angle triple-points (Lyall and Paterson, 1966).

Stanton (1964, p. 72) has shown that the apparent paragenetic sequence (i.e. chronological order of crystallization of successive minerals) in many ore mineral assemblages may be explained (at least in part) by a grain boundary energy control in mutual crystallization textures. Textures in metamorphosed ores, as in metamorphosed rocks, are the result of the interaction of deformation, crystallization and recrystallization (Richards, 1966).

PLUTONIC METAMORPHISM

No clear distinction can be made between high-grade regional metamorphism and what has been called plutonic metamorphism. It is convenient here, however, to distinguish a large group of rocks such as the granulites, charnockites, and eclogites which are characterized by granoblastic rather than gneissic textures and also by a distinctive mineralogy and field occurrence.

Granulites

The exact definition of the term *granulite* has been the subject of considerable debate (Scheumann, 1961; Scharbert, 1963) and its use is now so vague as to render it almost valueless as a petrographic name, e.g.:

1. *Granulite* (*sensu stricto*) is a rock which is texturally and mineralogically similar to the original Saxony granulite (Lehman, 1884; Tyrrell, 1926). This is not as clear as it first appears because rocks originally called granulites from this area include quartzites, schists and normal gneisses.

2. *Granulite* (*sensu lato*) refers to all rocks belonging to the "Granulite Facies", i.e. the products of high-grade regional or plutonic metamorphism as defined by mineralogical associations (Turner and Verhoogen, 1960,

p. 553). These rocks have a variety of compositions and textures and include granulites (*sensu stricto*), charnockites, marbles, anorthosites, etc. Eskola (1952) stated that "all the rocks of the granulite facies are granulites".

3. *Granulite* is used in a very loose way to refer to any *granular or granoblastic* metamorphic rock, e.g. "Moine granulite". There is little to recommend this practice but it is common.

4. The term is used in an unusual sense by French petrologists to refer to muscovite granite but this use is probably unknown in English.

Additional difficulties in terminology have appeared due to different use of the terms granulitic, granulose, etc., in that they are variously used to indicate granular, or granulite-like textures, e.g. *granulose, granular, granoblastic* and *granuloblastic* are approximate equivalents which mean a texture composed of interlocking xenoblastic crystals of the same size (Holmes, 1928). *Granulate* (granulated, granulation) means to "crush" in a dynamic or cataclastic manner (Daly, 1917; A.G.I. Glossary) and *granulitic* has been used by Holmes (1928) in the same way. *Granulitic* however has also been widely used as an adjective to mean a texture like that of a granulite but this is too vague to be useful. *Granulitization* (Harker, 1939) is a hypothetical process stated to be midway between cataclasis and protoclasis which produces a *granulitic* texture. It would appear that the meaning of *granoblastic* is clearly accepted to mean granular, that *granulated* approximately means cataclastic, but that *granulitic* is too vague and ambiguous to be acceptable.

Eskola (1952) gave the following classification of granulites (*sensu lato*):

(1) *Light, massive granulite:* a granoblastic quartz–feldspar rock with sparse, large garnets ("garnet–granite").

(2) *Light, foliated granulite:* a quartz–feldspar rock with a little garnet, the foliation being marked by quartz plates (e.g. Plate XXXIa). This resembles the original Saxony granulite and is granulite *sensu stricto.*

(3) *Field granulite:* a foliated quartz–feldspar rock with streaks of biotite; cordierite may be present and garnets range up to 2 cm. (This includes rocks which have been called gneiss, cordierite–granulite and kinzigite.)

(4) *Granitic granulite:* a coarse-grained, patchy granitic rock with biotite, cordierite and large garnets ("quartz–biotite–cordierite–garnet rock").

(5) *Noritic and quartz-noritic granulite:* a grey plagioclase-pyroxene rock lacking garnet, more or less foliated (pyroxene granulite or basic charnockite).

(6) *Pyroxene–diorite and granodioritic granulite:* granular pyroxene–plagioclase ($\pm$ quartz) rock (acid and intermediate charnockite).

(7) *Ultrabasic granulite:* Patchy or banded pyroxenite and peridotite (ultrabasic charnockite).

A useful classification involves subdivision according to texture into granoblastic, foliated and porphyroclastic (cataclastic) types; according to the original nature, into ortho- (igneous) and para- (sedimentary) types; and according to the composition, into acid, intermediate, basic, ultrabasic, etc., types.

Granoblastic Granulites

In the past, the term *granulite* has been applied in a wide sense to metamorphic rocks (mainly metasediments) with a granoblastic texture which may not belong to the Granulite Facies. Goldsmith (1959) has suggested that the non-genetic name *granofels* be given to such rocks. They range from medium to coarse-grained and are composed of minerals such as quartz, feldspar, garnet, scapolite, mica or pyroxene. Most contain banding which may be partly relict after bedding and partly metamorphic and many are very weakly foliated. The texture in thin section is dominated by granoblastic and polygonal quartz, feldspar, scapolite and ferromagnesian minerals. Thin sections cut normal to the lineation show differently oriented micas whose [001] may form a girdle normal to the lineation.

The granoblastic texture may arise for three reasons:

1. The regional metamorphism may be the low-stress type which tends not to give rise to dimensional or lattice preferred orientations. These conditions are most favourable for the preservation of original sedimentary textures.

2. The syntectonic phase of normal regional metamorphism may be followed by prolonged post-tectonic crystallization which obliterates a dimensional preferred orientation and replaces it by a surface-energy controlled granoblastic-polygonal aggregate. The lattice preferred orientation may also be destroyed or it may be modified by the annealing. Original sedimentary structures would probably not be preserved.

3. Certain minerals—mainly feldspar, but also quartz, scapolite and calcite—display negligible to slight tendencies to form elongate crystals, i.e. the crystals do not tend to grow much more quickly in one lattice direction than another. Pyroxene behaves similarly and although it forms somewhat elongate crystals the tendency is much less pronounced than in amphibole, mica, epidote, etc. Recrystallization is controlled by the ability to nucleate new crystals, thus instead of a quartz crystal being drawn out into an elongate grain or aggregate, it breaks down and forms several equant grains. Feldspar and pyroxene tend to be rather insensitive towards stress and lattice preferred orientations are not common; the dimensional preferred orientation in some rocks is probably a purely mechanical one due to the alignment of somewhat elongate pyroxene or feldspar by the matrix.

Granoblastic texture is also favoured by a lack of mica which tends to control quartz boundaries (so that quartz grains become elongate), which imparts an inhomogeneity to the rock, and which enhances slip along the foliation.

The Moine granulites (Plate XXXIb) consist of a granoblastic aggregate of quartz and feldspar with scattered sub-parallel mica flakes and mainly belong to the Amphibolite Facies. There is a dimensional and lattice preferred orientation of mica which is partly due to mimetic growth along bedding and which is partly tectonic. A lattice preferred orientation of quartz has a strong orthorhombic fabric (Phillips, 1937). The texture appears to be due to a combination of circumstances; low strain (sedimentary structures are commonly preserved), pronounced post-tectonic crystallization, and the abundance of feldspar and to a lesser extent of quartz (these meta-psammites might well have been called impure quartzites).

Edwards and Baker (1954) described banded scapolite–pyroxene "granulites" from the Cloncurry district in Queensland which are very similar to scapolite–diopside rocks from the Kanmantoo Group near Adelaide, South Australia (White, 1959) but which do not belong to the Granulite Facies. The texture is equigranular and polygonal with straight grain boundaries of scapolite and pyroxene meeting at equiangular triple-points and many grain boundaries parallel to rational lattice planes in both scapolite and pyroxene. The texture is similar to that of a hornfels with an equilibrium texture achieved by interfacial energy balance. The strong banding, giving layers rich in scapolite, pyroxene or amphibole, is mimetic after original bedding and appears to have been emphasized by metamorphic differentiation. There is little evidence of a dimensional preferred orientation but some scapolite has a lattice preferred orientation.

Charnockites belong to the group of granoblastic to foliated granulites. The original charnockite (named after Job Charnock, the founder of Calcutta, from whose tomb the rock was first described) is a particularly coarse-grained rather granular rock composed of xenoblastic to subidioblastic crystals of quartz, potash feldspar and hypersthene, equivalent in composition to a hypersthene granite. The rock is dark in colour, the quartz being blue and opalescent, and the feldspar brownish-grey. Holland (1900) extended his term to define a *Charnockite Series* which is a group of crystalline rocks consisting of hypersthene in addition to various proportions of quartz, potash–feldspar, plagioclase, garnet, etc., and analogous to the acid to ultrabasic igneous rocks, granite, granodiorite, diorite, norite and hypersthenite.

Rocks of the charnockite series throughout the world are not identical and probably have a number of different origins (Groves, 1935; Pichamuthu, 1953; Parras, 1958; Wilson, 1960; Howie, 1964) including: intrusion of a range of igneous rocks with concomitant protoclasis and later regional metamorphism, primary crystallization from a dry magma, plutonic metamorphism of intrusive igneous rocks, magmatic crystallization under plutonic conditions, high-grade regional metamorphism and metasomatism of sediments and igneous rocks, slight contamination and differentiation of dry magma, magmatic intrusion followed by consolidation and then regional metamorphism.

Although there is considerable doubt as to the origin of charnockites, at least some were igneous rocks whose present texture is due to an incomplete metamorphic process involving slight deformation and slight to possibly profound recrystallization under deep-seated (high-temperature and pressure) dry conditions. The textures of some charnockites still show their fundamentally igneous character (chilled margins, ophitic texture, zoned plagioclase) but most are acid to intermediate gneisses or granular rocks with a texture which could be interpreted either as entirely metamorphic or plutonic-igneous slightly modified. Many show extreme dynamic effects (crushing, granulation) and some have the typical elongate quartz ribbons of the foliated granulites.

The "granitic" charnockite is a coarse-grained massive or banded rock. In thin-section the texture is rather irregular with coarse xenoblastic feldspar, xenoblastic quartz or strongly elongate quartz-lenses cutting across other minerals. The potash feldspar is strongly perthitic (antiperthite is also common) and its relatively large size tends to give the rock a porphyroblastic aspect. The feldspar is undulose and marginally granulated and orthoclase appears to have been converted to microcline by deformation. Some large crystals are poikiloblastic and some are rimmed with or penetrated by "cauliflowers" of myrmekite. Plagioclase (generally oligoclase) which is abundant in granodioritic types (*farsundite*) is bent and a significant proportion is untwinned. Large shapeless crystals of garnet may be present. The ferromagnesian mineral is typically hypersthene, generally as clusters of somewhat prismatic crystals; alteration to a symplectic intergrowth of garnet and quartz is common.

The "syenitic" charnockite (*mangerite*) is a medium- to coarse-grained rock with an "allotriomorphic to hypidiomorphic granular" texture composed of an aggregate of coarsely perthitic potash feldspar and pyroxene. The simplest "dioritic" type (*enderbite*) is similar but is composed of plagioclase and a pyroxene which is typically hypersthene although a diopsidic variety may occur as well. The potash feldspar is microcline and has coarse perthitic structure. Both perthite and antiperthite occur; the regular varieties appearing to be due to exsolution although irregular vermicular intergrowths and "cauliflowers" of two feldspars or of feldspar and quartz may be due to replacement; much of the plagioclase is untwinned.

The texture is characteristically granoblastic with interlocking grain boundaries but the following variations are common: intergranular crushing and mortar structure accompanied by deformation twins in plagioclase; syntectonic crystallization and the development of a gneissic texture; intergranular graphic or granophyric quartz–feldspar intergrowths due to slight recrystallization or melting; intergranular reaction rims and coronas.

Pyroxene may be altered to hornblende or to garnet plus quartz; synantetic layered coronas form at some pyroxene–plagioclase contacts to give hornblende and quartz layers. Garnet not only occurs in symplectites, but forms pseudomorphs of hypersthene.

The basic charnockites range from noritic rocks of igneous appearance (i.e. ophitic texture) in which metamorphism has only caused bending of crystals, twinning in plagioclase and the formation of coronas, to coarse, banded, granoblastic aggregates of hypersthene (with clinopyroxene, etc.) and basic plagioclase representing complete recrystallization. Hypersthene-rich varieties are generally termed "basic charnockite" and garnet-rich varieties are termed "pyroxene granulite".

The ultrabasic charnockites are relatively rare. Almost unaltered peridotites and pyroxenites occur as masses within charnockite terrains but some completely recrystallized ultrabasic granulites occur as layers or patches in more acid charnockites or gneisses. Coarse granoblastic hypersthenite is typical (Joplin, 1968, Figs. 49, 50). It seems likely that at least some of this material results from metamorphic differentiation during the formation of banded charnockites from intermediate or basic igneous rocks.

It is convenient to discuss the *anorthosites* together with the charnockites and granulites. Anorthosite is generally classified as a basic igneous rock consisting essentially of large to very large crystals of plagioclase; if subsidiary orthopyroxene and clinopyroxene are present the rock grades into norite; if olivine is present then it grades into troctolite. Some anorthosites form part of stratiform igneous complexes, have igneous textures and contain a bytownite–labradorite plagioclase (e.g. Bushveld and Stillwater types). Other anorthosites occur in very large Precambrian batholithic bodies associated with rocks belonging to the Granulite or Amphibolite Facies. Anorthosites range in texture from granoblastic to foliated (parallel, elongate plagioclase prisms) to cataclastic (bent crystals, mortar texture, deformation twins). Intergranular crystallization results in coronas and graphic or myrmekitic intergrowths.

Igneous plagioclase has a tabular shape (the laths may be parallel), a composition of labradorite to bytownite, large size, is clouded or poikilitic, has albite and albite–Carlsbad twinning and compositional zoning. Metamorphic plagioclase has the composition of andesine, is unzoned, free of inclusions, twinned on albite and pericline laws, and forms medium grained interlocking granoblastic aggregates.

The origin of anorthosites is not clear (Turner and Verhoogen, 1960, p. 230) and the cataclastic textures may be attributed to protoclasis, to dynamic or to regional metamorphism of solid anorthosite of igneous origin. The association of anorthosites with other rocks showing obvious metamorphic features favours the third explanation.

Foliated Granulites

The original granulites from the "granulitgebirge" of Saxony (Lehmann, 1884) are light-coloured, medium- to rather fine-grained rocks consisting chiefly of shapeless quartz and feldspar, with possible pyroxene, kyanite and garnet and accessory sillimanite, garnet, spinel and rutile. The texture is characteristic; the rocks are strongly banded due to the streaky elongation of grains or aggregates of quartz in rods or ribbons. The rocks are associated with granular pyroxene-rich types, cordierite gneiss, quartzites, etc. The original granulite belongs to the modern Granulite Facies and contains a typical high-temperature, moderate- to high-pressure, anhydrous mineral assemblage. The term *leptynite* has been used for granulites of the Saxony type with strong banding and *leptite* for fine-grained rocks consisting chiefly of quartz and feldspar (but ranging almost from quartzites to amphibolites) associated with granulites.

The texture of the true granulite contains three elements (Plate XXXIa):

(1) It is dominated by parallel, rod-like, lenticular aggregates of coarse-grained quartz with the individual grains elongated in the same direction. These aggregates have a ribbon-like appearance in the two dimensions of a thin section and will be referred to here as *ribbons*.

(2) The ribbons are set in a rather fine-grained polygonal matrix of quartz and feldspar.

(3) Isolated crystals of garnet and kyanite occur in the fine-grained quartz matrix.

There is a strong lattice preferred orientation of quartz and a simple orthorhombic symmetry of both mesoscopic structures and quartz fabric in some granulites. Sahama (1936) considered that the texture of most granulites developed in two phases; the megascopic foliation and lineation in the first, and the quartz preferred orientation in the second. The work of Eskola (1952) shows that the main deformation was precrystalline (ibid., p. 149), that there is evidence of polymetamorphism (garnet with pressure shadows, wrapped around by biotite) and that at least some of the kyanite, sillimanite and garnet is post-tectonic with respect to the formation of the quartz ribbons.

No satisfactory explanation of the granulite texture has been advanced but two things seem clear: first, the rocks have been deformed, the inhomogeneity being probably at least partly due to the deformation, and second, the grain shapes and boundary relations indicate considerable post-tectonic crystallization. The texture is due to some combination of deformation and crystallization but differs markedly from that of compositionally similar quartzites and gneisses in the contrast between the quartz of the two main textural elements. The time relations of the two main elements are not known and a number of alternatives are possible. It is suggested that the texture is due to strong deformation and cataclasis followed by post-tectonic crystallization and that the rock is a kind of blasto-mylonite, i.e. the rock was first mylonitized to an inhomogeneous deformed aggregate and then annealed at high temperatures and confining pressures. The ribbons are relics of the mylonite structure but the dimensional and lattice preferred orientations are due mainly to annealing of the "cold-worked" aggregate.

The position and orientation of the grain boundaries (and hence the shapes of the grains) in the ribbons is strongly controlled by the outer boundary of the ribbon itself (Plate XXXIa). Quartz–quartz boundaries tend to be perpendicular to the ribbon boundary; the ribbons consist of a few very large crystals which are necessarily elongate; the quartz crystals in adjacent ribbons have similar orientations and ribbons appear to have coalesced to give large single crystals.

It is suggested that the quartz in the ribbons was crystallized post-tectonically and that single crystals or coarse aggregates replaced some kind of finer-grained quartz. Crystals in the ribbon grew by moving their grain boundaries into adjacent crystals and as the ribbon boundary prevented growth laterally, all movements took place parallel to the ribbon so that all boundaries tend to be perpendicular to the ribbon boundary. Lattice preferred orientations could be produced during this annealing by the higher mobility of certain boundaries causing the disappearance of unwanted crystal orientations.

Eclogites

Eclogite (*sensu stricto*) is a rock composed of the pyroxene omphacite (a jadeite–diopside mixture) and a pyrope-rich garnet. Plagioclase is typically absent (all of the soda and much of the alumina being in the pyroxene). Similar rocks consisting of garnet plus a pyroxene which is not omphacite (diopside, chrome-diopside, clinopyroxene, enstatite) are called *garnet pyroxenites*.

The presence of significant amounts of other minerals allows subdivision into varieties (Coleman *et al.*, 1965), e.g. kyanite–eclogite and amphibole–

eclogite. Accessories include rutile, quartz, amphibole, clinozoite, muscovite, diamond, etc.

Eclogites have the bulk chemical composition of a basic igneous rock and are the main representatives of the much-debated Eclogite Facies. They have four principal modes of occurrence (Eskola, 1921, 1939; Yoder, 1950; Turner and Verhoogen, 1960; Coleman *et al.*, 1965):

Group I (a) Nodules, xenoliths or inclusions in kimberlite, tuff, breccia, etc.
(b) Lenses or layers in peridotite.
Group II Lenses, inclusions or layers in glaucophane schists.
Group III Lenses or layers in regionally metamorphosed schists, gneisses and granulites.

Eclogites from different environments differ in mineralogy, bulk composition and type of secondary alteration. They show a wide range of textures (Eskola, 1921).

(1) *Granoblastic.* Massive or granular eclogites have granoblastic texture, possible strong compositional layering, a tendency for garnet to be larger than pyroxene and a tendency, in some rocks, for pyroxene, amphibole, zoisite and mica to be slightly elongate and parallel. This is by far the most common texture in eclogites (Glenelg, Norway, Tasmania, Spain, etc.).

(2) *Foliated.* Accentuation of the dimensional preferred orientation gives elongate, parallel laths of pyroxene, amphibole (zoisite, kyanite and mica), lenses of quartz and garnet aggregates (California, New Caledonia, Alps, Colombia).

(3) *Cataclastic.* Uncommon very fine-grained rock containing porphyroclasts in a very fine-grained xenoblastic matrix with possible fine compositional layering (Saltkjael Type, Eskola, 1921).

The rare "Gryting type pegmatitic eclogite" of Eskola (1921, p. 49) differs considerably in mineralogy and texture. It is a very coarse (up to 5 mm crystals), irregularly textured aggregate of enstatite, diopside and garnet with carbonate and mica. Its genetic relationships to other eclogites are not clear.

The most common (group III above) are massive, medium-grained rocks. Many have a vague compositional layering (apparently due to metamorphic differentiation) and many unaltered eclogites show few signs of deformation. The texture is a simple granoblastic polygonal aggregate of interlocking pyroxene and garnet. Grain boundaries between crystals of the same species tend to be straight, parallel to rational lattice directions, and to meet in equi-angle triple-points. Garnet differs from pyroxene in that it has a slightly greater tendency to form crystal faces. Primary amphibole, where present, tends to form large poikiloblastic crystals almost twice the size of pyroxene and garnet. Mica forms distinct flakes which may give a weak foliation by their parallelism. Quartz typically forms xenoblastic intergranular crystals, rutile occurs as small subidioblastic to idioblastic crystals and kyanite as randomly oriented subidioblastic laths.

Some eclogites are porphyroblastic with large crystals of garnet, pyroxene or hornblende. Some show a strong dimensional (and no doubt lattice) preferred orientation of pyroxene and/or mica, and the foliation may be marked by long grains of pyroxene or lenticular aggregates or layers of garnet grains.

Eclogites from glaucophane-schist terrains are different from "normal" eclogites in many ways. They do not have the attractive appearance of some European eclogites with their large red garnets in a matrix of coarse apple-green pyroxene and the Californian eclogites tend to be dark, massive, rather fine-grained rocks. The microscopic texture is somewhat irregular with garnet as porphyroblasts (xenoblastic to idioblastic) in a foliated equi-granular aggregate of pyroxene with lesser glaucophane, lawsonite, epidote, clinozoisite, muscovite and various secondary minerals. Some rocks are coarsely banded with compositional layering.

Eclogite textures are not as simple as they first appear and it is difficult to decide which minerals are primary and which are secondary. Associated glaucophane schists, pelitic schists or gneisses may show evidence of poly-metamorphism, and although the evidence is not clear, all "primary" eclogite minerals do not appear to have crystallized simultaneously during one phase of metamorphism (contrary to the views of Coleman *et al.*, 1965, p. 491). Some garnet and glaucophane appear to have crystallized after the finer-grained pyroxene-rich matrix. Quartz-filled pressure shadows around garnet (ibid., p. 491) in an eclogite from New Caledonia strongly suggest that this garnet is older than the surrounding omphacite–glaucophane aggregate. Idioblastic garnet in some Norwegian eclogites appears to have formed later than the pyroxene. Garnet commonly has a spongy core with numerous inclusions and a massive idioblastic rim (suggesting two stages of growth).

The alteration of eclogite is characteristic and partial or total retrograde conversion into amphibolite or greenschist is almost universal (Eskola, 1921). Four kinds of alteration have been recognized.

(1) A superficial alteration (mainly of pyroxene to finely fibrous amphi-bole) along widely spaced fractures.

(2) A characteristic kelyphitic alteration and amphibolization which is probably universal for eclogites associated with mica-schists and migmatites. *Amphibolization* is the conversion of some mineral (generally pyroxene) to amphibole. *Amphibolitization* is the conversion of some rock (e.g. eclogite)

to amphibolite. The common alteration first appears as rim alteration of pyroxene which develops fringes of fibrous brown amphibole, while muscovite–quartz boundaries develop biotite fringes. The pyroxene develops coronas of "kelyphite" and alters more quickly than garnet which may persist as relics. Pyroxene and garnet appear to react with each other and, with the addition of water of presumably external origin, give reaction rims, synantetic aggregates and irregular symplectic intergrowths of plagioclase, hornblende and garnet. The regular multi-layer complex coronas of the Granulite Facies do not appear to be formed in eclogites. Kyanite may be sheathed by finely fibrous, indefinite kelyphite or replaced by granular aggregates of plagioclase, zoisite or clinozoisite. Secondary plagioclase, hornblende, garnet and quartz form fine-grained, irregularly textured, almost unrecognizable growths or at medium grades crystallize into a medium-grained, foliated aggregate to give a normal-looking garnet–amphibolite. Low-grade retrograde metamorphism converts the eclogite into a fine-grained foliated aggregate of albite–epidote–actinolite–chlorite, etc., i.e. a greenschist, or at high confining pressures into a glaucophane schist.

The textural relationships show the relative times of formation of omphacite, garnet, secondary hornblende, chlorite, etc. and indicate that eclogite was converted to amphibolite, not that a basic rock was converted partly to eclogite and partly to amphibolite simultaneously.

(3) Eclogites associated with glaucophane schists lack coronas and kelyphitic growths. Garnet alters to chlorite, lawsonite, glaucophane and sphene. Pyroxene alters to glaucophane. Secondary minerals form by rim alteration but also as pseudomorphs and separate large cross-cutting crystals. Eclogite changes to glaucophane schist.

(4) Eclogite and kyanite–eclogite associated with Lewisian gneisses and schists at Glenelg (Alderman, 1936; Tilley, 1936a, b) have undergone extensive retrograde metamorphism at moderate grades. In addition to the development of coronas and kelyphite which are typical of amphibolitization, these rocks have been converted to "streaked eclogites". The eclogites have been deformed to produce a foliation, and migration of quartzo-feldspathic material has produced streaks, layers or "venules". The amphibolitized eclogite is cut by parallel layers of finely granular albite, oligoclase and quartz with accessory biotite, hornblende, kyanite and clinozoisite.

CHAPTER 11

Textures of Polymetamorphism

MANY rocks contain evidence of one episode of metamorphism following another, e.g. schist which has been intruded by a granite shows the effects of thermal metamorphism imprinted on those of regional metamorphism. *Repeated* metamorphism or *polymetamorphism* refers to several episodes of the same kind of metamorphism. *Retrograde* metamorphism is a lower grade process affecting a higher grade rock, e.g. a pelite is converted to a sillimanite gneiss during one phase then retrogressively altered to a sericite–chlorite phyllonite during a later phase.

THERMAL FOLLOWING DYNAMIC OR REGIONAL

This is a common occurrence as many regionally or dynamically altered terrains have been intruded by later igneous masses. The process is analogous to annealing, in that heating affects rocks composed of strained crystals. The net effect is to replace strained crystals by unstrained aggregates, to replace a directed or foliated texture by a random one, and to replace the earlier metamorphic minerals by an assemblage stable under the new conditions.

A reduction in grain size is common because polygonization of strained crystals produces a granular aggregate of unstrained ones (Plate XXVIc). A foliation may be preserved as compositional banding and there may be mimetic crystallization of micas along an old foliation to produce a "schistose hornfels".

The most conspicuous effect of thermal metamorphism of schist (Plate XXIVa) is the "complete recrystallization of the fine-grained schist to a compact hornfelsic mosaic, with the destruction of all schistosity" (Chapman, 1950, p. 206). Porphyroblasts grow across the foliation and produce a knotted (maculose) appearance. The foliation is rarely preserved helicitically and tends to be destroyed by the development of a granular or decussate texture.

Knotted slates, *garbenschiefer*, chiastolite slates, etc., are the result of thermal metamorphism of foliated pelites. The randomly oriented porphyroblasts have a post-tectonic relationship to the foliation. Thermal metamorphism converts foliated metaquartzite into coarse granular quartz–hornfels (quartzite). All traces of dimensional preferred orientation of quartz may be obliterated but a lattice preferred orientation may be retained. Greenschists and amphibolites are converted to massive dark hornfelses with the foliation preserved only as layers of granular pyroxene or amphibole.

Harker (1939) considered that one effect of thermal metamorphism superimposed on regional was the replacement of "stress" by "anti-stress" minerals. The stress–anti-stress distinction appears to have little validity but low-pressure (and high-temperature?) thermal minerals may replace high-pressure (and lower-temperature?) regional minerals. Thus sillimanite or kyanite may be pseudomorphed by andalusite while porphyroblasts of cordierite, garnet, etc., develop.

An extreme example of metamorphism of this kind is the thermal metamorphism of mylonite, e.g. the "sheared basic hornfels" of Spry (1955, p. 162). Sutton and Watson (1959, p. 10) described mylonites in which the matrix is fine-grained and granular with a texture which is "often more like that of a hornfels than of a rock formed by cataclasis" suggesting that "a period of mechanical granulation may have been followed by crystallization whose final stages took place under more or less static conditions". This post-tectonic crystallization is not a contact effect but seems to be in some way related to the formation of shear belts.

REGIONAL METAMORPHISM SUPERIMPOSED ON THERMAL

The dynamic aspects of regional metamorphism and the formation of a foliation tend to destroy the granular thermal-metamorphic textures just as any pre-metamorphic texture is destroyed, and examples of rocks which can be shown to have undergone regional metamorphism after thermal are rare. The classic example quoted by Harker (1939, pp. 341–3) is the aureole of the Carn Chuinneag granite in Ross and Cromarty. He considered that the intrusion of the granite produced a mass of hornfels which resisted the later regional metamorphism so that some rocks retained their hornfelsic character while others were partly or wholly converted into schist. Kyanite aggregates form pseudomorphs after chiastolite and cordierite. Later work (Harker, 1962; Long and Lambert, 1963) now suggests that in the hornfels zone there was first regional metamorphism to garnet–mica schist, then intrusion of the granite giving thermal metamorphism and producing andalusite and cordierite-bearing hornfelses, finally, regional metamorphism to kyanite grade.

RETROGRADE METAMORPHISM

Many metamorphic rocks contain evidence of retrograde mineral changes, i.e. alteration of higher "grade" minerals to lower ones. Many of these changes involve hydration and are considered to be the result of a decrease in temperature, increase in P_{H_2O}, or both. Examples include the alteration of garnet or biotite to chlorite, feldspar or various aluminium silicates to white mica (sericite, pinite), pyroxene to amphibole, olivine to serpentine, and plagioclase to albite plus epidote, zoisite or calcite. The secondary minerals form fibrous fringes around, inclusions within, and pseudomorphs of, the primary minerals. Many *coronas* or *reaction rims* in metamorphic rocks are formed by retrograde processes.

Retrograde metamorphism is normally a special (and common) case of repeated regional metamorphism where a lower-grade episode is superimposed on a higher grade, or where dynamic metamorphism is superimposed on regional (Knopf, 1931, p. 7). Many of the textural changes are discussed in sections on dynamic metamorphism and repeated regional metamorphism. Becke (1909) used the term *diapthoresis* (derived from the Greek root meaning "to destroy") more or less synonomously with retrograde metamorphism and the rocks formed by this process have been called *diapthorites.*

Metamorphism may convert coarse- or medium-grained, granular, gneissic or schistose rocks into fine-grained foliated aggregates to produce the *phyllonite.* The rock has the appearance of a phyllite but its texture is due to processes somewhat related to mylonitization but involving some recrystallization at low temperatures (Knopf, 1931). Although many phyllonites are the products of retrograde metamorphism this is not always strictly true because some phyllonites have been derived from igneous rocks; hence the phyllonitization is not really retrograde because the original coarse grain is not due to early metamorphism.

The textures of most phyllonites are complex, are dominated by mechanical effects and resemble those of mylonites. The rocks are foliated and thinly banded, rich in parallel mica flakes, and may contain several foliations of different ages. Knopf (1931, p. 18) considered that phyllonitization does not produce a new foliation but simply transposes and disrupts earlier foliations but there is no reason why this is necessarily so and many phyllonites contain a late growth foliation as well as old relics. A lenticular texture is typical; adjacent lenses are commonly of different sizes although the constituent quartz grains may be all of a similar size. The individual quartz crystals in a lens commonly show a sub-parallel optic orientation. Quartz occurs as elongate, lenticular, undulose crystals, along with fractured and sericitized feldspar, saussuritized plagioclase, sericitized sillimanite or kyanite, chloritized biotite, and ilmenite converted to leucoxene.

Retrograde metamorphism may take place under conditions of low directed stresses to produce rather massive rocks rich in small non-parallel mica flakes.

REPEATED REGIONAL METAMORPHISM (POLYMETAMORPHISM)

The textures of many metamorphic rocks are the result of the complex interaction of deformation and crystallization. Deformation acts in a negative way to break the rock down and reduce its grain size whereas crystallization acts in a positive way to build up new crystals and increase the grain size. The two processes may operate quite independently, for example when recrystallization follows deformation in the thermal metamorphism of a mylonite, or deformation follows crystallization when a gneiss is mylonitized. The two are considered to act together in the classical view of regional (dynamo-thermal) metamorphism, in the aureoles of synchronous granites, and in the protoclasis of moving, partly crystallized magma.

Modern work on a variety of regionally metamorphic provinces shows that the classical view of the simultaneous interaction of crystallization and deformation in regional metamorphism is an over-simplification. Many regional metamorphic provinces are not simple but have undergone several episodes of metamorphism (*polymetamorphism*) in which deformation and crystallization have acted independently. Some metamorphic events can be divided into separate phases or episodes. A rock may be deformed in several separate phases while it is held at an elevated temperature so that the duration of a *deformation phase* (*tectonic episode*) is much shorter than that of the whole *metamorphic event*.

The structures of these rocks are complicated because they contain folds, lineations and foliations of different ages. The microscopic textures of the rocks indicate the same metamorphic history as the mesoscopic structures. The techniques for analysing the textures of these rocks were first outlined by Sander (1930) but early attempts are of doubtful validity because of unreliable criteria. Little interest appears to have been aroused at the time by the important work of Turner and Hutton (1941) and Read (1949 in 1957) but chronological analysis was established as an invaluable tool by the work of Rast and Zwart.

Recognition of the importance of this approach is shown by the number of papers in the last ten years. These include Rast (1958, 1963, 1965); Rast and Sturt (1957); Harris and Rast (1960a); Zwart (1960a, b, 1963); Johnson (1961, 1962, 1963); Sturt and Harris (1961); Chatterjee (1961); Spry (1963a); Ramsay (1963) and Binns (1964).

The chronological analysis of crystallization and deformation consists of the interpretation of the time relations of various aspects of deformation (such as the formation of a foliation, fold, lineation or of a bent, crushed or twinned crystal) and of various aspects of crystallization (such as the growth of new minerals or the recrystallization of existing ones). This depends on the determination of:

(1) The time relations between different aspects of deformation, e.g. between different foliations or folds.

(2) The time relations between the crystallization of each mineral and some evidence of strain, generally a foliation but also a lineation, fold, or some pattern of preferred orientation,

(3) The relative time of crystallization of each mineral.

TIME RELATIONS BETWEEN DIFFERENT DEFORMATION EPISODES

The existence of a tectonic episode is recognized by its strain effects. On the large scale these will constitute macroscopic fold systems, foliations and lineations. In a region where a number of episodes have occurred they are labelled F_1, F_2, F_3, etc., in order of occurrence. An axial surface foliation and axial lineation are generally formed at the same time as the fold in which they lie and a series of foliations can be labelled S_1, S_2, S_3 and lineations L_1, L_2, L_3.

A common practice (Turner and Weiss, 1963, p. 86) is to label bedding S_1 so that the oldest foliation recognized (formed possibly during F_1) is S_2 and the oldest lineation (possibly formed during F_1 but also possibly during F_2) is L_1. This has the advantages of being simple and descriptive. It is preferred here to take a strongly genetic approach and to emphasize the chronological aspect of the nomenclature by giving the same numerical suffix to all structures formed during one tectonic episode. Thus S_0 is bedding and L_0 any sedimentary lineation. S_1 and L_1 are formed during the first tectonic episode F_1; S_2 and L_2 are formed during the second tectonic episode F_2 and so on. As a number of different kinds of lineation can be formed in one episode, e.g. mineral elongation, intersection of bedding with axial plane cleavage, rodding, mullions, etc., these can be given an additional suffix, 1, 2, 3, etc., after that denoting the tectonic episode. Thus lineations formed during F_1 can be denoted L_{11}, L_{12}, etc. Similarly different kinds of foliation formed simultaneously can be denoted S_{11}, S_{12}, S_{13}, etc.

The metamorphic phases may be designated M_1, M_2, etc., in order of occurrence and each may be further subdivided e.g. *MS*1 for the syntectonic part of the first metamorphic phase, *MP*2 for the post-tectonic part of the second metamorphic phase. It may be possible to equate M_1 to F_1 (e.g.

Sturt and Harris, 1961, p. 691) but in some areas the first metamorphism occurred during the second tectonic event (i.e. $M_1 = F_2$).

Distinguishing between the microscopic expression of several tectonic episodes may be difficult or even impossible and may involve detailed petrofabric analysis which will not be discussed here. It is best to cut thin

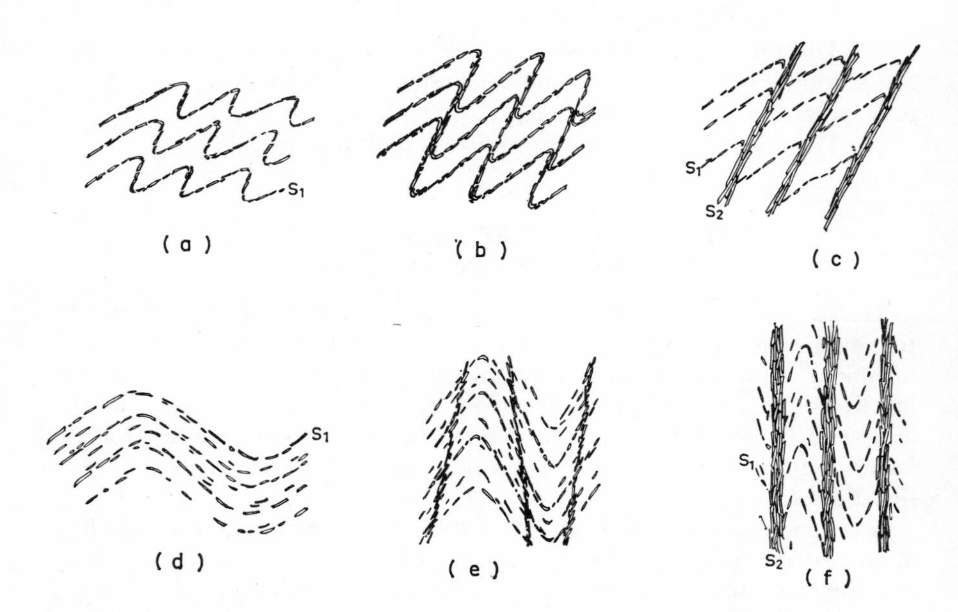

FIG. 62. Two foliations. Development of a second planar foliation S_2 during the deformation of the first foliation S_1. (a), (b) and (c) successive stages in asymmetrical folding. (d), (e) and (f) successive stages in symmetrical folding

sections so that structures whose relations are known macroscopically can be recognized under the microscope to allow all information to be integrated. As a general rule, later structures are less deformed than earlier ones and cut across them (Plate XXIXc). Common relations between two foliations of different ages are shown in Fig. 62.

Some such structures (Voll, 1960, pp. 544–64) are thought to result from the formation of two foliations simultaneously; continued deformation is supposed to leave one essentially planar while the other is rotated and becomes folded or dismembered. This process is probably very uncommon; the foliations S_1, S_2, S_3, etc., in Fig. 64 differ petrologically in the minerals developed.

TIME RELATIONS OF MINERAL CRYSTALLIZATION TO SOME EVIDENCE OF DEFORMATION

The crystallization of a given mineral is said to be *pre-*, *syn-*, or *post-tectonic* according to whether it took place before, during, or after some tectonic event. The terms *pre-*, *para-*, and *post-crystalline deformation* have a similar meaning but different emphasis. Criteria for recognizing these relations are given on pp. 251–9 and Figs. 58–61.

Time Relations of Minerals to Each Other

It is difficult to find unequivocal evidence of the relative ages of minerals in metamorphic rocks from their mutual relations alone.

Minerals which crystallized at the same time *may*:

(1) Show no alteration or reaction at mutual contacts.
(2) Be intimately intergrown.
(3) Have related preferred orientation fabrics.
(4) Show similar alteration to another mineral or minerals.
(5) Have some special chemical relationship.

On the other hand, minerals which crystallized at different times *may*:

(6) Show replacement, reaction or some special mutual texture.
(7) Have significant differences in fabric of preferred orientation.
(8) Be known to be stable under different physical conditions.
(9) Be enclosed, one within the other.

However, the first five features can also be shown by minerals which formed at different times and (6) and (7) by those formed at the same time. Inclusions of one mineral in another may be of:

(1) An older mineral engulfed by the growth of a younger one (helicitic texture).
(2) A phase exsolved by the host (exsolution texture).
(3) A younger phase growing in the host (replacement texture).

Inclusions are probably older than their host where they are idioblastic, consist of a variety of species or have a preferred orientation of the helicitic type.

Replacements may be simple to interpret, e.g. a six-sided aggregate of randomly oriented chlorite is commonly a pseudomorph of chlorite after garnet (similarly chlorite or actinolite after olivine or pyroxene, sericite after feldspar or aluminium silicates). Workman and Cowperthwaite (1963) described pseudomorphs of chiastolite by kyanite in which the idioblastic outlines, twins and inclusion patterns of the chiastolite have been retained.

Similarly, fine-grained veins or networks with non-dilational boundaries crossing crystals have a replacing aspect.

Fringes or borders of one mineral around another commonly are younger than the central core, e.g. chlorite around pyrite (Plate XVIIIa). *Atoll* structure consists of a core of one mineral surrounded by a single-crystal rim of a second phase. Garnet surrounding pyroxene in eclogite is commonly *younger* than the pyroxene and grew around it but Rast (1965) described *atoll* structure in which replacement took place from the centre outwards, the rim thus being *older* than the core.

The time relations of fine- and coarse-grained aggregates of the same mineral are particularly difficult to interpret. The replacement of a large idioblastic crystal by randomly oriented smaller crystals can be recognized where the original shape of the large crystal is retained but in a mixture of large and small crystals of the same species, relations are rarely clear. A disparity in grain size in one mineral does not necessarily imply an age difference between large and small crystals because a considerable range of sizes can be produced during normal recrystallization due to differences in nucleation or supply of material and large crystals could arise by abnormal or secondary annealing. It is generally found that crystals of a similar age are similar in appearance, i.e. size, shape, degree of strain, number of inclusions, etc.

A texture consisting of large crystals of quartz with mosaic substructure surrounded by a fine-grained aggregate of quartz can be a *mortar texture* in which the large crystals are old relics surrounded by new quartz, the substructure and undulose extinction being due to strain. Alternatively it has been postulated that it could be a *coalescence* texture in which small crystals have merged together to give a large superindividual.

Recrystallization in Successive Deformation Phases

Polymetamorphism produces departures from simple pre-, syn- and post-tectonic crystallization textures (Fig. 63). A mineral formed in an early phase may persist metastably through later phases and a crystal which is post-tectonic with regard to an early phase will be pre-tectonic to all later phases. *Interkinematic* crystals form in the quiet or static period between deformation phases.

Multiple-phase metamorphism (polymetamorphism) is suggested by multiple foliations and by crystals which have complex time relations, e.g. have discordant S_i and S_e, are post-tectonic to one foliation but pre-tectonic to another, or by a mineral species which has different forms or time relations in different crystals.

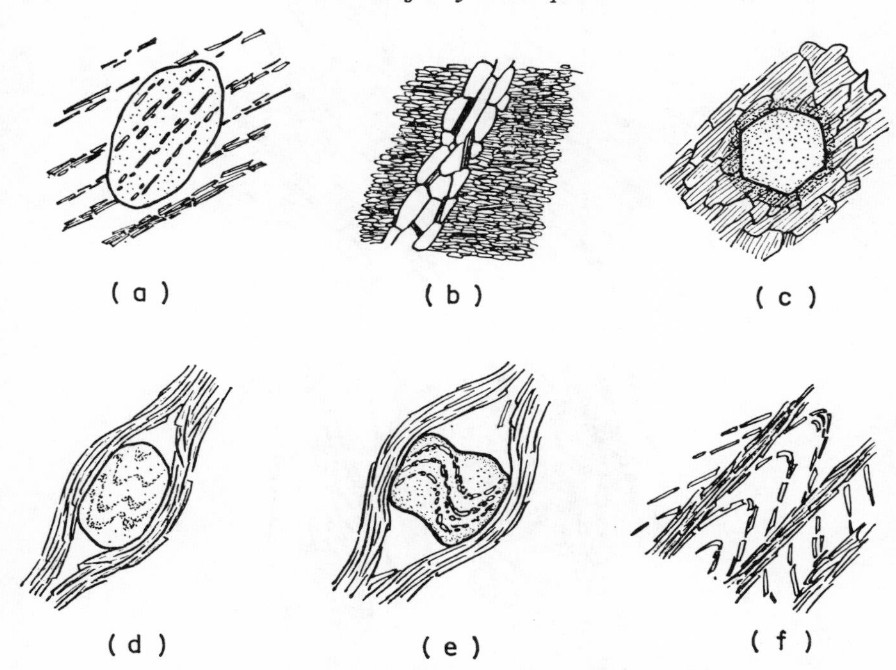

FIG. 63. Characteristic relations in polymetamorphic rocks. (a) Helicitic crystal, S_i discordant with S_e. (b) Layer of coarse micas cutting a fine foliation. (c) Blue-green coloration in actinolite in a rim around garnet. The actinolite formed after the garnet and derived certain constituents from it. (d) Helicitic crystal enclosing folded *S*-surface wrapped around by a later foliation. (e) Snowball garnet with S_1 as rotational S_i wrapped around by S_2. (f) Older folded foliation S_1 cut by planar S_2

Many polymetamorphic schists contain three foliations which represent three phases of deformation and which may be distinguished mesoscopically by their geometry. They may be separated microscopically on their degree of contortion, their degree of development (thickness, mineral orientation) or the presence or absence of minerals of certain kinds crystallizing along the foliation. The minerals constituting a foliation probably formed syntectonically with the foliation but those in an old foliation may have been wholly or partially recrystallized during later episodes.

The textural evolution of a typical polymetamorphic garnet–albite schist (Howell Group, Tasmania, Spry, 1963a, d) is shown in Fig. 64. The schist contains a single strong mesoscopic foliation which is marked by a fissility and thin compositional layering and which is parallel to interbedded quartzites and to the axial surfaces of isoclinal F_2 folds. In thin section this foliation is marked by layers alternately richer in quartz or in mica, both of which are

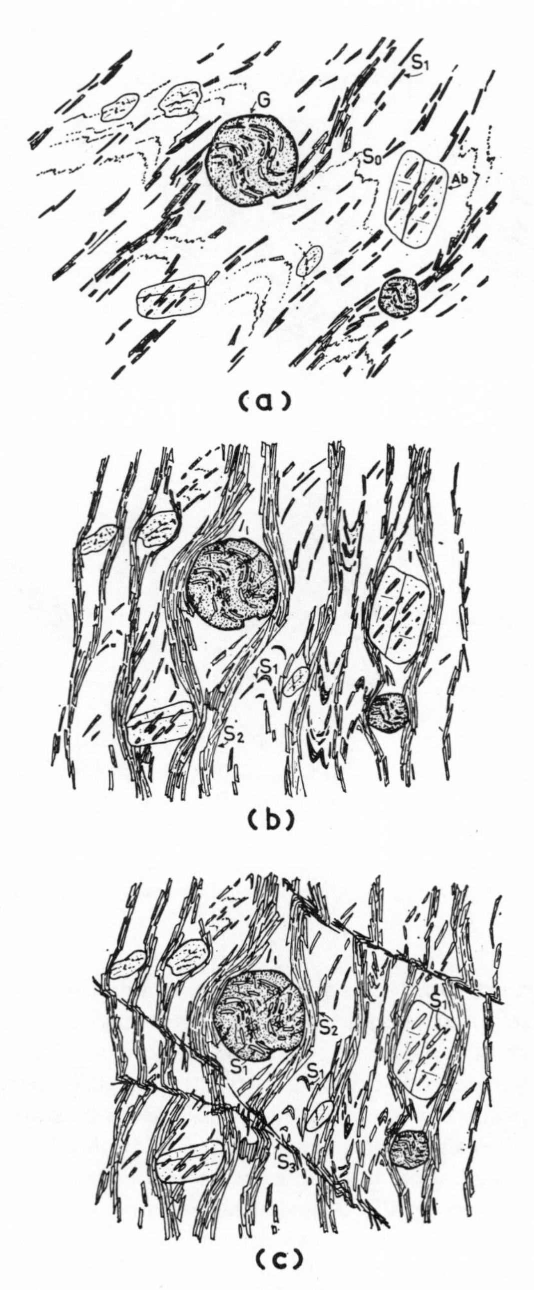

FIG. 64. Textural evolution of a polymetamorphic schist. (a) Bedding (S_0) is folded and a planar foliation S_1 is formed during F_1. Syntectonic crystallization of mica, quartz and rotational garnet (G). Post-tectonic crystallization of helicitic albite (Ab) (S_i concordant with S_e). (b) A second phase F_2 produces a strong foliation S_2, leaves S_1 as folded relics and obliterates S_0. S_2 wraps around garnet and albite giving pressure shadows; S_i is discordant with S_e. Mica and quartz are syntectonic. (c) A third minor phase folds S_2 and produces a strain slip cleavage S_3; there is no crystal growth

elongate parallel to the foliation. The foliation is not bedding but is S_2 (formed during F_2) and is younger than another foliation S_1 (formed during F_1) which is visible as contorted relics between layers of S_2. A still younger foliation S_3 occurs as sporadic fractures at about 15° to S_2. S_1 is outlined by tiny flakes of parallel muscovite and brown biotite, S_2 is outlined by large flakes of muscovite with very little biotite, and S_3 is marked by a few flakes of green biotite and chlorite. Micas of S_2 wrap around crystals of garnet (with pressure shadows) which are thus pretectonic to F_2. The garnets contain centres with snowball structure (syntectonic to F_1) and idioblastic massive rims (post-tectonic to F_1).

Albite forms large porphyroblasts with simple and multiple twinning and helicitic texture wrapped around by S_2. Like the garnet, the albite is thus pretectonic to F_2. The helicitic structure is of two types. One is outlined by contorted lines of tiny opaque granules which may represent bedding but the more common consists of slightly curved lines of elongate quartz, parallel mica flakes (S_1), and small idioblastic garnets and tourmalines. The albite is later than S_1 which is retained helicitically and is also later than the small garnets and tourmalines which are thus early post-tectonic to F_1. Some large helicitic tourmalines are randomly oriented and enclose S_2; they are thus post-tectonic to F_2.

Chlorite occurs as envelopes to garnet where it is drawn out along S_2 (syntectonic to F_2); some is randomly oriented within S_2 (post-tectonic to F_2) some garnets have been replaced pseudomorphously by randomly oriented chlorite and as the garnet shape has been retained, this chlorite is post-tectonic to F_2; some chlorite is randomly oriented along S_3 (post-tectonic to F_3).

The history may be summarized as shown in Table 10.

The associated amphibolites have a similar textural history. They consist of small, parallel actinolite prisms which are parallel to the S_2 of the enclosing schists, and wrap around large hornblende and garnet crystals. These porphyroblasts are undulose, shattered and granulated and have pressure shadows and are thus pretectonic to F_2 and belong to F_1. The small actinolite crystals are partially syntectonic to F_2 but most are poorly oriented and post-tectonic to F_2. Some garnet has a snowballed centre and massive idioblastic rim and thus is partly syntectonic, partly post-tectonic to F_1. Biotite forms large rusty-brown twisted or kinked flakes which wrap around the garnet and are parallel to S_2. These are probably old (F_1) relics which are pretectonic to F_2 but which have been rotated mechanically into parallellism with S_2. Andesine occurs as large, ragged dusty crystals and probably belongs to F_1; albite forms aggregates of clear elongate crystals along S_2 and belongs to F_2. The amphibolite was thus a coarse almandine–biotite–hornblende–andesine rock belonging to the Amphibolite Facies during F_1 but the

TABLE 10

	F_1			F_2		F_3	
Quartz	——	——	——	——			
Muscovite	——		——	——			
Garnet	——	——					
Biotite	——	——			——		——
Albite	——	——			——		
Chlorite			——	——	——		——
Tourmaline			——		——		
Deformation stage	pre-	syn- tectonic	post-	syn-	post- tectonic	syn-	post- tectonic
S formed		S_1		S_2		S_3	
S folded		S_0		S_0, S_1		S_0, S_1, S_2	

assemblage changed to albite–epidote–actinolite (Greenschist Facies) in F_2. The rock does not appear to have been affected by F_3.

The polymetamorphic history of the Buchan schists has been discussed by Johnson (1962, 1963). These differ from the examples above (which are more like the products of Barrovian metamorphism) in that they contain andalusite and cordierite and thus belong to a lower pressure metamorphism. Cordierite forms helicitic porphyroblasts which contain S_i discordant with S_e which is S_2. The crystals are thus post-tectonic to F_1. Andalusite has grown with random orientations across F_2 folds; S_i is concordant with S_e which is S_2.

The polymetamorphic history of monomineralic rocks such as pure quartzites or marbles may be impossible to recognize from the texture alone. Some quartzites associated with the Tasmanian polymetamorphic schists and amphibolites described above have mortar texture in which the large quartz porphyroclasts are relict from a coarse-grained rock belonging to F_1 and the fine-grained matrix is syntectonic to F_2. Thin sections of isoclinally folded micaceous quartzites may contain S_1 outlined by mica and elongate quartz; S_1 and S_2 may be recognized as discordant on the hinges of folds but merge to give a compound foliation on the limbs.

Repeated metamorphism at high grade has been described by Sutton and Watson (1950) in the Lewisian of the North-West Highlands. A range of sediments and igneous rocks was metamorphosed to the granulite facies in the Scourian Orogeny to produce charnockites, gneisses, etc. Uplift and dolerite intrusion followed. The later Laxfordian metamorphism took place at Amphibolite Facies and was accompanied by deformation, migmatization and introduction of microcline. The Laxfordian was followed by a period of mylonitization and production of pseudotachylyte.

Polymetamorphic Regions

The polymetamorphic nature of regionally metamorphosed areas and the textural evolution of the rocks has been studied in detail in the Scottish Highlands, the Pyrenees and in Tasmania.

The grade of metamorphism during each episode is judged by the presence of index minerals such as almandine, sillimanite, kyanite, biotite, chlorite, plagioclase, etc., taking into consideration at the same time the chemical composition and the grain size.

The amount of deformation is estimated in a very approximate way from the size, number and types of folds or foliations produced. Tight isoclinal folds or nappes indicate one extreme and strain slip cleavage and kinks the other. The separate tectonic events are recognized mainly from mesoscopic structural evidence such as folded lineations, folds and foliations.

The relative roles of crystallization and deformation as depicted by Rast (1958) at Schichallion, by Ramsay (1963) in the Moine, by Johnson (1963) in the Buchan Dalradian and Spry (1963a) in the Tasmanian Precambrian are shown in Fig. 65. Such diagrams are constructed by determining the grade of metamorphism during, and between, successive deformation episodes which are arbitrarily placed at equal intervals on the horizontal time axis. Neither axes have true scales, the shapes of the curves are not controlled at all closely, and so not too much should be read into such diagrams. Eventually it is to be hoped that real values of time, temperatures and various pressures can be placed on the axes. However the diagrams demonstrate the independent nature of the roles of temperature and deformation in regional metamorphism.

As far as can be seen from the limited evidence available, the classical view of regional metamorphism being due to the simultaneous action of increased temperature and directed pressures is true in that both generally begin together; however, crystallization due to elevation of temperature or other factors is not necessarily most intense at the height of tectonic activity; in fact, the highest temperatures (plus considerable metasomatism) may be reached in the interkinematic periods between successive tectonic phases. In the Tasmanian Precambrian it appears that the highest temperatures (garnet, kyanite, etc.) were reached during F_1 but that most folding (with

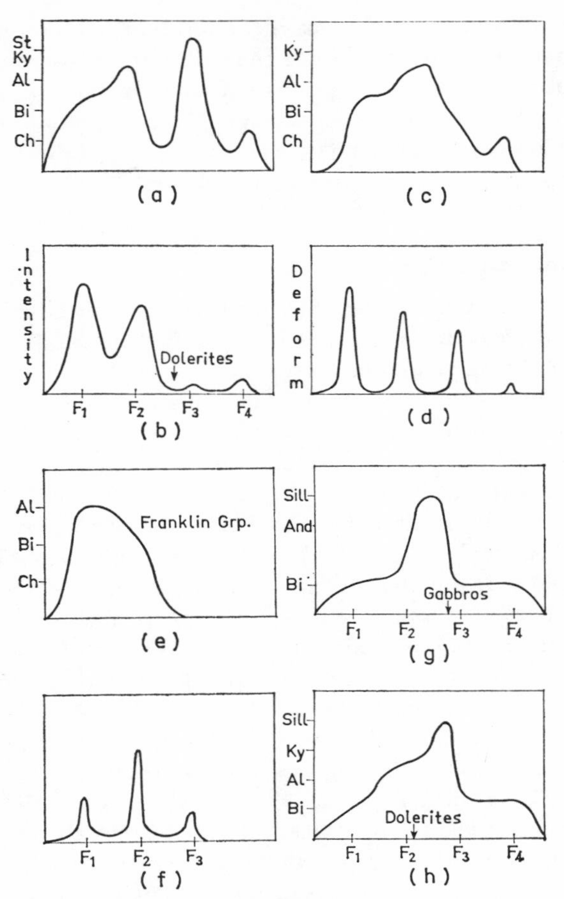

FIG. 65. Time relations of deformation and metamorphism in various regional metamorphic provinces. The first three regions are represented by a pair of diagrams, the lower showing the variation of deformation with time, culminating in tectonic episodes named consecutively F_1, F_2, F_3 and F_4, and the upper diagram showing the intensity of metamorphism. The horizontal time scale is the same in both members of the pairs. (a) and (b) Schichallion (Rast, 1958); (c) and (d) Moine Nappe, western margin (Ramsay, 1963); (e) and (f) Frenchman's Cap (Spry, 1963); (g) and (h) show the variation of grade with time but no attempt is made to estimate the degree of deformation during each deformation phase (Johnson, 1963). (g) Buchan Dalradian; (h) Barrovian Dalradian

possibly large recumbent nappes) took place in F_2. The converse is found elsewhere. Very large structures such as nappes in the Alps, Pyrenees and Caledonides were formed first; regional metamorphism followed in several phases which were independent of the large structures and possibly related to later cross-folding. If the nappes are truly large sedimentary structures, as has been postulated, then most deformation took place before recrystallization. There is some suggestion (pretectonic cordierite, etc.) in parts of the Pyrenees that the temperature had risen with resultant metamorphic crystallization, before the first deformation.

The exact dating of the various episodes is not easy and it is possible in some cases that episodes such as F_1 and F_2 are so closely spaced in time that they can be considered to be the same. The division into *episodes* (minor) and *events* (major) is not a simple one and the decision as to whether F_1 and F_2 represent two episodes separated by a million years or less, or two events occurring in the Precambrian and Ordovician depends on independent geochronological evidence.

The Lewisian rocks of the North-West Highlands have been affected by the Scourian Orogeny (2500 m.y.) followed by the Laxfordian (1600 m.y.) and have then been deformed and metamorphosed during what has been called the Caledonian Orogeny. This orogeny has been divided into a number of tectonic, metamorphic and intrusive episodes which range from Precambrian to Silurian. In a very generalized way it is possible that there was early folding and metamorphism in the late Precambrian (730 m.y.) followed by granitic intrusion (Carn Chuinneag, 530 m.y.), a second recrystallization period at 475–500 m.y. and more intrusives at 365 m.y. (Sutton, 1963, 1965). Which of these later episodes are F_2, F_3, F_4, etc., is not yet clear; in fact it is not necessary that phases named F_2 in adjacent areas are actually concurrent. The F_2 of one area may be F_3 of another in abolute chronological terms.

Binns and Miller (1963) and Binns (1963) have dated isotopically a series of metamorphic and igneous events at Broken Hill. The first (Willyama, M_1) metamorphism was 1600 m.y. or older; metamorphisms M_2 and M_3 belong to 1345 m.y., and were closely followed by the Mundi-Mundi Granite at 1307 m.y. with the metamorphism and folding M_4 dated at 916 m.y.

The duration of an episode, phase or event is not clear. Orogeny may affect an area intermittently over hundreds of millions of years or may be completed in as little as 30 m.y. Metamorphism in the Alps (Steiger, 1963) lasted for 30 m.y. and deformation is divisible into 3 or 4 episodes, each of which lasted between 2 and 10 m.y. Rocks which were regionally metamorphosed only 15 m.y. ago are now exposed in the Alps and it would appear that the interval between alteration and exposure may be less than 10 m.y. in many places (Sutton, 1965).

This information, though scanty and very preliminary in nature, suggests that the various times of recrystallization in a single regional metamorphic province may be separated by appreciable spans of time and should not be considered merely as minor parts of a single metamorphism.

References

ACKERMANN, P. B. and WALKER, F., Vitrification of arkose by Karoo dolerite near Heilberon, Orange Free State, *Q. Jl. geol. Soc. Lond.*, **116**, 239 (1960).

AGRON, S. L., Structure and petrology of the Peach Bottom slate, *Bull. geol. Soc. Am.*, **61**, 1265 (1950).

AHRENS, T. J., and GREGSON, V. G., Shock compression of crustal rocks, *J. Geophys. Res.*, **69**, 4839 (1964).

AKIZUKI, M., Dislocations in muscovite, *Sci. Rep. Univ. Tohoku*, **3**, 9, 485 (1966).

ALDERMAN, A. R., Eclogites in the vicinity of Glenelg, Inverness-shire, *Q. Jl. geol. Soc. Lond.*, **92**, 488 (1936).

ALEXANDER, B. H., Discussion in *Phase Transformations in Solids*, pp. 582–7 (Eds. R. Smoluchowski, J. E. Mayer and W. A. Weyl). Wiley, New York, 1951.

ALLING, A. L., *Interpretive Petrology*. McGraw-Hill, New York, 1936.

ALLING, A. L., Plutonic perthites, *J. Geol.*, **46**, 142 (1938).

ALMOND, D. C., Metamorphism of Tertiary lavas in Strathaird, Skye, *Trans. R. Soc. Edinb.*, **65**, 16, 413 (1964).

AMELINCKX, S., *Direct Observation of Dislocations*. Academic Press, New York, 1964.

ANDERSON, G. H., Pseudo-cataclastic texture of replacement origin in igneous rocks, *Am. Miner.*, **19**, 5, 185 (1934).

ANDERSON, O., Genesis of some types of feldspar from granite pegmatites, *Norsk. geol. Tidsskr.*, **10**, 116 (1928).

ARMSTRONG, E. J., Mylonization of hybrid rocks near Philadalphia, Penn., *Bull. geol. Soc. Am.*, **52**, 667 (1941).

ARNOLD, G. W., Defects in natural and synthetic quartz, *Physics Chem. Solids*, **13**, 306 (1960).

ATHERTON, M. P., Chemical significance of isograds. In *Controls of Metamorphism* (Eds. W. S. Pitcher and G. S. Flinn). Oliver & Boyd, Edinburgh, 1965.

AUGUSTITHIS, S. S., Non-eutectic, graphic, micrographic and graphic-like myrmekite structures and textures, *Beitr. Miner. Petrogr.*, **8**, 491 (1962).

AUST, K. T., Annealing twins and coincidence-site boundaries in zone-refined aluminium, *Trans. Am. Inst. Min. Metall. Engrs.*, **221**, 758 (1961).

AUST, K. T. and RUTTER, J. W., Annealing twins and coincidence-site boundaries in high purity lead, *Trans. Am. Inst. Min. metall. Engrs.*, **218**, 1023 (1960).

AUST, K. T. and RUTTER, J. W., Effects of grain-boundary mobility and energy on preferred orientation in annealed high purity lead, *Trans. Am. Inst. Min. metall. Engrs.*, **224**, 111 (1962).

AZAROFF, L. V., *Introduction to Solids*. McGraw-Hill, New York, 1960.

BAILEY, E. B., Geology of Ben Nevis and Glen Coe, *Mem. geol. Surv. U.K.* (1916).

BAILEY, E. B., Metamorphism of the South West Highlands, *Geol. Mag.*, **60**, 137 (1923).

BAILEY, E. B., *Contribution to* Sutton, J., and Watson, J., 1950. Pre-Torridonian metamorphic history of Loch Torridon and Scourie Areas, *Q. Jl. geol. Soc. Lond.*, **106**, 241 (1950).

BAILEY, E. B., and MCCALLIEN, W. J., Metamorphic rocks in North-East Antrim, *Trans. R. Soc. Edinb.*, **58**, 1 (1934).

BAILEY, S. W., BELL, R. A. and PENG, C. J., Plastic deformation of quartz in nature, *Bull. geol. Soc. Am.*, **69**, 1452 (1958).

BAIN, G. W., Skeleton quartz crystals, *Am. Mineralog.*, **10**, 435 (1925).
BAIN, G. W., Wall-rock alteration along Ontario gold deposits, *Econ. Geol.*, **28**, 705 (1933).
BALK, R., Structural geology of the Adirondack anorthosite, *J. Geol.*, **38**, 289 (1931).
BARRETT, C. S., *Structure of Metals*. McGraw-Hill, New York, 1952.
BARTH, T. F. W., Polymorphic phenomena and crystal structure, *Am. J. Sci.*, **27**, 273 (1934).
BARTH, T. F. W., *Theoretical Petrology*. Wiley, New York, 1952.
BARTH, T. F. W., Zonal structure in feldspars of crystalline schists, *Reun. Int. React. Solid. III, Madrid*, 363 (1956).
BASTIN, E. S., Interpretation of ore textures, *Mem. geol. Soc. Am.*, **45**, (1950).
BATHURST, R. G. C., Diagenetic fabrics in some British Dinantian Limestones, *Lpool. Manch. geol. J.*, **2**, 1, 11 (1958).
BATTEY, M. H., Alkali metasomatism and the petrology of some keratophyres, *Geol. Mag.*, **92**, 104 (1955).
BEAVIS, F. C., Mylonites of the Upper Kiewa Valley, *Proc. R. Soc. Vic.*, **74**, 55 (1961).
BECK, P. A., Annealing of cold-worked metals, *Adv. in Phys.*, **3**, 11, 245 (1954).
BECK, P. A. and SPERRY, P. R., Strain induced grain-boundary migration in high purity aluminium, *J. appl. Phys.*, **21**, 150 (1950).
BECK, P. A., SPERRY, P. R. and HU, H., The orientation dependence of the rate of grain-boundary migration, *J. appl. Phys.*, **21**, 420 (1950).
BECK, P. A., KREMER, J. C., DEMER, L. J. and HOLZWORTH, M. L., Grain growth in high purity aluminium in an aluminium–magnesium alloy, *Trans. Am. Inst. Min. metall. Engrs.*, **175**, 372 (1948).
BECKE, F., Über Zonenstruktur an Feldspaten, *Sver. dt. naturiv-med. Ver.*, **3** (1897).
BECKE, F., Über Mineralbestand und Struktur der Kristallinen Schiefer, *Denkschr. Akad. Wiss. Wien*, **75**, 1 (1903); also *Congr. geol. Internat. 9th Vienna*, 553 (1903).
BECKE, F., Über diapthorite, *Tscherm. Min. Pet. Mitt.*, **28**, 369 (1909).
BECKE, F., Zur physiographie der Gemengenteile der Krystallinen Schiefer, *Denkschr. Akad. Wiss. Wien*, **75**, 97 (1913a).
BECKE, F., Über Mineralbestand und Struktur der Kristallinen Schiefer, *Denskchr. Akad. Wiss. Wien*, **75**, 1 (1913b).
BECKE, F., Struktur und Kluftung, *Fortschr. Miner., Kristallogr. Petrogr.*, **9**, 185 (1924).
BECKER, G. F. and DAY, A. L., Linear force of growing crystals, *Washington Acad. Sci. Proc.*, **7**, 283 (1905).
BECKER, G. F. and DAY, A. L., Note on the linear force of growing crystals, *J. Geol.*, **24**, 313 (1916).
BELL, J. F., Morphology of mechanical twinning in crystals, *Am. Miner.*, **26**, 247 (1941).
BELL, R. L. and CAHN, R. W., Dynamics of twinning and the inter-relation of twinning and slip in zinc crystals, *Proc. R. Soc., Lond.*, **A239**, 494, (1957).
BENNINGTON, K. O., Role of shearing stress and pressure in differentiation, *J. Geol.*, **64**, 558 (1965).
BENSON, G. C., and DEMPSEY, E., Caloric and surface energies of some crystals possessing the fluorite structure, *Proc. R. Soc., Lond.*, **A266**, 344 (1962).
BENSON, G. C., FREEMAN, P. I. and DEMPSEY, E., Calculation of cohesive and surface energies of thorium and uranium oxides, *J. Am. Ceram. Soc.*, **46**, 43 (1963).
BERNAL, J. D., DASGUPTA, D. H. and MACKAY, A. L., Oriented transformations in iron oxides and hydroxides, *Nature*, **180**, 645 (1957).
BHATTACHARJEE, C. C., Late structural and petrological history of the Lewisian rocks of the Meall Deise area north of Gairloch, Ross-shire, *Trans. geol. Soc. Glasg.*, **25**, 1, 31 (1964).
BIKERMAN, J. J., *Surface Chemistry, Theory and Applications*. Academic, New York, 1958.
BILBY, B. A. and CHRISTIAN, J. W., *Mechanism of the Phase Transformations in Metals*. Inst. Metals, London, 1961.
BILBY, B. A., COTTRELL, A. H. and SWINDON, K. H., The spread of plastic yield from a notch, *Proc. R. Soc., Lond.*, **A272**, 304 (1963).
BINNS, R. A., Some observations on metamorphism at Broken Hill, N.S.W., *Proc. Australas. Inst. Min. Metall.*, **207**, 239 (1963).

BINNS, R. A., Zones of progressive regional metamorphism in the Willyama complex, Broken Hill district, N.S.W., *J. geol. Soc. Aust.*, **11**, 283 (1964).

BINNS, R. A., Granitic intrusions and regional metamorphic rocks of Permian age from the Wongwibinda district, N.S.W., *J. R. Soc. N.S.W.*, **99**, 5 (1966).

BINNS, R. A. and MILLER, J. A., Potassium–Argon age determinations on some rocks from the Broken Hill region of New South Wales, *Nature*, **199**, 4890, 274 (1963).

BLATT, H. and CHRISTIE, J. M., Undulatory extinction in quartz etc., *J. sedim. Petrol.*, **33**, 559 (1963).

BLOOMFIELD, K., Nepheline gneisses of northern Nyasaland, *Int. Geol. Congress XX, Mexico, Assoc. de Serv. Geol. Africanos*, 291 (1959).

BLOXAM, T. W., Jadeite-bearing metagreywackes in California, *Am. Miner.*, **41**, 489 (1956).

BLOXAM, T. W., Jadeite-rocks and glaucophane schists from Angel Island, San Francisco Bay, *Am. J. Sci.*, **258**, 555 (1960).

BOLLING, G. F. and RICHMAN, R. H., Continuous mechanical twinning, *Acta metall.* **13**, I, 707; II, 723; III, 745 (1965).

BOLLMAN, W., Atom movements and dislocation structures in some common minerals, *Acta metall.*, **9**, 971 (1961).

BONET, F., La facies Urgoniana del Cretacico Medio de la region de Tampico, *Boln Assoc. Mex. geol. Petro.*, **4**, 153 (1952).

BORCHERT, H. and MUIR, R. D., *Metamorphism of Salt Deposits*. Van Nostrand, London, 1964.

BOSE, M. K., Garnet coronites from Koraput, Orissa, *Geol. Mag.*, **98**, 409 (1961).

BOSMAN, W., Spots in the spotted slates of Steige (Vosges) and Vogtland (Saxony), *Geologie Mijnb.*, **43e**, 476 (1964).

BOWDEN, F. P. and COOPER, R. E., Velocity of twin propagation in crystals, *Nature*, **195**, 1091 (1962).

BOWES, D. and CONVERY, H. J. E., Composition of some Ben Ledi grits and its bearing on the origin of albite schists in the south west Highlands, *Scott. Jnl. Geol.*, **2**, 1, 67 (1966).

BRACE, W. F., Orientation of anisotropic minerals in a stress field, *Mem. geol. Soc. Am.*, **79**, 9, (1960).

BRACE, W. F., and WALSH, J. B., Some direct measurements of the surface energy of quartz and orthoclase, *Am. Miner.*, **47**, 1111 (1962).

BRADLEY, J., Geology of the West Coast Range, III. *Pap. Proc. R. Soc. Tasm.*, **91**, 163 (1957).

BRAITSCH, O., Entstehung und Stoffbestand der Salzlagerstätten, *Miner. u Petrog. in Einzeldarstellungen*, Bd. 3 (1962).

BRAY, J. M. Ilmenite–hematite–magnetite relations in some emery ores, *Am. Miner.*, **24**, 162 (1939).

BRETT, R., Experimental data from the system Cu-Fe-S and their bearing on exsolution textures in ores, *Econ. Geol.*, **59**, 1241 (1964).

BRIDGEWATER, D., Clouded feldspars, *Geol. Mag.*, **103**, 284 (1966).

BRINDLEY, G. W., Structural control in some minerals. In *Kinetics of High Temperature Processes* (Ed. W. D. Kingery). Tech. Press, M.I.T. and Wiley, New York, 1959.

BRINDLEY, G. W., Crystallographic aspects of some decomposition and recrystallization reactions. In *Progress in Ceramic Science*, vol. III (Ed. J. E. Burke). Pergamon Press, Oxford, 1963.

BRINDLEY, G. W. and HIYAMI, R., Mechanism of formation of forsterite and enstatite from serpentine, *Mineralog. Mag.*, **35**, 189 (1965).

BRINDLEY, G. W. and NAKAHIRA, M., The kaolinite-mullite series, *J. Am. Ceram. Soc.*, **42**, II–311, III–319 (1959).

BRINDLEY, G. W. and ZUSSMAN, J., A structural study of the transformation of serpentine minerals to forsterite, *Am. Miner.*, **46**, 461 (1957).

BRÖGGER, W. C., Die Mineralen der Syentpegmatitgänge der Südnorwegischen Augit und Nephelinsyenite, *Z. Kristallogr. Miner.*, 16 (1890).

BRÖGGER, W. C., On several Archean rocks from the south coast of Norway II. The South Norwegian hyperites and their metamorphism, *Skr. norske Vidensk-Akad.*, **1**, 1 (1934).

BROWN, G. and STEPHEN, I., Structural study of iddingsite, *Am. Miner.*, **44**, 251 (1959).

BROWN, W. H., FYFE, W. S. and TURNER, F. J., Aragonite in Californian glaucophane schists, and the kinetics of the aragonite-calcite transformation, *J. Petrology*, **3**, 566 (1962).

BROWN, W. L., Peristerite unmixing in the plagioclases and metamorphic facies series, *Norsk. geol. Tidsskr*, **42**, 354 (1962).

BROWN, W. L., Crystallographic aspects of feldspars in metamorphism. In *Controls of Metamorphism*, p. 342 (Eds. W. S. Pitcher and G. W. Flinn). Oliver & Boyd, Edinburgh, 1965.

BRUNNER, H. and GRANT, N. J., Measurement of deformation resulting from grain boundary sliding in Al and Al-Mg from 410° to 940°F. *Inst. Min. Metall. Engrs. G.B.*, **218**, 122 (1960).

BUCKLEY, H. E., Crystallization of potassium alum and effect of impurities on its habit, *Z. Kristallogr. Miner.*, **73**, 5 (1930).

BUCKLEY, H. E., Oriented inclusions in crystals, *Z. Kristallogr. Miner.*, **88**, 248 (1934).

BUCKLEY, H. E., *Crystal Growth*. Wiley, New York, 1951.

BUDDINGTON, A. F., Adirondack igneous rocks and their metamorphism, *Mem. geol. Soc. Am.*, **7** (1939).

BUDDINGTON, A. F., Chemical petrology of some metamorphosed Adirondack gabbroic, syenitic and quartz-syenitic rocks, *Am. J. Sci.*, *Bowen Vol.*, 37 (1952).

BUEREN, H. G. VAN, *Imperfections in Crystals*. North-Holland, Amsterdam, 1961.

BUERGER, M. J., Plastic deformation of ore minerals, *Am. Miner.*, **13**, 35 (1928).

BUERGER, M. J., Translation-gliding in crystals, *Am. Miner.*, **15**, 45 (1930a).

BUERGER, M. J., Translation-gliding in crystals of the NaCl structure type, *Am. Miner.*, **15**, 174, 226 (1930b).

BUERGER, M. J., Lineage structure in crystals, *Z. Kristallogr. Miner.*, **87** (1934).

BUERGER, M. J., Genesis of twin crystals, *Am. Miner.*, **30**, 469 (1945).

BUERGER, M. J., Relative importance of the several faces of a crystal, *Am. Miner.*, **32**, 593 (1947).

BUERGER, M. J., Role of temperature in mineralogy, *Am. Miner.*, **33**, 101 (1948).

BUERGER, M. J., Crystallographic aspects of phase transformations. In *Phase Transformations in Solids*, p. 183 (Eds. R. Smoluchowski, J. E. Mayer and W. A. Weyl). Wiley, New York, 1951.

BUERGER, M. J. and WASHKEN, E., Metamorphism of minerals, *Am. Miner.*, **32**, 296 (1947).

BUERGER, N. W. and BUERGER, M. J., Crystallographic relations between cubanite segregation plates, chalcopyrite matrix and secondary chalcopyrite twins, *Am. Miner.*, **19**, 289 (1934).

BUNCH, T. E. and COHEN, A. J., Shock deformation of quartz from two meteorite craters. *Bull. geol. Soc. Am.*, **75**, 1263 (1964).

BUNN, C. W., Adsorption, oriented overgrowth and mixed crystal formation, *Proc. R. Soc. Lond.*, **A141**, 567 (1933).

BURGERS, W. G., Principles of recrystallization. In *Art and Science of Growing Crystals* (Ed. J. J. Gilman). Wiley, New York, 1963.

BURGERS, W. G. and LOUWERSE, P. C., Über den Zusammenhang zwischen Deformationsvorgang und Rekrystallisationstextur bei Aluminium, *Z. Phys.*, **67**, 605 (1931).

BURGERS, W. G., MEIJS, J. C. and TIEDEMA, T. J., Frequency of annealing twins in copper crystals grown by recrystallization, *Acta metall.*, **1**, 75 (1953).

BURKE, J. E., Some factors affecting the rate of grain growth in metals, *Trans. Am. Inst. Min. metall. Engrs.*, **180**, 73 (1949).

BURKE, J. E., The formation of annealing twins, *Trans. Am. Inst. Min. metall. Engrs.*, **188**, 1324 (1950).

BURKE, J. E., Grain growth in ceramics. In *Kinetics of High Temperature Processes* (Ed. W. D. Kingery). Tech. Press, M.I.T., and Wiley, 1959.

BURKE, J. E. and SHIAU, Y. G., The effect of mechanical deformation on grain growth in α brass, *Trans. Am. Inst. Min. metall. Engrs.*, **175**, 141 (1948).

BURKE, J. E. and TURNBULL, D., Recrystallization and grain growth, *Prog. Metal Phys.*, **3**, 220 (1952).

BURN, R. and MURRAY, G. T., Plasticity and dislocation etch pits in CaF_2, *J. Am. Ceram. Soc.*, **45**, 251 (1962).

BURNS, D. J., Chemical and mineralogical changes associated with the Laxford metamorphism of dolerite dykes in the Scourie–Loch Laxford area, Sutherland, Scotland, *Geol. Mag.*, **103**, 19 (1966).

BURSILL, L. A. and MCLAREN, A. C., Transmission electron microscope observations of fracture in quartz and zircon, *J. appl. Phys.*, **36**, 2084 (1964).

BUSH, W. F. and WILLIAMSON, W. O., Enhanced reactivity of magnesian silicates from strained magnesia, *Mineralog. Mag.*, **35**, 177 (1965).

BUTLER, B. C. M., Metamorphism and metasomatism of rocks of the Moine Series by a dolerite plug in Glenmore, Ardnamurchan, *Mineralog. Mag.*, **32**, 866 (1961).

BUTLER, B. C. M., The dehydration of chrystotile in air and under hydrothermal conditions, *Mineralog. Mag.*, **33**, 261, 467 (1963).

BYRNE, J. G., *Recovery, Recrystallization and Grain Growth*, McMillan, New York, 1965.

CAHN, R. W., Recrystallization of single crystals after plastic bending, *J. Inst. Metals*, **76**, 121 (1949).

CAHN, R. W., A new theory of recrystallization nucleii, *Proc. Phys. Soc. Lond.*, **63**, 323 (1950).

CAHN, R. W., Plastic deformation of α-uranium; twinning and twinned crystals, *Acta metall.*, **1**, 49 (1953).

CAHN, R. W., Twinned crystals, *Adv. Phys.*, **3**, 363 (1954).

CAHN, R. W., Theory of crystal growth and interface motion in crystalline substances, *Acta metall.*, **8**, 554 (1960).

CANN, J. R., Metamorphism of amygdales at 'Airde Beinn, N. Mull., *Mineralog. Mag., Tilley Vol.*, 92 (1965).

CANNON, R. T., Plagioclase zoning and twinning in relation to the metamorphic history of some amphibolites and granulites. *Am. J. Sci.*, **264**, 526 (1966).

CAROZZI, A. V., *Microscopic Sedimentary Petrology*. Wiley, New York, 1960.

CARTER, N. L., Basal quartz deformation lamellae—a criterion for recognition of impactites. *Am. J. Sci.*, **263**, 786 (1965).

CARTER, N. L., CHRISTIE, J. M. and GRIGGS, D. T., Experimental deformation and recrystallization of quartz, *J. Geol.*, **72**, 687 (1964).

CARTER, N. L. and FRIEDMAN, M., Dynamic analysis of deformed quartz and calcite from the Dry Creek Ridge Anticline, Montana, *Am. J. Sci.*, **263**, 747 (1965).

CHAO, E. C. T., Shock effects in certain rock-forming minerals, *Science*, **156**, 192 (1967).

CHAPMAN, R. W., Contact-metamorphic effects of Triassic dibase at Safe Harbour, Pennsylvania, *Bull. geol. Soc. Am.*, **61**, 911 (1950).

CHATTERJEE, N. D., Alpine metamorphism in the Simplon area, Switzerland and Italy, *Geol. Rdsch.*, **51**, 1, (1961).

CHAYES, F., On the association of perthitic microcline with highly undulant and granular quartz in some calc-alkaline granites, *Am. J. Sci.*, **250**, 281 (1952).

CHELIUS, C., Das Granitmassiv des Melibocus und seine Ganggesteine, *Notizblatt des Vereins für Erdkunde und der Groszh. Geol. Land zu Darmstadt*, **4**, 13, 1 (1892).

CHEN, F. P. H., Kinetic studies of crystallization of synthetic mica glass, *J. Am. Ceram. Soc.*, **46**, 476 (1963).

CHINNER, G. A., Pelitic gneisses with varying ferrous/ferric ratios from Glen Clova, Angus, Scotland, *J. Petrology*, **1**, 178 (1960).

CHINNER, G. A., The origin of sillimanite in Glen Clova, *J. Petrology*, **2**, 312 (1961).

CHINNER, G. A., Almandine in thermal aureoles, *J. Petrology*, **3**, 316 (1962).

CHRISTIAN, J. W., *Theory of Transformations in Metals and Alloys*. Pergamon Press, Oxford, 1965.

CHRISTIE, J. M., Mylonitic Rocks of the Moine Thrust Zone in the Assynt District, North West Scotland, *Trans. geol. Soc. Edinb.*, **18**, 79 (1960).

CHRISTIE, J. M., Moine Thrust Zone in the Assynt region, North West Scotland, *Univ. Calif. Publ. geol. Sci.*, **40**, 6, 345 (1963).

CHRISTIE, J. M. and RALEIGH, C. B., Origin of deformation lamellae in quartz, *Am. J. Sci.*, **257**, 385 (1959).

CHRISTIE, J. M., GRIGGS, D. T. and CARTER, N. L., Experimental evidence of basal slip in quartz, *J. Geol.*, **72**, 734 (1964).

CLARK, R. and CRAIG, C. B., Twinning, *Prog. Metal Phys.*, **3**, 115 (1952).

Cloos, E., Lineation, *Mem. geol. Soc. Am.*, **18** (1946).

Clough, C. T., Geology of the Cheviot Hills, *Mem. geol. Surv. U.K.* (1888).

Clough, C. T., In W. Gunn *et al.*, Geology of Cowal, *Mem. Geol. Surv. Scotland* (1897).

Clough, C. T., Maufe, H. B. and Bailey, E. B., Cauldron Subsidence of Glen-Coe, *Q. Jl. geol. Soc. Lond.*, **65**, 611 (1909).

Cohen, M., Martensite transformation. In *Phase Transformation in Solids*, p. 588 (Eds. R. Smoluchowski *et al.*) Wiley, New York, 1951.

Coleman, R. G. and Lee, D. E., Metamorphic aragonite in the glaucophane schists of Cazadero Calif., *Am. J. Sci.*, **260**, 577 (1962).

Coleman, R. G. and Lee, D. E., Glaucophane-bearing metamorphic rock types of the Cazadero area, California, *J. Petrology*, **4**, 260 (1963).

Coleman, R. G., Lee, D. E., Beatty, L. B. and Brannock, W. W., Eclogites and eclogites, their similarities and differences, *Bull. geol. Soc. Am.*, **76**, 483 (1965).

Collette, B. J., On helicitic structures and the occurrence of elongate crystals in the direction of the axis of a fold, *Proc. K. ned. Akad. Wet., Ser. B*, **62**, 3, 161 (1959).

Comagni, P., Chianotti, G. and Manara, A., Generation of vacancies during plastic deformation of KCl crystals, *Physics Chem. Solids*, **17**, 165 (1960).

Conrad, H., Mechanical behavior of sapphire, *J. Amer. Ceram. Soc.*, **48**, 195 (1965).

Conrad, H., Stone, G. and Janowski, K., Yielding and flow of sapphire (alpha-Al_2O_3) crystals in tension and compression, *Trans. Am. Inst. metall. Engrs.*, **233**, 889 (1964).

Coombs, D. S., Nature and alteration of some Triassic sediments from Southland, New Zealand, *Trans. R. Soc. N.Z.*, **82**, 65 (1954).

Coombs, D. S., Lower grade mineral facies in New Zealand, *Rept. Int. Geol. Congress XXI, Norden*, **13**, 339 (1960).

Coombs, D. S., Some recent work on the lower grades of metamorphism, *Aust. J. Sci.*, **24**, 5, 203 (1961).

Coombs, D. S., Ellis, A. J., Fyfe, W. S. and Taylor, A. M., Zeolite facies with comments on the interpretation of hydrothermal synthesis, *Geochim. cosmochim. Acta*, **17**, 53 (1959).

Correns, C. W., Explanation of the so-called force of crystallization, *Sber. Akad. Wiss. math-physik. Klasse*, **11**, 81, 8 (1926).

Correns, C. W., *Einführung in die Mineralogie* (*Kristallographie und Petrologie*). Springer, Berlin, 1949.

Cottrell, A. H., *Dislocations and Plastic Flow in Crystals*. Clarendon Press, Oxford, 1953.

Cottrell, A. H., *Theory of Crystal Dislocations*. Blackie, London, 1964a.

Cottrell, A. H., *Mechanical Properties of Matter*. Wiley, New York, 1964b.

Crampton, C. B., Quartz fabric reorientation in the region of Ben More Assynt, North West Highlands of Scotland, *Geol. Mag.*, **100**, 361 (1963).

Crook, A. W. K., Petrology of Parry Group, Upper Devonian-Lower Carboniferous, Tamworth-Nundle District, New South Wales, *J. sedim. Petrol.*, **30**, 538 (1960).

Crook, A. W. K., Diagenesis in the Wahgi Valley sequence, New Guinea, *Proc. R. Soc. Vict.*, **74**, 77 (1961).

Crook, A. W. K., Burial metamorphic rocks from Fiji, *N.Z. Jl. Geol. Geophys.*, **6**, 681 (1963).

Crook, A. W. K. and Crook, P. J., Gosses Bluff-Diapir, Crypto-volcanic Structure or Astrobleme? *J. geol. Soc. Aust.*, **13**, 495 (1966).

Crowell, J. C. and Walker, J. W. R., Anorthosite and related rocks along the San Andreas Fault, Calif., *Univ. Calif. Pub. Geol. Sci.*, **40**, 219 (1962).

Cunningham-Craig, E. H., Metamorphism in the Loch Lomond district, *Q. Jl. geol. Soc. Lond.*, **60**, 10 (1904).

Currie, K. L., Shock metamorphism in the Carswell Circular Structure, Saskatchewan, *Nature*, **213**, 56 (1967).

Cutler, I. B., Nucleation and nucleii growth in sintered alumina. In *Kinetics of High Temperature Processes*, p. 120 (Ed. W. D. Kingery). Tech. Press, M.I.T. and Wiley, New York, 1959.

Daly, R. A., Metamorphism and its phases, *Bull. geol. Soc. Am.*, **28**, 400 (1917).

DASGUPTA, D. R., Oriented transformation of aragonite into calcite, *Mineralog. Mag.*, **33**, 265, 924 (1964).

DASGUPTA, D. R., Transformation of ankerite during heat treatment, *Mineralog. Mag.*, **35**, 634 (1965).

DASH, M. J. and BROWN, N., Investigation of the origin and growth of annealing twins, *Acta metall.*, **11**, 1067 (1963).

DAVIES, J. T. and RIDEAL, A. K., *Interfacial Phenomena*. Academic Press, New York and London, 1961.

DAY, R. B. and STOKES, J. A., Mechanical behaviour of polycrystalline magnesium oxide at high temperatures, *J. Am. Ceram. Soc.*, **49**, 345 (1966).

DE CARLI, P. S. and JAMIESON, J. C., Formation of an amorphous form of quartz under shock conditions, *J. Chem. Phys.*, **31**, 1675 (1959).

DEER, W. A., HOWIE, R. A. and ZUSSMAN, J., *Rock-forming Minerals*. Vols. 1–5. Longmans, London, 1962–4.

DENT, B. M., Effect of boundary distortion on the surface energy of a crystal, *Phil. Mag.*, **8**, 530 (1929).

DENT GLASSER, L. S. D., GLASSER, F. P. and TAYLOR, H. F., Topotactic reactions in inorganic oxy-compounds, *Quart. Rev.*, **16**, 343 (1962).

DESCH, C. H., Crystallization of alloys, *Proc. R. Inst. G. B.*, 1934.

DE VORE, G. W., Role of absorption in the fractionation and distribution of sediments, *J. Geol.*, **63**, 153 (1955a).

DE VORE, G. W., Crystal growth and the distribution of elements, *J. Geol.*, **63**, 471 (1955b).

DE VORE, G. W., Role of minimum interfacial free energy in determining the macroscopic features of mineral assemblages, I, the model, *J. Geol.*, **67**, 211 (1959).

DE VORE, G. W., Compositions of silicate surfaces and surface phenomena, *Wyoming Uni. Contrib. geol.*, **2**, 21 (1963).

DIENES, C. J., Introduction to defects in silicas, *Physics Chem. Solids*, **13**, 272 (1960).

DIETZ R. S., Cryptoexplosion structures, *Am. J. Sci.*, **261**, 650 (1963).

DILLAMORE, I. C., Factors affecting the rolling recrystallization textures in face-centred-cubic crystals, *Acta metall.*, **12**, 1005 (1964).

DOHERTY, P. E. and CHALMERS, B., Origin of lineage substructure in aluminium, *Trans. Am. Inst. Min. metall. Engrs.*, **224**, 1124 (1962).

DONNAY, J. H. D., Width of albite-twinning lamellae, *Am. Miner.*, **25**, 578 (1940).

DONNAY, J. H. D., Plagioclase twinning, *Bull. geol. Soc. Am.*, **54**, 1645 (1943).

DONNELLY, I. W., Kinetic considerations in the genesis of growth twinning, *Am. Mineralog.* **52**, 1 (1967).

DOUGLAS, R. W., Recent studies in the physics and chemistry of glass, *Sci. Prog., Wash.*, **53**, 61 (1965).

DREVER, H. I., Symplectite-bearing nodules from Argyllshire, *Geol. Mag.*, **73**, 448 (1936).

DREVER, H. I., Pseudotachylite of the Gairloch district, Ross-shire, *Am. J. Sci.*, **259**, 542 (1961).

DROMSKY, J. A., LORD, F. V. and ANSELL, G. S., Growth of aluminium oxide particles in a nickel matrix, *Trans. Am. Inst. Min. metall. Engrs.*, **224**, 236 (1962).

DUNN, C. G., DANIELS, F. W. and BOLTON, M. J., Relative energies of grain boundaries in silicon–iron, *Trans. Am. Inst. Min. Metall. Engrs.*, **188**, 1245 (1950).

DUSCHATKO, R. W. and POLDERVAART, A., Spilitic intrusion near Ladron Peak, Socorro Co., New Mexico, *Bull. geol. Soc. Am.*, **66**, 1097 (1955).

ECKERMAN, H. VON, Anorthosite and Kenningite of the Nordingra-Rodo region, *Medd. Stockh. Högsk Min. Inst.*, **101**, 283 (1938).

EDWARDS, A. B., *Textures of the Ore Minerals*, 2nd ed. Aust. Inst. Min. Met. Engrs., Melbourne. 1960.

EDWARDS, A. B. and BAKER, G., Contact phenomena in the Morang Hills, Victoria, *Proc. R. Soc. Vict.*, **56**, 19 (1944).

EDWARDS, A. B. and BAKER, G., Scapolitization in the Cloncurry district of Northwestern Queensland, *J. geol. Soc. Aust.*, **1**, 1 (1954).

EITEL, W., *Physical Chemistry of the Silicates*. Uni. Chicago Press, 1954.

EITEL, W., Structural conversions in crystalline systems, *Spec. Pap. geol. Soc. Am.*, **66**, (1958).
EITEL, W., *Silicate Science*, Vols. 1 and 2. Academic Press, New York, 1965.
ELLISTON, J. N., Sediments of the Warramunga Geosyncline. In *Syntaphral Tectonics and Diagenesis*, p. li (Ed. S. W. Carey) Univ. Tasmania. Publ.
EMBURY, J. D. and NICHOLSON, R. B., The system Al-Zn-Mg, *Acta metall.*, **13**, 403 (1965).
EMMONS, R. C. and GATES, R. M., Plagioclase twinning, *Bull. geol. Soc. Am.*, **54**, 287 (1943).
EMMONS, R. C. and MANN, V., A twin-zone relationship in plagioclase feldspar, *Mem. geol. Soc., Am.*, **52**, 41 (1953).
ERNSBERGER, F. M., Structural effects in the chemical reactivity of silica and the silicates, *Physics Chem. Solids*, **13**, 347 (1952).
ERNST, W. G., Petrogenesis of glaucophane schists, *J. Petrology*, **4**, 1 (1963).
ERNST, W. G. and BLATT, H., Experimental study of quartz overgrowths and synthetic quartzites, *J. Geol.*, **72**, 461 (1964).
ESKOLA, P., On the eclogites of Norway, *Vidensk. Meddr. dansk naturh. Foren. Kl.* 8 (1921).
ESKOLA, P., On the principles of metamorphic differentiation, *Bull. Comm. geol. Finl.*, **97**, 68 (1932b).
ESKOLA, P., In Barth, T. F. W., Correns, C. W., and Eskola, P., *Die Enstehung der Gesteine*. Springer, Berlin, 1939.
ESKOLA, P., On the granulites of Lapland, *Am. J. Sci., Bowen Vol.*, 133 (1952).
EVANS, B. W. and LEAKE, B. E., Composition and origin of striped amphibolites of Connemara, Ireland, *J. Petrology*, **1**, 337 (1960).
FAIRBAIRN, H. W., Correlation of quartz deformation with its crystal structure, *Am. Miner.*, **24**, 351 (1939).
FAIRBAIRN, H. W., *Structural Petrology of Deformed Rocks*, 2nd ed. Addison-Wesley, Reading. 1949.
FAIRBAIRN, H. W., Pressure shadows and relative movements in a shear zone, *Trans. Am. geophys. Un.*, **31**, 914 (1950).
FELLOWS, R. E., Recrystallization and flowage in Appalachian quartzite, *Bull. goel. Soc. Am.*, **54**, 1399 (1943).
FLETT, J. H., Geology of Ben Wyvis, Carn Chuinneag, etc., *Mem. geol. Surv. Scotland*, (1912).
FLINN, D., Deformation in metamorphism. In *Controls of Metamorphism* (Eds. W. S. Pitcher and G. S. Flinn). Oliver & Boyd, Edinburgh, 1965.
FLINN, P. A., Theory of deformation in superlattices, *Trans. Am. Inst. Min. metall. Engrs.*, **218**, 145 (1960).
FOSTER, R. J., Origin of embayed quartz crystals in acidic volcanic rocks, *Am. Miner.*, **45**, 892 (1960).
FOURIE, J. T., WEINBERG, F. and BOSWELL, F. W. C., Growth of twins in single crystals as observed by transmission electron microscopy, *Acta metall.*, **8**, 851 (1960).
FRANCIS, G. H., Facies boundaries in pelites at the middle grades of regional metamorphism, *Geol. Mag.*, **93**, 353 (1956).
FRANKEL, J. J., Abraded pyrite crystals from the Witwatersrand gold mines, *Mineralog. Mag.*, **31**, 392 (1957).
FREDERICKSON, A. F., Mosaic structure in quartz, *Am. Miner.*, **40**, 1 (1955).
FRIEDEL, G., *Leçons de crystallographie professées à la Faculté des Sciences de Strasbourg*. Berger Lerrault, Paris, 1926.
FRIEDEL, G., *Dislocations*. Pergamon Press, Oxford, 1964.
FRIEDMAN, G. M., Terminology of crystallization textures and fabrics in sedimentary rocks, *J. sedim. Petrol.*, **35**, 643 (1965).
FRIEDMAN, M., Petrofabric techniques for the determination of principle stress directions in rocks. In *State of Stress in the Earth's Crust*, p. 451 (Ed. W. R. Judd). Elsevier, New York, 1964.
FRONDEL, C., Selective incrustation of crystal forms, *Am. Miner.*, **19**, 316 (1934).
FRONDEL, C., Oriented inclusions of staurolite, zircon and garnet in muscovite; skating crystals and their significance, *Am. Miner.*, **25**, 69 (1940).
FRONDEL, C., Secondary Dauphiné twinning in silica, *Am. Miner.*, **30**, 447 (1945).

FRONDEL, C. and ASHBY, G. E., Oriented inclusions of magnetite and hematite in muscovite, *Am. Miner.*, **22**, 104 (1937).

FRONDELL, C., NEWHOUSE, W. R. and JARRELL, R. F., Spatial distribution of minor elements in single-crystals, *Am. Miner.*, **27**, 726 (1942).

FROST, A. A. and PEARSON, R. G., *Kinetics and Mechanism*. John Wiley, New York and London, 1961.

FUKUTA, N. and MASON, B. J., Epitaxial growth of ice on organic crystals, *Physics Chem. Solids*, **24**, 715 (1963).

FULLMAN, R. L. and FISHER, J. L., Formation of annealing twins and coincidence-site boundaries in high purity lead, *Trans. Am. Inst. Min. metall. Engrs.*, **218**, 1023 (1951a).

FULLMAN, R. L. and FISHER, J. L., Formation of annealing twins during grain growth, *J. appl. Phys.*, **22**, 1350 (1951b).

FYFE, W. S., *Geochemistry of Solids*. McGraw-Hill, New York, 1964.

FYFE, W. S., TURNER, F. J. and VERHOOGEN, J., Metamorphic reactions and metamorphic facies, *Mem. geol. Soc. Am.*, **73** (1958).

GAERTNER, H. R. VON, Petrographie . . . der Gipse vom Suedrande des Harzes, *Jb. preuss. geol. Landesanst. Berg. Akad.*, **53**, 655 (1932).

GALWEY, A. K. and JONES, K. A., Attempt to determine the mechanism of a natural mineral-forming reaction, *J. chem. Soc.*, **1084**, 5681 (1963).

GANSSER, A., Der Nortrand der Tambodecke, *Schweiz. min. pet. Mitt.* **17**, 291 (1937).

GARNICK, F., Initiation of spherulitic growth, *J. appl. Phys.*, **36**, 3012 (1965).

GARRIDO, J., Development of twinning in the dehydration of brucite, *Am. Miner.*, **36**, 773 (1951).

GATES, R. M., Petrogenetic significance of perthite, *Mem. geol. Soc. Am.*, **53**, 55 (1953).

GATES, R. M., Amphibolites: syntectonic intrusives. *Am. J. Sci.*, **265**, 118 (1967).

GAY, P., Note on albite twinning in feldspars, *Mineralog. Mag.*, **31**, 301 (1956).

GAY, P., Subsolidus relations in the plagioclase feldspars, *Norsk. geol. Tidsskr.*, **42**, 37 (1962).

GEISLER, A. H., Precipitation from solid solutions of metals, In *Phase Transformations in Solids*, p. 387 (Eds. R. Smoluchowski, J. E. Mayer and W. A. Weyl). Wiley, New York, 1951.

GHENT, E. D., Glaucophane schist facies metamorphism in the Black Butte area, Northern Coast Ranges, California, *Am. J. Sci.*, **263**, 345 (1965).

GIBBS, P., Imperfection interactions in aluminium oxide. In *Kinetics of High Temperature Processes* (Ed. W. D. Kingery). Technical Press M.I.T., and Wiley, New York, 1959.

GIFKINS, R. C. and LANGDON, T. G., On the question of low-temperature gliding at grain boundaries, *J. Inst. Metals*, **93**, 347 (1965).

GILMAN, J. J., Cleavage, ductility and tenacity in crystals. In *Fracture. Proc. Conference Swampscott*. Wiley, New York, 1959.

GILMAN, J. J., Direct measurement of the surface energies of crystals, *J. app. Phys.*, **31**, 2208 (1960).

GILMAN, J. J., Mechanical behavior of ionic crystals. In *Progress in Ceramic Science*, vol. 1, p. 146. Pergamon Press, Oxford, 1961.

GILMAN, J. J. (Ed.) *Art and Science of Growing Crystals*. Wiley, New York, 1963.

GIRIFALCO, L. A., *Atomic Migration in Crystals*. Blaisdall, New York and London, 1964.

GJELSVIK, T., Metamorphosed dolerites in the gneiss area of Sunnmore on the West Coast of Southern Norway, *Norsk. geol. Tidsskr.*, **30**, 33 (1952).

GLOVER, E. D. and SIPPEL, R. F., Experimental pseudomorphs: a replacement of calcite by fluorite, *Am. Miner.*, **47**, 1156 (1962).

GOGUEL, J., *Tectonics*. Freeman, San Francisco, 1960.

GOLDICH, S. S. and KINSER, J. H., Perthite from Toby Hill, Ontario, *Am. Miner.*, **24**, 407 (1939).

GOLDMAN, M. I., Deformation, metamorphism and mineralization in gypsum–anhydrite cap rock, Sulphur Salt Dome, Louisiana, *Mem. geol. Soc. Am.*, **50** (1952).

GOLDSCHMIDT, V. M., Die Kontaktmetamorphose im Kristianagebiet, *Vidensk. Meddr. dansk naturh. Foren.*, **11** (1911).

GOLDSMITH, R., Granofels, a new metamorphic rock. *J. Geol.*, **67**, 109 (1959).

GOMER, R. and SMITH, C. S., *Structure and Properties of Solid Surfaces*. Univ. Chicago Press, 1952.
GOODSPEED, G. E., Development of quartz porphyroblasts in siliceous hornfels, *Am. Miner.*, **22**, 133 (1937a).
GOODSPEED, G. E., Development of plagioclase porphyroblasts, *Am. Miner.*, **22**, 1135 (1937b).
GOODSPEED, G. E., Some textural features of magmatic and metasomatic rocks, *Am. Miner.*, **44**, 211 (1959).
GORAI, M., Petrological studies of plagioclase twins, *Am. Miner.*, **36**, 884 (1951).
GRAY, T. J., Diffusion in ionic lattices, sintering and reaction in the solid state. In *Defect Solid State*. Interscience, New York, 1957.
GRAY, T. J., DETWILER, D. P., RASE, D. E., LAWRENCE, W. G., WEST, R. R. and JENNINGS, T. J., *Defect Solid State*. Interscience, New York, 1957.
GREENWOOD, J. W., Syntheses and stability of anthophyllite, *J. Petrology*, **4**, 317 (1963).
GREENWOOD, J. W., and MCTAGGART, K. C., Correlation of zones in plagioclase, *Am. J. Sci.*, **255**, 656 (1957).
GRIGGS, D. T. and BLASIC, J. D., Quartz; anomalous weakness of synthetic crystals, *Science*, **147**, 3655, 292 (1965).
GRIGGS, D. T., PATERSON, M. S., HEARD, H. C. and TURNER, F. J., Annealing recrystallization in calcite crystals and aggregates, *Mem. geol. Soc. Am.*, **79**, 21 (1960).
GRIGGS, D. T., TURNER, F. J. and HEARD, H., Deformation of rocks at 500° to 800°C., *Mem. geol. Soc. Am.*, **79**, 60 (1960).
GRIGOREV, D. P., *Ontogeny of Minerals*. Israel Program for Scientific Translations, Jerusalem, 1965.
GRIM, R. E., *Clay Mineralogy*. McGraw-Hill, New York, 1953.
GROSS, K. A., X-ray line broadening and stored energy in deformed and annealed calcite, *Phil. Mag.*, **12**, 118, 801 (1965).
GROUT, F. F., *Petrography and Petrology*. McGraw-Hill, New York, 1932.
GROUT, F. F., Formation of igneous-looking rocks by metasomatism. A critical review and suggested research, *Bull. geol. Soc. Am.*, **52**, 1595 (1941).
GROVES, A. W., The charnockite series of Uganda, British East Africa, *Q. Jl. geol. Soc. Lond.*, **19**, 150 (1935).
GRUBENMANN, U., *Die Kristallinen Schiefer*. Berlin, 1904.
GRUBENMANN, U. and NIGGLI, P., *Die Gesteinmetamorphose*. Borntraeger, Berlin, 1924.
GRUNER, J. W., Structural reasons for oriented intergrowths in some minerals, *Am. Miner.*, **14**, 227 (1929).
GUY, A. G. and PHILIBERT, J., Effect of strain on diffusion in metals, *Trans. Am. Inst. Min. metall. Engrs.*, **221**, 1174 (1961).
HALL, G. M., Zoisite and other minerals included in mica from Spruce Pine, N. Carolina, *Am. Miner.*, **19**, 76 (1937).
HAMILTON, J., Banded olivine in some Scottish Carboniferous olivine-basalts, *Geol. Mag.*, **94**, 135 (1957).
HARKER, A., *Metamorphism*. Methuen, London, 1939.
HARKER, D. and PARKER, E. R., Grain shape and grain growth, *Trans. Am. Soc. Metals*, **34**, 156 (1945).
HARKER, R. I., Older orthogneisses of Carn Chuinneag and Inchbae, *J. Petrology*, **3**, 215 (1962).
HARRIS, A. L. and RAST, N., Oriented quartz inclusions in garnets, *Nature*, **185**, 448 (1960a).
HARRIS, A. L. and RAST, N., The evolution of quartz fabrics in the metamorphic rocks of Central Perthshire, *Trans. geol. Soc. Edinb.*, **18**, 51 (1960b).
HARTE, B. and HENLEY, K. J., Occurrence of compositionally zoned almandite garnets in regionally metamorphosed rocks, *Nature*, **210**, 5037, 689 (1966).
HARTMANN, P., On the morphology of growth twins, *Z. Kristallogr.* **107**, 225 (1956).
HATCH, F. H., WELLS, A. K. and WELLS, M. K., *Petrology of the Igneous Rocks*. Murby, London, 1961.
HAUFFE, K., Investigations on the oxidation of metals as a tool for the exploration of the movement of ions in solid oxides. In *Kinetics of High Temperature Processes*, p. 283 (Ed. W. D. Kingery). Technical Press, M.I.T. and Wiley, New York, 1959.

HEAD, R. B., Ice nucleation by some cyclic compounds, *Physics Chem. Solids*, **23**, 1371 (1962).
HEALD, M. T., Stylolites in sandstones, *J. Geol.*, **63**, 101 (1955).
HEIER, K. S., Amphibolite-granulite facies transition reflected in the mineralogy of potassium feldspars, *Estudos Geologicos, Cursillos y Conferencias*, **8**, 131 (1961).
HERRING, C., Diffusional viscosity of a polycrystalline solid, *J. appl. Phys.*, **21**, 437 (1950).
HERRING, C., Theorems on the free energies of crystal surfaces, *Phys. Rev.*, **82**, 87 (1951).
HIETANEN, A., Petrology of Finnish quartzites, *Geol. Comm. Finlande Bull.*, **122** (1938).
HIETANEN, A., Kyanite, andalusite and sillimanite in the schist in Boehls Butte Quadrangle, Idaho, *Am. Miner.*, **41**, 1 (1956).
HIGGS, D. V. and HANDIN, J. W., Experimental deformation of single dolomite crystals, *Bull. geol. Soc. Am.*, **70**, 247 (1959).
HILLS, E. S., *Elements of Structural Geology*. Methuen, London, 1963.
HIRTHE, W. M. and BRITTAIN, J. O., Dislocations in rutile as revealed by the etch pit technique, *J. Am. Ceram. Soc.*, **45**, 546 (1962).
HIRTHE, W. M. and BRITTAIN, J. O., High temperature steady state creep in rutile, *J. Am. Ceram. Soc.*, **46**, 411 (1963).
HOBBS, B. E., Microfabric of tectonites from the Wyangala Dam area, N.S.W., *Bull. geol. Soc. Am.*, **77**, 685 (1966).
HODGSON, A. A., FREEMAN, A. G. and TAYLOR, H. F. W., Thermal decomposition of crocidolite from Koegos, S. Africa, *Mineralog. Mag.*, **35**, 3 (1965).
HOGAN, L. M., Nucleation and growth of the Al-$CuAl_2$ eutectic, *J. Aust. Inst. Metals*, **10**, 78 (1965).
HOLLAND, T., On the origin and growth of garnets, *Rec. geol. Surv. India*, **29**, 20 (1896).
HOLLAND, T., The charnockite series—a group of Archean hypersthene rocks in Peninsular India, *Mem. geol. Surv. India*, **28**, 2, 119 (1900).
HOLLISTER, L. S., Garnet zoning: an interpretation based on the Raleigh fractionation model, *Sci.*, **154**, 1647 (1966).
HOLMES, A., *Nomenclature of Petrology*. Murby, London, 1928.
HOLMQUIST, P. J., Die Hochgebirgsbildungen am Torne Träsk in Lappland, *Geol. För. Stockh. Förh.*, **32**, 913 (1910).
HOLSER, W. T., Metamorphism and associated mineralization in the Phillipsburg region, Montana, *Bull. geol. Soc. Am.*, **61**, 1053 (1950).
HORNBOGEN, E., Dynamic effects during twinning in alpha-iron, *Trans. Am. Inst. Min. metall. Engrs.*, **221**, 711 (1961).
HORNSTRA, J., Dislocations, stacking faults and twins in the spinel structure, *Physics Chem. Solids*, **15**, 311 (1960).
HORVAY, G. and CAHN, J. W., Dendritic and spheroidal growth, *Acta metall.*, **9**, 695 (1961).
HOWIE, R. A., Charnockites, *Sci. Progr.*, **62**, 208, 628 (1964).
HOWKINS, J. B., Helicitic textures in garnets from Moine rocks of Moidart, *Trans. geol. Soc. Edinb.*, **18**, 315 (1961).
HSU, H., Granulites and mylonites of the . . . San Gabriel Mtns., *Univ. Calif. Publ. Geol. Sci.*, **30**, 44 (1955).
HU, H., Direct observations on the annealing of a Si-Fe crystal in the electron microscope, *Trans. Am. Inst. Min. metall. Engrs.*, **224**, 75 (1962).
HU, H. and SMITH, C. S., Formation of low-energy interfaces during grain growth in alpha and beta brasses, *Acta. metall.*, **4**, 638 (1956).
HUANG, W. T. and MERRIT, C. A., Petrography of the troctolite of the Wichita Mountains, Oklahoma, *Am. Miner.*, **39**, 549 (1952).
HUBBARD, F. H., Antiperthite and mantled feldspar textures in charnockite (enderbite) from S.W. Nigeria, *Am. Miner.*, **50**, 2040 (1965).
HULSE, C. O., COPLEY, G. M. O. and PASK, J. A., Effect of crystal orientation on plastic deformation of magnesium oxide, *J. Am. Ceram. Soc.*, **46**, 417 (1963).
HUTTON, C. O. and TURNER, F. J., Metamorphic zones in North West Otago, *Trans. R. Soc. N.Z.*, **65**, 405 (1936).
INGERSON, E. and BARKSDALE, J. D., Iridescent garnet from the Adelaide mining district, Nevada, *Am. Miner.*, **28**, 303 (1943).

INMAN, M. C. and TIPPLER, H. R., Interfacial energy and composition in metals and alloys, *Metals Rev.*, **8**, 30, 105 (1963).
IRVING, C., MIODOWNIK, A. P. and TOWNER, J. M., Twin abundance of some copper alloys and its relation to stacking fault energy, *J. Inst. Metals.*, **10**, 360 (1965).
JAEGER, J. C., Temperatures outside of a cooling intrusive sheet, *Am. J. Sci.*, **257**, 44 (1957).
JAFFE, H. W., Postanorthosite gabbro near Avalanche Lake in Essex County, New York, *J. Geol.*, **54**, 105 (1946).
JAGITSCH, R. and MATS-GORAN, D., Geologische Diffusionen in Kristallierten Phasen, *Proc. Internat. Symp. Reactivity of Solids, Gothenburg*, p. 463 (1954).
JOHANNSEN, A., *Descriptive Petrography of the Igneous Rocks*, vol. 1. Univ. Chicago Press, 1939.
JOHNS, W. D., Review of topotactic development of high temperature phases from two-layer silicates, *J. Am. Ceram. Soc.*, **44**, 682 (1965).
JOHNSON, A., Biegungen und Translationen, *Neues Jb. Miner. Geol. Paläont.*, Beil Bd., **2**, 133 (1902).
JOHNSON, M. R. W., Polymetamorphism in movement zones in the Caledonian thrust belt of Northwest Scotland, *J. Geol.*, **69**, 417 (1961).
JOHNSON, M. R. W., Relations of movement and metamorphism in the Dalradian of Banffshire, *Trans. geol. Soc. Edinb.*, **19**, 29 (1962).
JOHNSON, M. R. W., Some time-relations of movement and metamorphism in the Scottish Highlands, *Geologie Mijnb.*, **5**, 121 (1963).
JONES, K. A., Origin of albite porphyroblasts in rocks of the Ben More-Am Binnein area, Western Perthshire, *Geol. Mag.*, **98**, 41 (1961).
JONES, K. A. and GALWEY, A. K., Study of possible factors concerning garnet formation in rocks from Ardara, Co. Donegal, *Geol. Mag.*, **101**, 79 (1964).
JONES, K. A. and GALWEY, A. K., Size distribution, composition and growth kinetics of garnet crystals in some metamorphic rocks from the West of Ireland, *Q. Jl. geol. Soc. Lond.*, **122**, 29 (1966).
JOPLIN, G. A., Note on the origin of basic xenoliths in plutonic rocks, *Geol. Mag.*, **72**, 227 (1935).
JOPLIN, G. A., *A Petrography of Australian Metamorphic Rocks*. Angus & Roberton, Sydney, 1968.
JOSHI, M. S. and VAGH, A. S., Role of spirals in the growth of prism faces on cultured quartz, *J. app. Phys.*, **37**, 315 (1966).
JOYCE, B. A., BENNETT, R. J., BICKNELL, R. W. and ETTER, P. J., Epitaxial deposition of silicon on quartz and alumina, *Trans. Am. Inst. Min. metall. Engrs.*, **233**, 556 (1965).
KALLIOKOSKI, J., Metamorphic features in North American massive sulphide deposits, *Econ. Geol.*, **60**, 485 (1965).
KAMB, W. B., Theory of preferred orientation developed by crystallization under stress, *Mem. geol. Soc. Am.*, **79**, 9 (1959).
KANTER, M. A., Mechanism for atom motion in graphite crystals, In *Kinetics of High Temperature Processes* (Ed. W. D. Kingery). Technical Press M.I.T., and Wiley, New York, 1959.
KEITH, R. E. and GILMAN, J. J., Dislocation etch pits and plastic deformation in calcite, *Acta metall.*, **8**, 1 (1960).
KELLER, A., Morphology of crystalline polymers. In *Growth and Perfection of Crystals*, p. 499 (Eds. R. H. Doremus, B. W. Roberts and D. Turnbull). Wiley, New York, 1958.
KEUNEN, PH. H., Pitted pebbles, *Leid. geol. Meded.*, **13**, 1, 189 (1943).
KINGERY, W. D., Sintering in the presence of a liquid phase. In *Kinetics of High Temperature Processes*, p. 187 (Ed. W. D. Kingery). Technical Press M.I.T., and Wiley, New York, 1959.
KINGERY, W. D., *Introduction to Ceramics*. Wiley, New York, 1960.
KLASSEN-NEKLYUDOVA, M. V., *Mechanical Twinning of Crystals*. Am. Consultants Bur. (Transl. J. E. S. Bradley), 1964.
KNAPP, O., *Devitrification of Glasses*. Hungarian Acad. Science, Budapest, 1965.
KNOPF, A., Partial fusion of granodiorite by intrusive basalt, Owens Valley, California, *Am. J. Sci.*, **36**, 373 (1938).

KNOPF, E. B., Retrogressive metamorphism and phyllonitization, *Am. J. Sci.*, **21**, 1 (1931).

KNOPF, E. B. and INGERSON, E., Structural Petrology, *Mem. geol. Soc. Am.*, 6 (1938).

KOKORSCH, R., Zur Kenntnis von Genesis, Metamorphose und Faziesverhälten des Stassfurtlagers im Grubenfeld Hildesia—Mahildenhall, Dieckholzen bei Hildescheim, *Beih. geol. Jb.*, **41**, 1 (1960).

KOPPENOAK, T. J., PARTHASARATHI, M. N. and BECK, P. A., Effect of grain growth on the formation of cube-texture in an Al-Mn alloy, *Trans. Am. Inst. Min. metall. Engrs.*, **218**, 98 (1960).

KOZLOWSKI, K., On the eclogite-like rocks of Stary Gieraltow, *Bull. Acad. pol. Sci. cl III, Math.*, **6**, 723 (1958).

KRANK, E. H. and OJA, R. V., Experimental studies of anatexis, *Int. geol. Congress, Copenhagen*, **14**, 16 (1960).

KRETZ, R., Interpretation of the shape of mineral grains in metamorphic rocks, *J. Petrology*, **7**, 68 (1966a).

KRETZ, R., Grain size distribution for certain metamorphic minerals in relation to nucleation and growth, *J. Geol.*, **74**, 147 (1966b).

KRONBERG, M. L. and WILSON, F. H., Dynamical flow properties of single crystals of sapphire, *J. Am. Ceram. Soc.*, **45**, 274 (1962).

KUNO, H., Petrology of Hakone Volcano, Japan, *Bull. geol. Soc. Am.*, **61**, 95 (1950).

KUZNETZOV, V. D., *Surface Energy of Solids*. Trans. from Russian by D.S.I.R. H.M. Stationery Office, London, 1957.

LACEY, E. D., Configuration change in silicates with particular reference to network structure, *Acta crystallogr.*, **18**, 141 (1965a).

LACEY, E. D., Factors in the study of metamorphic reaction rates. In *Controls of Metamorphism* (Eds. W. D. Pitcher and G. S. Flinn). Oliver & Boyd, Edinburgh, 1965(b).

LACROIX, A., *Les enclaves des roches volcaniques*. Protat, Mason, 1893.

LAPWORTH, C., The Highland controversy in British geology, *Nature*, **32**, 558 (1885).

LARSEN, E. S. and SWITZER, G., An obsidian-like rock formed from the melting of granodiorite, *Am. J. Sci.*, **237**, 62 (1939).

LAUDER, W. R., Reaction of crystal structures and reaction fabric, *Am. Miner.*, **46**, 1317 (1961).

LAUDER, W. R., Geology of Dun Mtn., Nelson, New Zealand, *N.Z. J. Geol. Geophys.*, **8**, 475 (1965).

LAVES, F., The lattice and twinning of microcline, *J. Geol.*, **58**, 548 (1950).

LAVES, F., Relationship between exsolved plagioclase and its host, *Am. Cryst. Assn. Wash. Meeting* (1951).

LAVES, F., Phase relations of the alkali feldspars, *J. Geol.*, **60**, (1), 436, (2), 549 (1952a).

LAVES, F., Über die Einfluss von Ordnung und Unordnung auf mechanische Zwillingsbildung, *Naturwissenschaften*, **39**, 346 (1952b).

LAVES, F., Mechanische Zwillingsbildung in Feldspaten in Abhängigkeit von Ordnung-Unordnung der Si/Al Verteilung innerhalb des $(Si,Al)_4O_8$ Gerüstes, *Naturwissenschaften*, **39**, 546 (1952c).

LAVES, F., Mechanical twinning in acid plagioclases, *Am. Miner.*, **50**, 511 (1965).

LE COMTE, P., Creep in rock salt, *J. Geol.*, **73**, 469 (1965).

LEFEVER, R. A. and CHASE, A. B., Analysis of surface features of single crystals of synthetic garnets, *J. Am. Ceram. Soc.*, **45**, 32 (1962).

LEFEVER, R. A., CHASE, A. B. and TORPHY, J. W., Characteristic imperfections in flux-grown crystals of yttrium iron garnet, *J. Am. Ceram. Soc.*, **44**, 141 (1961).

LEHMANN, J., *Untersuchungen über die Entstehung der allkristallinen Schiefergesteine im Sächsischen Granulitgebirge*. Bonn, 1884.

LI, J. C. M., Possibility of sub-grain rotation during recrystallization, *J. app. Phys.*, **33**, 2958 (1962).

LIMA-DE-FARIA, J. and LOPES-VIERIRA, A., Transformation of groutite (alpha MnOOH) into pyrolusite (MnO_2), *Mineralog. Mag.*, **33**, 1024 (1964).

LONG, L. E. and LAMBERT, R. St. J., Rb–Sr isotope ages from the Moine Series. In *The British Caledonides* (Eds. M. R. W. Johnson and F. H. Stewart). Oliver & Boyd, Edinburgh, 1963.

LOVERING, J. F. and WHITE, A. J. R., Significance of primary scapolite in granulitic inclusions from deep-seated pipes, *J. Petrology*, **5**, 195 (1964).

LYALL, K. D., Origin of mechanical twinning in galena, *Am. Miner.*, **51**, 243 (1966).

LYALL, K. D. and PATERSON, M. S., Plastic deformation of galena, *Acta metall.*, **14**, 371 (1966).

MCCALL, G. J. H., Dalradian geology of the Creeslough area, Co. Donegal, *Q. Jl. geol. Soc. Lond.*, **110**, 153 (1954).

MCCALLIEN, W. J., Metamorphic rocks of Kintyre, *Trans. R. Soc. Edinb.*, **58**, 1, 163 (1929).

MCCONNELL, J. D. C., Electron optical effects associated with partial inversion in a silicate phase, *Phil. Mag.*, **11**, 114, 1289 (1965).

MACDONALD, G. J. F., Thermodynamics of solids under non-hydrostatic stress with geologic applications, *Am. J. Sci.*, **255**, 266 (1957).

MACDONALD, G. J. F., Orientation of anisotropic minerals in a stress field, *Mem. geol. Soc. Am.*, **79**, 1 (1960).

MACGREGOR, A. G., Clouded feldspars and thermal metamorphism, *Mineralog. Mag.*, **27**, 524 (1931a).

MACGREGOR, A. G., Scottish pyroxene-granulite hornfelses and Ostenwald beerbachites, *Geol. Mag.*, **68**, 506 (1931b).

MACGREGOR, A. J., Ice crystals in glaciers compared with quartz crystals in dynamically metamorphosed sandstones, *J. Glaciol.*, **1**, 564 (1950).

MACKENZIE, J. K., Stresses and energies associated with inter-crystalline boundaries, *Proc. phys. Soc. Lond.*, **A63**, 1370 (1950).

MACKENZIE, J. K., MOORE, A. J. W. and NICHOLAS, J. F., Bonds broken at atomically flat crystal surfaces, *Physics Chem. Solids*, **23**, 185 (1962).

MACKENZIE, W. S., Kyanite gneiss within a thermal aureole, *Geol. Mag.*, **86**, 251 (1949).

MACKENZIE, W. S., Orthoclase-microcline inversion, *Mineralog. Mag.*, **30**, 354 (1954).

MACKENZIE, W. S., Some comments on the application of experimental results to the study of metamorphism. In *Controls of Metamorphism* (Eds. W. S. Pitcher and G. S. Flinn). Oliver & Boyd, Edinburgh, 1965.

MCLAREN, A. C. and PHAKEY, P. P., Transmission electron microscope study of amethyst and citrine, *Aust. J. Phys.*, **18**, 135 (1965a).

MCLAREN, A. C. and PHAKEY, P. P., Dislocations in quartz observed by transmission electron microscopy, *J. appl. Phys.*, **36**, 3244 (1965b).

MCLAREN, A. C., RETCHFORD, J. A., GRIGGS, D. T. and CHRISTIE, J. M., Transmission electron microscope study of Brazil twins and dislocations experimentally produced in natural quartz, *Phys. stat. sol.*, **19**, 631 (1967).

MCLEAN, D., *Grain Boundaries in Metals*. Clarendon Press, Oxford, 1957.

MCLEAN, D., Science of metamorphism in metals. In *Controls of Metamorphism* (Eds. W. S. Pitcher and G. S. Flinn). Oliver & Boyd, Edinburgh, 1965.

MAGNÉE, I. D., Observations sur l'origine des gisements de pyrite du sud de l'Espagne et du Portugal, *Rept. Int. Geol. Congrs. VII, Paris*, 95 (1935).

MANASEVIT, H. M., MILLER, A., MORRITZ, F. L. and NOLDER, R., Heteroepitaxial silicon-aluminium oxide interface, Part I, *Trans. Am. Inst. Min. metall. Engrs.*, **233**, 540 (1965).

MANDELKERN, L., Crystallization kinetics in polymeric systems. In *Growth and Perfection of Crystals*, p. 467 (Eds. R. H. Doremus, B. W. Roberts and D. Turnbull). Wiley, New York, 1958.

MARMO, V., On the microcline of the granitic rocks of central Sierra Leone, *Schweiz. miner. petrogr. Mitt.*, **35**, 105 (1955a).

MARMO, V., Reply to the remarks of Prof. F. Laves, *Schweiz. miner. petrogr. Mitt.*, **83**, 23 (Abstract in English in *Min. Abstr.*) (1955b).

MARSHALL, R. R., Devitrification of natural glasses, *Bull. geol. Soc. Am.*, **72**, 1493 (1961).

MAY, J. E., Polygonization of sapphire. In *Kinetics of High Temperature Processes* (Ed. W. D. Kingery). Technical Press M.I.T., and Wiley, New York, 1959.

MEANS, W. D. and PATERSON, M. S., Experiments on preferred orientation of platy minerals, *Cont. Min. Petr.*, **13**, 108 (1966).

MEGAW, H. D., Order and disorder in feldspars, *Norsk. geol. Tidsskr.*, **42**, 104 (1962).

MERWE, VAN DER, J. H., On the stresses associated with inter-crystalline boundaries, *Proc. phys. Soc. Lond.*, **63A**, 616 (1950).

MICHOT, P., Essai sur la géologie de la catazone, *Bull. Acad. r. Belg. Cl. Sci. 5. Sér.*, **37**, 271 (1951).

MILES, K. R., Metamorphism of the jasper bars of Western Australia, *Q. Jl. geol. Soc. Lond.*, **52**, 115 (1945).

MILLER, W. J., Origin of foliation in the Precambrian rocks of northern New York, *J. Geol.*, **24**, 567 (1916).

MISCH, P., Metasomatic granitization of batholithic dimensions, *Am. J. Sci.*, **247**, 372 (1949).

MISCH, P., Zoned plagioclase in metamorphic rocks, *Abstract Am. Miner.*, **40**, 327 (1955).

MISCH, P., and HAZZARD, J. C., Stratigraphy and metamorphism of late Precambrian rocks of Central North-eastern Nevada and adjacent Utah, *Bull. Am. Ass. Petrol. Geol.*, **46**, 3, 289 (1962).

MITCHELL, R. S. and COREY, A. S., Coalescence of hexagonal and cubic polymorphs in tetrahedral structures as illustrated by some wurtzite-sphalerite crystal groups, *Am. Miner.*, **39**, 773 (1954).

MIYASHIRO, A., Evolution of metamorphic belts, *J. Petrology*, **2**, 277 (1961).

MOMENT, R. L. and GORDON, R. B., Energy of grain boundaries in halite, *J. Am. Ceram. Soc.*, **47**, 570 (1964).

MONDOLFO, L. F., Nucleation in eutectic alloys, *J. Aust. Inst. Metals.*, **10**, 169 (1965).

MOORHOUSE, W. W., *Study of Rocks in Thin Section.* Harper, New York, 1959.

MOREY, G. W., *Properties of Glass.* Rheinhold, New York, 1938.

MOTT, N. F., Slip at grain boundaries in metals, *Proc. phys. Soc. Lond.*, **60**, 391 (1948).

MÜGGE, O., Beiträge für Kentniss der Strukturflächen des Kalkspathes und über die Beziechungen derselben untereinander und der Zwillingsbilden an Kalkspath und einigen anderen Mineralien. *Neues Jb. Miner. Geol. Paläont.*, Beil Bd., **1**, 32 (1883).

MÜGGE, O., Zur Kentniss der durch sekundäre Zwillingsbildung bewirkten Fächer-verschiebungen, *Neues Jb. Miner. Geol. Paläont.*, Beil Bd., **2**, 44 (1885).

MÜGGE, O., Ueber Translationen und verwandte Erscheinungen in Kristallen, *Neues Jb. Miner. Geol. Paläont.*, Beil Bd., **1**, 1 (1898).

MÜGGE, O., Bewegungen von Porphyroblasten, *Neues Jb. Miner. Geol. Paläont.*, Biel Bd. **3**, 469 (1930).

MÜGGE, O. and HEIDE, F., Einfache Schiebungen am Anorthosit, *Neues Jb. Miner. Geol. Paläont.*, Biel Bd. **64**, 163 (1931).

MURTHY, M. V. N., Coronites from India and their bearing on the origin of coronas, *Bull. geol. Soc. Am.*, **68**, 23 (1958).

NABARRO, F. R. N., Deformation of crystals by the motion of single ions. In *Report of a Conference on the Strength of Solids. Proc. phys. Soc. Lond.*, **75**, 1948.

NEWHOUSE, W. H. and FLAHERTY, C. F., Texture and origin of some banded and schistose ores, *Econ. Geol.*, **25**, 600 (1930).

NEWTON, R. C., The kyanite-sillimanite equilibrium at 750° C, *Science*, **151**, 1222 (1966).

NICHOLAS, J. F., Rearrangement of a flat surface without altering the number of broken bonds, *Physics. Chem. Solids.*, **24**, 1279 (1963).

NIGGLI, P., Die Chloritoid Schiefer und die sedimentäre Zone am Nordostrand des Gotthard massivs, *Beitr. geol. Karte Schweiz.*, **36** (1912).

NIGGLI, P., *Lehrbuch der Mineralogie.* Gebrüder Borntraeger, Berlin, 1920.

NIGGLI, P., *Rocks and Mineral Deposits.* Freeman, San Francisco, 1954.

NIXON, P. H., KNORRING, O. and ROOKE, J. M., Kimberlite and associated inclusions in Basutoland, *Am. Miner.*, **48**, 1090 (1963).

NORTON, F. H., Crystal growth during calcination. In *Kinetics of High Temperature Processes*, p. 116 (Ed. W. D. Kingery). Wiley, New York, 1959.

OBREIMOFF, J. W., The splitting strength of mica, *Proc. R. Soc. Lond.*, **A127**, 290 (1930).

OROWAN, E., A type of plastic deformation new in metals, *Nature*, **149**, 643 (1942).

ORVILLE, P. M., Alkali metasomatism and feldspars, *Norsk. geol. Tidsskr.*, **42**, 354 (1962).

OSBORN, F. F. and ADAMS, F. D., Deformation of galena and pyrrhotite, *Econ. geol.*, **26**, 884 (1931).

PABST, A., Pressure shadows and the measurement of orientation of minerals in rocks, *Am. Miner.*, **16**, 55 (1931).
PABST, A., Large and small garnets from Fort Wrangell, Alaska, *Am. Miner.*, **28**, 233 (1943).
PABST, A., Transformation of indices in twin gliding, *Bull. geol. Soc. Am.*, **66**, 55 (1955).
PACKHAM, G. H. and CROOK, K. A. W., Principle of diagenetic facies and some of its implications, *J. Geol.*, **68**, 392 (1960).
PARK, R. G., Pseudotachylite of the Gairloch district, Ross-shire, *Am. J. Sci.*, **259**, 542 (1961).
PARRAS, K., On the charnockites in the light of a highly metamorphic rock complex in South-Western Finland, *Bull. Comm. geol. Finl.*, **181**, 1 (1958).
PARTHASARATHI, M. N. and BECK, P. A., The oriented-growth mechanism of the formation of recrystallization textures in Al, *Trans. Am. Inst. Min. metall. Engrs.*, **221**, 831 (1961).
PASHLEY, W. D., The study of epitaxy in thin surface films, *Adv. Phys.*, **5**, 173 (1956).
PATERSON, M. S., X-ray line broadening in plastically deformed calcite, *Phil. Mag.*, **4**, 451 (1959).
PATERSON, M. S. and WEISS, L. E., Symmetry concepts in the structural analysis of deformed rocks, *Bull. geol. Soc. Am.*, **72**, 849 (1961).
PATTERSON, J. H., Thermal decomposition of crocidolite in air and in a vacuum, *Mineralog. Mag.*, **35**, 31 (1965).
PAULING, L., Structure of the chlorites, *Proc. Natn. Acad. Sci., U.S.A.*, **16**, 578 (1930).
PETTIJOHN, F. J., *Sedimentary Rocks*. Harper, New York, 1957.
PHEMISTER, J., Geology of Strath Oykell and lower Loch Shin, *Mem. geol. Surv. Scotld.* (1926).
PHEMISTER, J., Zoning in plagioclase feldspar, *Mineralog. Mag.*, **23**, 541 (1934).
PHILLIPS, F. C., Some mineralogical and chemical changes induced by progressive metamorphism in the Green Bed Group of the Scottish Dalradians, *Mineralog. Mag.*, **22**, 239 (1930).
PHILPOTTS, A. R., Origin of pseudotachylites, *Am. J. Sci.*, **262**, 1008 (1964).
PICHAMUTHU, C. S., The charnockite problem, *Proc. Mysore Geol. Ass.* (1953).
PICHAMUTHU, C. S., The significance of clouded plagioclase in the basic dykes of Mysore State, India, *J. geol. Soc. India*, **1**, 68 (1959).
PITCHER, W. S., The aluminium silicate polymorphs. In *Controls of Metamorphism* (Eds. W. S. Pitcher and G. W. Flinn). Oliver & Boyd, Edinburgh, 1965.
PITCHER, W. S. and FLINN, G. W., (Eds.) *Controls of Metamorphism*. Oliver & Boyd, Edinburgh, 1965.
PITCHER, W. S. and READ, H. H., The aureole of the Main Donegal Granite, *Q. Jl. geol. Soc. Lond.*, **116**, 1 (1960).
PITCHER, W. S. and READ, H. H., Contact metamorphism in relation to manner of emplacement of the granites of Donegal, Ireland, *J. Geol.*, **71**, 261 (1963).
POLDERVAART, A. and ECKELMANN, F. D., Growth phenomena in zircon of autochthonous granites, *Bull. geol. Soc. Am.*, **66**, 947 (1955).
POLDERVAART, A. and GILKEY, A. K., On clouded plagioclase, *Am. Miner.*, **39**, 75 (1954).
POWELL, D., On the preferred crystallographic orientation of garnet in some metamorphic rocks, *Mineralog. Mag.*, **35**, 1094 (1966).
POWELL, D. and TREGUS, J. E., On the geometry of S-shaped inclusion trails in garnet porphyroblasts, *Mineralog. Mag.*, **36**, 453 (1967).
PRIESTNER, R. and LESLIE, W. C., Nucleation of deformation twins at slip-plane intersections in body-centred-cubic metals, *Phil. Mag.*, **11**, 895 (1965).
PRINZ, M. and POLDERVAART, A., Layered mylonite from Beartooth Mountains, Montana, *Bull. geol. Soc. Am.*, **75**, 741 (1964).
QUENSEL, P., Zur Kenntnis der Mylonitbildung, *Bull. geol. Instn. Univ. Upsala*, **55**, 91 (1916).
RAGAN, D., Emplacement of the Twin Sisters dunite, Washington, *Am. J. Sci.*, **261**, 549 (1963).
RALEIGH, C. B., Crystallization and recrystallization of quartz in a single piston-cylinder device, *J. Geol.*, **73**, 369 (1965a).
RALEIGH, C. B., Glide mechanism in experimentally deformed minerals, *Science*, **150**, 3697 (1965b).
RALEIGH, C. B. and TALBOT, J. L., Mechanical twinning in naturally and experimentally deformed diopside, *Am. J. Sci.*, **265**, 151 (1967).

RAMBERG, H., Force of crystallization as a well definable property of crystals, *Geol. För. Stockh. Förh*, **69**, 189 (1947).

RAMBERG, H., *Origin of igneous and metasomatic rocks*. Univ. Chicago Press (1952).

RAMBERG, H. and EKSTROM, T., Note on preferred orientation of pyrite cubes in grit layers in slates, *Neues Jb. Miner. Geol. Paläont.*, Biel Bd. **8**, 246 (1964).

RAMDOHR, P., Die Lagerstätte von Broken Hill in New South Wales im Lichte der neuen geologischen Erkenntnisse und Erzmikroscopischer Untersuchungen, *Heidelb. Beitr. Miner. Petrogr.*, **2**, 4, 291 (1920).

RAMSAY, J. G., The geometry and mechanics of formation of "similar" type folds, *J. Geol.*, **70**, 309 (1962).

RAMSAY, J. G., Structure and metamorphism of the Moine and Lewisian rocks of the northwest Caledonides. In *The British Caledonides* (Eds. M. R. W. Johnson and F. H. Stewart). Oliver & Boyd, Edinburgh, 1963.

RAO, A. B. and RAO, M. S., Some observations on the plagioclase twinning in charnockite rocks, *Proc. nat. Inst. Sci. India*, **19**, 501 (1953).

RAO, A. B. and RAO, G. V. V., Intergrowth in ilmenite of the beach sands of Kerala, *Mineralog. Mag.*, **35**, 118 (1965).

RAST, N., Metamorphic history of the Schichallion Complex, *Trans. R. Soc. Edinb.*, **63**, 2 413 (1958).

RAST, N., Structure and metamorphism of the Dalradian Rocks of Scotland. In *The British Caledonides* (Eds. M. R. W. Johnson and F. H. Stewart). Oliver & Boyd, Edinburgh, 1963.

RAST, N., Nucleation and growth of metamorphic minerals. In *Controls of Metamorphism* (Eds. W. S. Pitcher and G. S. Flinn). Oliver & Boyd, Edinburgh, 1965.

RAST, N. and STURT, B. A., Crystallographic and geological factors in the growth of garnets from Central Perthshire, *Nature Lond.*, **179**, 215 (1957).

READ, H. H., Metamorphic geology of Unst in the Shetland Islands, *Q. Jl. geol. Soc. Lond.*, **90**, 637 (1934).

READ, H. H., Stratigraphical order of the Dalradian rocks of the Banffshire Coast, *Geol. Mag.*, **73**, 468 (1936).

READ, H. H., Mylonitisation and cataclasis in acidic dykes in the Insch (Aberdeenshire) Gabbro and its aureole, *Proc. Geol. Ass. Lond.*, **62**, 4, 237 (1951).

READ, H. H., *The Granite Controversy*. Murby, London, 1957.

READ, H. H. and WATSON, J., *Introduction to Geology*. Macmillan, London, 1962.

READ, W. T., *Dislocations in Crystals*. McGraw-Hill, New York, 1953.

READ, W. T. and SCHOCKLEY, W., Dislocation models of crystal grain boundaries, *Phys. Rev.*, **78**, 275 (1950).

REYNOLDS, D. L., Demonstrations in petrogenesis from Kiloran Bay, Colonsay, *Mineralog. Mag.*, **24**, 367 (1936).

REYNOLDS, D. L., The albite schists of Antrim and their petrogenetic relationships to Caledonian orogenesis, *Proc. R. Irish Acad.*, **48**, 8, 43 (1942).

REYNOLDS, D. L., Fluidisation as a geological process, *Am. J. Sci.*, **252**, 577 (1954).

RICHARDS, D. A., Magnetite inclusions in mica, *Proc. phys. Soc. Lond.*, **A63**, 852 (1950).

RICHARDS, S. M., Mineragraphy of fault-zone sulphides, Broken Hill, N.S.W., *C.S.I.R.O. Tech. Pub.* **5** (1966).

RICHARZ, S., Some inclusions in basalt, *J. Geol.*, **32**, 685 (1924).

RIGGS, B. A., Consideration of diffusion in strained systems, *Acta metall.*, **12**, 952 (1964).

RINGWOOD, A. E. and MAJOR, A., Synthesis of Mg_2SiO_4-Fe_2SiO_4 spinel solid solutions. In *Petrology of the Upper Mantle*, Australian National Univ. Dept. Geophys. Pub. 444, 234 (1966).

RINNE, F., Über künstlich und natürlich umgeformtes Steinsalz und seine Rekristallisation, *Tschermaks miner petrog. M.H.*, **37**, 237 (1926).

RINNE, F., *Gesteinskunde*. Jänecke, Leipzig, 1928.

ROBERTS, J. L., Emplacement of the main Glencoe fault-intrusion at Stob Mhic Mhartuin, *Geol. Mag.*, **103**, 299 (1966).

ROBERTSON, F., Sphalerite-dolomite orientation relations at the Renfrew prospect, Ontario, *Am. Miner.*, **36**, 116 (1951).

ROBERTSON, F., Perthite formed by reorganization of albite from plagioclase during potash metasomatism, *Am. Miner.*, **44**, 603 (1959).

ROSENFELD, J. L. and CHASE, A. B., Pressure and temperature of crystallization from elastic effects around solid inclusions in a mineral, *Am. J. Sci.*, **259**, 519 (1961).

ROSENQVIST, I. TH., Some investigations in the crystal chemistry of silicates, II, Orientation of perthite lamellae in feldspars, *Norsk. geol. Tidsskr.*, **28**, 192 (1950).

ROSENQVIST, I. TH., Metamorphic facies and the feldspar minerals, *Univ. Bergen Arbok., Naturvit. rekke.*, **4**, 1 (1952).

ROSS, J. V., Combination twinning in plagioclase feldspars, *Am. J. Sci.*, **255**, 650 (1957).

ROY, D. M. and ROY, R., An experimental study in the formation and properties of synthetic serpentines and related layer silicates minerals, *Am. Miner.*, **39**, 957 (1954).

ROYER, L., Recherches expérimentales sur l'epitaxie ou orientation mutuelle des critaux d'éspèces différentes, *Bull. Soc. fr. Miner. Crystallogr.*, **51**, 7 (1928).

RUDENKO, S. A., Morphological-genetic classification of perthite intergrowths, *Mem. Soc. miner. Russe*, **83**, 23 (Abstract in English in *Miner. Abstr.*) (1954).

RUSSELL, G. S., Crystal growth in solution under local stress, *Am. Miner.*, **20**, 733 (1935).

RUTLAND, R. W. R., Tectonic overpressures. In *Controls of Metamorphism* (Eds. W. S. Pitcher and G. S. Flinn). Oliver & Boyd, Edinburgh, 1965.

RUTLAND, R. W. R., HOLMES, M. and JONES, M.A., Granites of the Glomfjord area, Northern Norway, *Internat. Geol. Congr., 21st, Copenhagen*, **19**, 43 (1960).

RUTTER, J. W. and AUST, K. T., Kinetics of grain boundary migration of high purity lead containing very small additions of silver and gold, *Trans. Am. Inst. Min. metall. Engrs.*, **218**, 682 (1960).

SAENZ DE, I. M., Origin of ternary film and string perthites from a Uruguayan migmatite, *Schweiz. miner. petrogr. Mitt.*, **45**, 103 (1965).

SAHAMA, T., Die Regelung von Quartz und Glimmer in den Gesteinen der Finnisch-Lappländischen Granulitformation, *Bull. Comm. geol. Finl.*, **113**, 1 (1936).

SANDER, B., Über Zusammenhänge zwischen Teilbewegung und Gesteinen, *Tschermaks miner. petrogr. Mitt.*, **30**, 281 (1911).

SANDER, B., *Einführung in die Gefügekunde der geologischen Körper*. Springer, Berlin, 1st ed. 1930, 2nd ed. 1950.

SARATOVKIN, D. D., *Dendritic Crystallization* (Trans. J. E. S. Bradley). Consultants Bureau, New York, 1959.

SARROT-REYNAUD, J., Le socle crystallophyllien du Dome de la Mure, *C. R. Acad. Sci.*, **246**, 2008 (1958).

SAUER, J. A., MORROW, D. R. and RICHARDSON, C. C., Morphology of solution-grown polypropylene crystal aggregates, *J. appl. Phys.*, **36**, 10, 3017 (1965).

SCHARBERT, H. G., Zur Nomenklatur der Gesteine im Granulitfazies, *Tschermaks miner. petrogr. Mitt.*, **8**, 591 (1963).

SCHEERER, P. E., MIKAMI, H. M. and TAUBER, J. A., Microstructure of chromite-periclase at 1650° to 2310°C., *J. Am. Ceram. Soc.*, **47**, 297 (1964).

SCHEUMANN, K. H., "Granulit" eine petrographische Definition. *Neues Jb. Miner. Geol. Paläont.*, **3**, 75 (1961).

SCHMIDT, W., Bewegungsspuren in Porphyroblasten Kristalliner Schiefer, *Sver. Akad. Wiss. Wien*, **1**, 127, 293 (1918).

SCHMIDT, W., *Tektonik und Verformungslehre*. Barnstraeger, Berlin, 1932.

SCHOFIELD, T. H. and BACON, A. E., Study of the recrystallization and grain growth of cold-rolled and annealed Ti-10% Mo alloy, *Acta metall.*, **9**, 653 (1961).

SCHRAUF, S., Beiträge zur Kenntnis des Assoziationskreises der Magnesiasilicate, *Z. Kristallogr. Miner.*, **6** (1882).

SCHUILING, R. D., Formation of pegmatitic carbonatite in a syenite-marble contact, *Nature Lond.*, **192**, 1280 (1961).

SCHUILING, R. D. and WENSINK, H., Porphyroblastic and poikiloblastic textures. The growth of large crystals in a solid medium, *Neues Jb. Miner. Geol. Palaeont. Beil Bd.*, 247 (1962).

SCHWERDTNER, W. M., Preferred orientation of hornblende in a banded hornblende gneiss, *Am. J. Sci.*, **262**, 1212 (1964).

Sclar, C. B., Layered mylonites and the process of metamorphic differentiation, *Bull. geol. Soc. Am.*, **76**, 611 (1965).

Scott, J. S. and Drever, H. I., Frictional fusion along the Himalayan Thrust, *Proc. R. Soc. Edinb.*, **65**, 121 (1954).

Sederholm, J. J., On synantetic minerals and related phenomena, *Bull. Comm. geol. Finl.*, **48** (1916).

Segnit, E., Oriented overgrowth of hematite on beta-alumina, *Mineralog. Mag. Tilley Vol.*, 416 (1965).

Seitz, F., *Physics of Metals*. McGraw-Hill, New York and London, 1943.

Seitz, F. and Read, T. A., Theory of the plastic flow of solids, *J. appl. Phys.*, **12**, I, 100; II, 170; III, 470; IV, 538 (1941).

Sen, S. K., Potassium content of natural plagioclase and the origin of antiperthites, *J. Geol.*, **67**, 479 (1959).

Sengupta, A., Preliminary results of the study of porphyroblasts from Sakoli metamorphites, Bhandara District, *J. geol. Soc. India*, **4**, 85 (1963).

Shand, S. J., Pseudotachylyte of Parijs (Orange Free State), *Q. Jl. geol. Soc. Lond.*, **72**, 198 (1916).

Shand, S. J., Coronas and coronites, *Bull. geol. Soc. Am.*, **56**, 247 (1945).

Shaub, B. M., Paragenesis of the garnet and associated minerals of the Barton Mine near North Creek, N.Y., *Am. Miner.*, **34**, 573 (1949).

Shelley, D., On myrmekite, *Am. Miner.*, **49**, 41 (1964).

Shewmon, P. C., New method of measuring surface energies and torques of solid surfaces, *Trans. Am. Inst. Min. metall. Engrs.*, **227**, 400 (1963).

Shockley, W. (Ed.) *Imperfections in Nearly Perfect Crystals*. Wiley, New York, 1952.

Short, N. M., Effects of shock pressures from a nuclear explosion on mechanical and optical properties of granodiorite, *J. geophys. Res.* **71**, 1195 (1966).

Shubnikov, A. V., Vorläufige Mitteilung über die Messung der sogenannten Kristallisation-kraft, *Z. Kristallogr. Miner.*, **88**, 466 (1934).

Shumskii, P. A., *Principles of Structural Glaciology*. Dover, New York, 1964.

Shuttleworth, J. K., Surface tension of solids, *Phys. Soc. Lond.*, **63A**, 444 (1950).

Siefert, K. E., The genesis of plagioclase twinning in the Nonewang granite, *Am. Miner.*, **49**, 297 (1964).

Siefert, K. E., Deformation bands in albite, *Am. Miner.*, **50**, 1469 (1965).

Siegle, S., Order–disorder transitions in metal alloys. In *Phase Transformations in Solids*, p. 366 (Eds. R. Smoluchowski, J. E. Mayer and W. A. Weyl). Wiley, New York, 1951.

Silk, E. C. H. and Barnes, R. S., Observations of dislocations in mica, *Acta metall.*, **9**, 558 (1961).

Simpson, A., The metamorphism of the Manx Slate Series, Isle of Man, *Geol. Mag.*, **101**, 20 (1964).

Singewald, J. T., Weathering and albitization of the Wissahickon Schist at the Pretty Bay Dam, Baltimore Co., Maryland, *Bull. geol. Soc. Am.*, **43**, 449 (1932).

Singh, S., Cordierite in the Precambrian rocks of the S. Savanna, Brit. Guiana, *Geol. Mag.*, **103**, 36 (1966).

Smallman, R. E., *Modern Physical Metallurgy*. Butterworths, London, 1963.

Smith, C. S., Grains, phases and interfaces: an interpretation of microstructure, *Trans. Am. Inst. Min. metall. Engrs.*, **175**, 15 (1948).

Smith, C. S., Microstructure, *Trans. Am. Soc. Metals*, **45**, 533 (1953a).

Smith, C. S., Further notes on the shape of metal grains, *Acta metall.*, **1**, 295 (1953b).

Smith, C. S., The shape of things, *Scient. American*, **190**, 58 (1954).

Smith, C. S., Some elementary principles of polycrystalline microstructure, *Met. Rev.*, **9**, 33, 1 (1964).

Smith, D. G. W., Chemistry and mineralogy of some emery-like rocks from Sithean Sluaigh, Argyllshire, *Am. Miner.*, **50**, 1982 (1965).

Smith, F. G., *Physical Geochemistry*. Addison-Wesley, Mass., 1963.

Smith, J. V., The effect of temperature, structural state and composition on albite, pericline and acline-A twins of plagioclase feldspar, *Am. Miner.*, **43**, 546 (1958).

SMITH, J. V., Genetic aspects of twinning in feldspar, *Norsk. geol. Tidsskr.*, **42**, 244 (1962).
SMITH, W. W., Pseudomorphs after olivine in the Markle basalt, *Mineralog. Mag.*, **32**, 324 (1959).
SMITH, W. W., Structural relationships within pseudomorphs after olivine, *Mineralog. Mag.*, **32**, 823 (1961).
SMOLUCHOWSKI, R., Nucleation theory. In *Phase Transformations in Solids*, p. 149 (Eds. R. Smoluchowski, J. E. Mayer and W. A. Weyl). Wiley, New York, 1951.
SORBY, H. C., On the origin of slaty cleavage, *Edinb. New Philos. J.*, **10**, 136 (1853).
SORBY, H. C., On the theory of slaty cleavage, *Phil. Mag.*, **12**, 127 (156).
SOSMAN, R. B., *The Properties of Silica.* Chemical Catalog Co., N.Y., 1927.
SPITZER, N. G., and LIGENZA, J. R., Oxygen exchange between silica and high pressure steam, *J. Phys. Chem. Solids*, **17**, 196 (1961).
SPRIGGS, R. M., BRISETTE, L. A. and VASILOS, T., Grain growth in fully dense magnesia *J. Am. Ceram. Soc.*, **47**, 417 (1964).
SPRY, A., Thermal metamorphism of portions of the Woolomin Group in the Armidale district, Pt. I, *Proc. R. Soc. N.S.W.*, **87**, 129 (1953).
SPRY, A., Thermal metamorphism of portions of the Woolomin Group in the Armidale district, Pt. II, *Proc. R. Soc. N.S.W.*, **89**, 157 (1955).
SPRY, A., Chronological analysis of crystallization and deformation of some Tasmanian Precambrian rocks, *J. geol. Soc. Aust.*, **10**, 1, 193 (1963a).
SPRY, A., Origin and significance of snowball structure in garnet, *J. Petrology*, **4**, 211 (1963b).
SPRY, A., Ripple marks and pseudo-ripple marks in deformed quartzite, *Am. J. Sci.*, **261**, 756 (1963c).
SPRY, A., Notes on the petrology and structure of the Precambrian metamorphic rocks of the Upper Mersey-Forth area, *Rec. Queen Vict. Mus.*, **16**, 1 (1963d).
SPRY, A. and SOLOMON, M., Columnar buchites at Apsley, Tasmania, *Q. Jl. geol. Soc. Lond.*, **120**, 519 (1964).
STANTON, R. L., General features of the conformable "pyritic" overbodies, *Trans. Can. Inst. Min. Metall.*, **63**, 22 (1960).
STANTON, R. L., Mineral interfaces in stratiform ores, *Bull. Instn. Min. Metall., Lond.*, **74**, 2, 45 (1964).
STARKEY, J., Chess-board albite from New Brunswick, Canada, *Geol. Mag.*, **96**, 141 (1959).
STAUB, R., Petrographische Untersuchungen im westlichen Berninagebirge, *Vjschr. naturf. Ges. Zurich*, **60**, 71 (1915).
STEIGER, R. H., Use of K-Ar ages of hornblende for dating phases of the Alpine Orogeny, *Program* 1963 *Meet. Geol. Soc. Am.*, **157A** (1963).
STEPHENS, D. L. and ALFORD, W. J., Dislocation structures in single-crystal Al_2O_3, *J. Am. Ceram. Soc.*, **47**, 81 (1964).
STILWELL, F. L., Metamorphic rocks of Adelie Land, *Aust. Antarctic Expedition, 1911–1914, Sci. Repts.*, **A**, **3**, 1 (1918).
STOOKEY, S. D. and MAURER, R. D., Catalyzed crystallization in glass—theory and practice. In *Progress in Ceramic Science*, **2**, 77 (1962).
STURT, B. A., Preferred orientation of nepheline in deformed nepheline syenite gneisses from Söroy, Northern Norway, *Geol. Mag.*, **98**, 464 (1961).
STURT, B. A. and HARRIS, A. L., Metamorphic history of the Loch Tummel area, Central Perthshire, *L'pool Manchr. geol. J.*, **2**, 689 (1961).
STURT, B. A. and RAMSAY, D. M., The alkaline complex of the Breivikbotn area, Söroy, N. Norway, *Norg. Geol. Unders.*, **231** (1965).
SUBRAMANIAM, A. P., Charnockites of the type area near Madras—a reinterpretation, *Am. J. Sci.*, **257**, 321 (1959).
SUNDIUS, N. and BYSTROM, A. M., Decomposition products of muscovite at temperatures between 1000° and 1200°C., *Trans. Br. Ceram. Soc.*, **53**, 632 (1953).
SUTTON, J., *Caledonides* (Eds. M. R. W. Johnson and F. H. Stewart). Oliver & Boyd, Edinburgh, 1963.
SUTTON, J., Some recent advances in our understanding of the controls of metamorphism. In *Controls of Metamorphism* (Eds. W. S. Pitcher and G. S. Flinn). Oliver & Boyd, Edinburgh, 1965.

SUTTON, J. and WATSON, J., Pre-Torridonian history of the Loch Torridon and Scourie areas in the North-West Highlands, *Q. Jl. geol. Soc. Lond.*, **106**, 241 (1950).

SUTTON, J. and WATSON, J., Varying trends in the metamorphism of dolerites, *Geol. Mag.*, **88**, 25 (1951).

SUTTON, J. and WATSON, J., Metamorphism in deep-seated zones of transcurrent movement at Kungwe Bay, Tanganyika Territory, *J. Geol.*, **67**, 1 (1959).

TABER, S., Growth of crystals under external pressure, *Am. J. Sci.*, **41**, 532 (1916).

TAYLOR, G. I., Plastic strain in metals, *J. Inst. Metals. Clay Mineralogy Bull.*, **62**, 307 (1938).

TAYLOR, W. H., Structure of the principal feldspars, *Norsk. geol. Tidsskr.*, **44**, 1 (1962).

TEALL, J. J. H., Dynamic metamorphism, *Proc. Geol. Ass. Lond.*, **29**, 1 (1918).

TEMPLE, A. K. and HEINRICH, E. W., Spurrite from northern Coahuila, Mexico, *Mineralog. Mag.*, **33**, 841 (1964).

THAYER, T. P., Serpentinization as a constant volume metasomatic process, *Am. Miner.*, **51**, 685 (1966).

THOMAS, H. H. and CAMPBELL-SMITH, W., Xenoliths of igneous origin in the Tregastel-Ploumanach granite, France, *Q. Jl. geol. Soc. Lond.*, **88**, 274 (1932).

THOMAS, J. M., Interaction of gases and solid surfaces, *Sci. Progr.*, **50**, 46 (1962).

THOMAS, L. A. and WOOSTER, W. A., Piezocrescence—the growth of Dauphiné twinning in quartz under stress, *Proc. R. Soc.* **A208**, 43 (1951).

TILLEY, C. E., Paragenesis of kyanite amphibolites, *Mineralog. Mag.*, **24**, 555 (1936a).

TILLEY, C. E., Paragenesis of kyanite eclogites, *Mineralog. Mag.*, **24**, 422 (1936b).

TOBI, A. C., Petrographical and geological investigations in the Merdaret-Lac Crop region, *Leid. geol. Meded.*, **24**, 181 (1959).

TOBI, A. C., Pattern of plagioclase twinning as a significant rock property, *Proc. K. ned. Akad. Wet.*, **B64**, 576 (1961).

TOBI, A. C., Characteristic patterns of plagioclase twinning, *Norsk. geol. Tidsskr.*, **42**, 264 (1962).

TÖRNEBOHM, A. E., Om Sveriges viktigare diabas-osh gabbro-arter, *K. svenska Vetensk-Handl. B.*, **4**, 13 (1877).

TOLANSKY, S., *Surface Microtopography*. Longmans, London, 1960.

TONEY, S. and AARONSON, H. I., Effects of grain boundary structure on precipitate morphology in an Fe-1·55% Si Alloy, *Trans. Am. Inst. Min. metall. Engrs.*, **224**, 909 (1962).

TOWNER, R. J. and BERGER, J. A., X-ray studies of polygonization and subgrain growth in aluminium, *Trans. Am. Inst. Min. metall. Engrs.*, **218**, 611 (1960).

TOZER, C. F., Mode of occurrence of sillimanite in the Glen District, Co. Donegal, *Geol. Mag.*, **92**, 310 (1955).

TREFFNER, W. S., Microstructure of periclase, *J. Am. Ceram. Soc.*, **47**, 410 (1964).

TROIANO, A. R. and GRENINGER, A. G., The martensite transformation, *Metal Prog.*, **50**, 303 (1946).

TURNBULL, D., Transient nucleation, *Trans. Am. Inst. Min. metall. Engrs.*, **175**, 774 (1948).

TURNBULL, D., Theory of grain boundary migration rates, *Trans. Am. Inst. Min. metall. Engrs.*, **191**, 661 (1951).

TURNER, F. J., Genesis of oligoclase in certain schists, *Geol. Mag.*, **60**, 529 (1933).

TURNER, F. J., Progressive regional metamorphism in southern New Zealand, *Geol. Mag.*, **75**, 160 (1938).

TURNER, F. J., Development of pseudo-stratification by metamorphic differentiation in the schists of Otago, *Am. J. Sci.*, **239**, 1 (1941).

TURNER, F. J., Mineralogical and structural evolution of the metamorphic rocks, *Mem. geol. Soc. Am.*, **30** (1948).

TURNER, F. J., Observations of twinning in metamorphic rocks, *Am. Miner.*, **36**, 581 (1951).

TURNER, F. J., Rotation of the crystal lattice in deformation bands and twin lamellae of strained crystals, *Proc. natn. Acad. Sci. U.S.A.*, **48**, 955 (1952).

TURNER, F. J., GRIGGS, D. T. and HEARD, H., Experimental deformation of calcite crystals, *Bull. geol. Soc. Am.*, **65**, 883 (1954).

TURNER, F. J., HEARD, H. and GRIGGS, D. J., Experimental deformation of enstatite and accompanying inversion to clinoenstatite, *Int. Geol. Congr. Copenhagen*, **18**, 399 (1960).

Turner, F. J. and Hutton, C. O., Some porphyroblastic albite schists from Waikouaiti River, Otago, *Trans. R. Soc. N.Z.*, **71**, 223 (1941).

Turner, F. J. and Verhoogen, J., *Igneous and Metamorphic Petrology*, 2nd Edn. McGraw-Hill, New York, 1960.

Turner, F. J. and Weiss, L. E., *Structural Analysis of Metamorphic Tectonites*. McGraw-Hill, New York, 1963.

Tuttle, O. F., Origin of the contrasting mineralogy of extrusive and plutonic salic rocks, *J. Geol.*, **60**, 107 (1952).

Tuttle, O. F. and Smith, J. V., The nepheline-kalsilite system II., *Am. J. Sci.*, **256**, 571 (1958).

Tyrrell, G. W., *Principles of Petrology*. Methuen & Co., London, 1926.

Uglow, W. L., Gneissic galena ore from the Slocan district, British Columbia, *Econ. Geol.*, **12**, 643 (1917).

Urusovskaya, A. A., Formation of regions with a reoriented lattice as a result of deformation of mono- and polycrystals. In *Plasticity of Crystals* (Ed. M. V. Klassen-Neklyudova). Consultants Bureau, New York, 1962.

Van Diver, B. B., Contemporaneous faulting—metamorphism in Wenatchee Ridge area, Northern Cascades, Washington, *Am. J. Sci.*, **265**, 132 (1967).

Vance, J. A., Polysynthetic twinning in plagioclase, *Am. Miner.*, **46**, 1097 (1961).

Vance, J. A., Zoning in igneous plagioclase: normal and oscillatory, *Am. J. Sci.*, **260**, 746 (1962).

Vance, J. A., Zoning in igneous plagioclase: patchy zoning, *J. Geol.*, **73**, 636 (1965).

Vasilos, R., Mitchell, J. B. and Spriggs, R. M., Creep of polycrystalline magnesia, *J. Am. Ceram. Soc.*, **47**, 203 (1964).

Veit, K., Künstliche Schiebungen und Translationen in Mineralien, *Neues Jb. Miner. Geol. Paläont.*, Beil Bd. **45**, 121 (1922).

Venables, J. A., The electron microscopy of deformation twinning, *Physics Chem. Solids*, **25**, 685 (1964a).

Venables, J. A., Nucleation and propagation of deformation twins, *Physics Chem. Solids*, **25**, 693 (1964b).

Venkatesh, V., Development and growth of cordierite in para-lavas, *Am. Miner.*, **37**, 831 (1952).

Venkatesh, V., Twinning in cordierite, *Am. Miner.*, **39**, 636 (1954).

Verhoogen, J., Geological significance of surface tension, *J. Geol.*, **56**, 210 (1948).

Vernon, R., Plagioclase twins in some mafic gneisses from Broken Hill, Australia, *Mineralog. Mag.*, **35**, 488 (1965).

Vernon, R., Intergranular microstructures of high grade metamorphic rocks at Broken Hill, Australia, *J. Petrology*, **9**, 1 (1968).

Vesasalo, A., Talc schists and soapstone occurrences in Finland, *Bull. Comm. geol. Finl.*, **216**, 9 (1962).

Vogel, D. E. and Bahezre, C., Composition of partially zoned garnet and zoisite from Cabo Ortegal, N. W. Spain, *Neues Jb. Miner. Geol. Palaeont. Beil Bd.*, **5**, 140 (1965).

Vogel, Th. A., Optical-crystallographic scatter in plagioclase, *Am. Miner.*, **49**, 614 (1964).

Vogel, Th. A., and Seifert, K. E., Deformation twins in ordered plagioclase, *Am. Miner.*, **50**, 511 (1965).

Voll, G., New work on petrofabrics, *Lpool. Manch. geol. J.*, **2**, 503 (1960).

Voruz, T. A., Jewett, R. P. and Accountius, O. E., Direct observations of dislocations in sapphire, *J. Am. Ceram. Soc.*, **46**, 9, 459 (1963).

Wackerle, J., Shock-wave compression of quartz, *J. Appl. Phys.*, **33**, 922 (1962).

Wahl, F. M., High temperature phase of three-layer clay minerals and their interactions with common ceramic materials, *Bull. Am. Ceram. Soc.*, **44**, 676 (1965).

Waldbaum, D. R., Thermodynamic properties of mullite, andalusite, kyanite and sillimanite, *Am. Miner.*, **50**, 186 (1965).

Walker, H. L., Discussion of effect of mechanical deformation on grain growth in alpha brass, *Trans. Am. Inst. Min. metall. Engrs.*, **75**, 157 (1948).

Walker, K. R., Joplin, G., Lovering, J. F. and Green, R., Metamorphic and metasomatic

convergence of basic igneous rocks and lime-magnesia sediments of the Precambrian of North-western Queensland, *J. geol. Soc. Aust.*, **6**, 2, 149 (1959).
WALTON, A. G., Calculation of ionic crystal surface energies from thermodynamic data, *J. Am. Ceram. Soc.*, **48**, 151 (1965a).
WALTON, A. G., Nucleation of crystals from solution, *Science*, **148**, 3670 (1965b).
WALTON, M., HILLS, A. and HANSEN, E., Compositionally zoned granitic pebbles in three metamorphosed conglomerates, *Am. J. Sci.*, **262**, 1 (1964).
WATERS, A. C. and CAMPBELL, C. D. Mylonites from the San Andreas Fault Zone, *Am. J. Sci.*, **29**, 473 (1935).
WATERS, A. C. and KRAUSKOFF, J., Protoclastic border of the Colville Batholith, *Bull. geol. Soc. Am.*, **52**, 1355 (1941).
WATSON, J., Late sillimanite in the migmatites of Sutherland, *Geol. Mag.*, **85**, 149 (1948).
WATT, W. S., Textural and field relationships of basement granitic rocks, Quaersuarssuk, Greenland, *Meddr. Gronland*, **179**, 8, 134 (1965).
WAYMAN, C. M., *Introduction to the Crystallography of Martensitic Transformations*. Collier-McMillan, London, 1964.
WEGMANN, C. E., Zur Deutung der Migmatite, *Geol. Rdsh.*, **26**, 305 (1935).
WELLS, A. K., Crystal habit and internal structure, *Phil. Mag.*, **37**, I, 189; II, 217 (1946).
WENK, E., Beiträge zur Petrographie und Geologie des Silvrettkristallins, *Schweiz. miner. petrogr. Mitt.*, **14**, 196 (1934).
WENK, E., Zur Definition von Schiefer und Gneiss, *Neues. Jb. Miner. Geol. Paläont.*, Beil Bd., **5**, 97 (1963).
WESTBROOK, J. H., Segregation at grain boundaries, *Metals Rev.*, **9**, 36, 415 (1964).
WEYL, W. A., Transitions in glass. In *Phase Transformations in Solids*, p. 296 (Eds. R. Smoluchowski, J. E. Mayer and W. A. Weyl). Wiley, New York, 1951.
WHITE, A. J. R., Scapolite-bearing marbles and calc-silicate rocks from Tungkillo and Milendella, S. Australia, *Geol. Mag.*, **96**, 285 (1959).
WILLEMSE, J., On the granite of the Vredefort region and some of its associated rocks, *Trans. geol. Soc. S. Africa*, **40**, 43 (1937).
WILLIAMS, H., TURNER, F. J. and GILBERT, C. M., *Petrography*. Freeman, San Francisco, 1954.
WILLIS, B. T. M., Screw dislocation in quartz, *Nature Lond.*, **170**, 1115 (1952).
WILLOWS, R. S. and HATSCHCK, E., *Surface Tension and Surface Energy*. Churchill, London, 1923.
WILSON, A. F., Metamorphism of granite rocks by olivine dolerite in Central Australia, *Geol. Mag.*, **89**, 73 (1952).
WILSON, A. F., Charnockitic granites and associated granites of Central Australia, *Trans. R. Soc. S. Aust.*, **83**, 37 (1960).
WILSON, G. B., Tectonic significance of small scale structures and their importance in the field, *Annal. Soc. geol. Belgium*, **74**, 423 (1961).
WISEMAN, J. D. H., Central and South-west Highland epidiorites, *Q. Jl. geol. Soc.*, **90**, 354 (1934).
WORKMAN, D. R. and COWPERTHWAITE, I. A., An occurrence of kyanite pseudomorphing andalusite from Southern Rhodesia, *Geol. Mag.*, **100**, 456 (1963).
WYLLIE, P. J., Fusion of Torridonian arkose by a picrite sill in Soay (Hebrides), *J. Petrology*, **2**, 1 (1961).
WYLLIE, P. J. and HAAS, J. L., The system $CaO–SiO_2–CO_2–H_2O$. II., *Geochim. cosmochim. Acta*, **30**, 525 (1966).
YODER, H. S., The jadeite problem, *Am. J. Sci.*, **248**, 225, 312 (1950).
ZEGGEREN, F. VAN and BENSON, G. C., Calculation of the surface energy of alkali halide crystals, *J. chem. Phys.*, **26**, 1077 (1957).
ZWART, H. J., Relations between folding and metamorphism in the Central Pyrenees, *Geologie Mijnb.*, **39**e, 163 (1960a).
ZWART, H. J., Chronological succession of folding and metamorphism in the Central Pyrenees, *Geol. Rdsch.*, **50**, 203 (1960b).
ZWART, H. J., Metamorphic history of the Central Pyrenees, II, *Leid. geol. Meded.*, **28**, 321 (1963).

Author Index

Subject Index